BASIC SPACE
PLASMA PHYSICS

T0325153

Wolfgang Baumjohann

Max-Planck-Institut für extraterrestrische Physik, Garching &
Institut für Geophysik, Ludwig-Maximilians-Universität, München

Rudolf A. Treumann

Max-Planck-Institut für extraterrestrische Physik, Garching &
Institut für Geophysik, Ludwig-Maximilians-Universität, München

BASIC SPACE
PLASMA PHYSICS

Imperial College Press

ICP

Published by

Imperial College Press
57 Shelton Street
Covent Garden
London WC2H 9HE

Distributed by

World Scientific Publishing Co. Pte. Ltd.
P O Box 128, Farrer Road, Singapore 912805
USA office: Suite 1B, 1060 Main Street, River Edge, NJ 07661
UK office: 57 Shelton Street, Covent Garden, London WC2H 9HE

British Library Cataloguing-in-Publication Data
A catalogue record for this book is available from the British Library.

First published 1996
Reprinted 1999

BASIC SPACE PLASMA PHYSICS

ISBN   1-86094-017-X

Printed in Singapore.

Contents

Preface

One more textbook on plasma physics? Indeed, there are a number of excellent textbooks on the market, like the incomparable book *Introduction to Plasma Physics and Controlled Fusion* by Francis F. Chen. It is impossible to compete with a book of this clarity, or some of the other texts which have been around for longer or shorter. However, we found most of the books not well-suited for a course on space plasma physics. Some are directed more toward the interests of laboratory plasma physics, like Chen's book, others are highly mathematical, such that it would have required an additional course in applied mathematics to make them accessible to the students. The vast majority of books in the field of space plasma physics, however, are collections of review articles, like the recent *Introduction to Space Physics* edited by Margaret G. Kivelson and Christopher T. Russell. These books require that the reader already has quite some knowledge of the field.

The only textbook specifically addressed to the needs of space plasma physics is *Physics of Space Plasmas* by George K. Parks. This book covers many aspects of space plasma physics, but is ordered in terms of phenomena rather than with respect to plasma theory. To give the students a feeling for the coherency of our field, we felt the need to find a compromise between classical plasma physics textbooks and the books by Parks and Kivelson & Russell. We tried to achieve this goal during a third-year space plasma physics course, which we gave regularly at the University of Munich since 1988 for undergraduate and graduate students of geophysics, who had an average knowledge of fluid dynamics and electromagnetism.

This textbook collects and expands lecture notes from these two-semester courses. However, the first part can also be used for a one-semester undergraduate course and research scientists may find the later chapters of the second part helpful. The book is written in a self-contained way and most of the material is presented including the basic steps of derivation so that the reader can follow without need to consult original sources. Some of the more involved mathematical derivations are given in the Appendix. Special emphasis has been placed on providing instructive figures. Figures containing original measurements are scarce and have mostly been redrawn in a more schematic way.

The first five chapters provide an introduction into space physics, based on a mixture of simple theory and a description of the wealth of space plasma phenomena. A concise description of the Earth's plasma environment is followed by a derivation of single particle motion in electromagnetic fields, adiabatic invariants, and applications to the Earth's magnetosphere and ring current. Then the origin and effects of collisions and conductivities and the formation of the ionosphere are discussed. Ohm's law and the frozen-in concept are introduced on a somewhat heuristic basis. The first part ends with an introduction into magnetospheric dynamics, including convection electric fields, current systems, substorms, and other macroscopic aspects of solar wind-magnetosphere and magnetosphere-ionosphere coupling.

The second part of the book presents a more rigorous theoretical foundation of space plasma physics, yet still contains many applications to space physics. It starts from kinetic theory, which is built on the Klimontovich approach. Introducing moments of the distribution function allows the derivation of the single and multi-fluid equations, followed by a discussion of fluid boundaries and shocks, with the Earth's magnetopause and bow shock as examples. Both, fluid and kinetic theory are then applied to derive the relevant wave modes in a plasma, again with applications from space physics.

The material presented in the present book is extended in *Advanced Space Plasma Physics*, written by the same authors. This companion textbook gives a representative selection of the many macro- and microinstabilities in a plasma, from the Rayleigh-Taylor and Kelvin-Helmholtz to the electrostatic and electromagnetic instabilities, and a comprehensive overview on the nonlinear aspects relevant for space plasma physics, e.g., wave-particle interaction, solitons, and anomalous transport.

We are grateful to Rosmarie Mayr-Ihbe for turning our often rough sketches into the figures contained in this book. It is also a pleasure to thank Jim LaBelle for valuable contributions, Anja Czaykowska and Thomas Bauer for careful reading of the manuscript and many suggestions, and Karl-Heinz Mühlhäuser and Patrick Daly for helping us with LaTeX. We gratefully acknowledge the support of Heinrich Soffel, Gerhard Haerendel and Gregor Morfill, and acknowledge the patience of our colleagues at MPE, when we worked on this book instead of finishing other projects in time. Both of us owe deep respect to our teachers who introduced us into geophysics, Jürgen Untiedt and the late Gerhard Fanselau.

Last but not least, we would like to mention that we have profited from many books and reviews on plasma and space physics. References to most of them have been included into the suggestions for further reading at the ends of the chapters. These suggestions, however, do not include the very large number of original papers, which we made use of and are indebted to.

Needless to say, we have made all efforts to make the text error-free. However, this is an unsurmountable task. We hope that the readers of this book will kindly inform us about misprints and errors they may find in here, preferably by electronic mail to *bj@mpe-garching.mpg.de*. We will be grateful for any hints and post them with other errors on *http://www.mpe-garching.mpg.de/bj/bspp.html*.

1. Introduction

The context of the term 'geophysics' has changed considerably during the second half of this century. Well into the fifties the key interest of geophysics was the interior of our planet, i.e., solid Earth geophysics covering seismology, rock physics, magnetic and electric properties of crust and mantle, etc. With the advent of the spaceflight era, the interests of geophysicists broadened and extended into the external neighborhood of our planet. It was realized that the extraterrestrial matter is in an ionized state, very different from the state of known matter near the ground of the Earth.

Matter of this kind behaves unexpected because of its sensitivity to electric and magnetic fields and its ability to carry electric currents. Within this context, the concept of a plasma became introduced and space plasma physics became a new and important branch of geophysics. Nowadays, methods of plasma physics are not only used in external geophysics, but are essential to understand the dynamics of the Earth's fluid core and the generation of the terrestrial magnetic field.

1.1. Definition of a Plasma

A *plasma* is a gas of charged particles, which consists of equal numbers of free positive and negative charge carriers. Having roughly the same number of charges with different signs in the same volume element guarantees that the plasma behaves *quasineutral* in the stationary state. On average a plasma looks electrically neutral to the outside, since the randomly distributed particle electric charge fields mutually cancel.

For a particle to be considered a free particle, its typical potential energy due to its nearest neighbor must be much smaller than its random kinetic (thermal) energy. Only then the particle's motion is practically free from the influence by other charged particles in its neighborhood as long as no direct collisions take place.

Since the particles in a plasma have to overcome the coupling with their neighbors, they must have thermal energies above some electronvolts. Thus a typical plasma is a hot and highly ionized gas. While only a few natural plasmas, such as flames or lightning strokes, can be found near the Earth's surface, plasmas are abundant in the universe. More than 99% of all known matter is in the plasma state.

Debye Shielding

For the plasma to behave quasineutral in the stationary state, it is necessary to have about equal numbers of positive and negative charges per volume element. Such a volume element must be large enough to contain a sufficient number of particles, yet small enough compared with the characteristic lengths for variations of macroscopic parameters such as density and temperature. In each volume element the microscopic space charge fields of the individual charge carriers must cancel each other to provide macroscopic charge neutrality.

To let the plasma appear electrically neutral, the electric *Coulomb potential* field of every charge, q

$$\phi_C = \frac{q}{4\pi\epsilon_0 r} \tag{1.1}$$

with ϵ_0 being the free space permittivity, is shielded by other charges in the plasma and assumes the *Debye potential* form

$$\phi_D = \frac{q}{4\pi\epsilon_0 r} \exp\left(-\frac{r}{\lambda_D}\right) \tag{1.2}$$

in which the exponential function cuts off the potential at distances $r > \lambda_D$. The characteristic length scale, λ_D, is called *Debye length* and is the distance, over which a balance is obtained between the thermal particle energy, which tends to perturb the electrical neutrality, and the electrostatic potential energy resulting from any charge separation, which tends to restore charge neutrality. Figure 1.1 shows the shielding effect.

In Sec. 9.1 we will show that the Debye length is a function of the electron and ion temperatures, T_e and T_i, and the plasma density, $n_e \simeq n_i$ (assuming singly charged ions)

$$\boxed{\lambda_D = \left(\frac{\epsilon_0 k_B T_e}{n_e e^2}\right)^{1/2}} \tag{1.3}$$

where we have assumed $T_e \simeq T_i$ and where k_B the Boltzmann constant and e the electron charge. We will give the exact definition for the temperature in Sec. 6.5. Until then we will use the terms temperature and average energy, $\langle W \rangle = k_B T$, as synonyms.

In order for a plasma to be quasineutral, the physical dimension of the system, L, must be large compared to λ_D

$$\lambda_D \ll L \tag{1.4}$$

Otherwise there is not enough space for the collective shielding effect to occur, and we have a simple ionized gas. This requirement is often called the first plasma criterion.

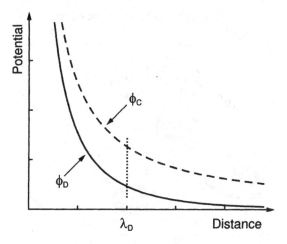

Fig. 1.1. Comparison of Debye and Coulomb potential.

Plasma Parameter

Since the shielding effect is the result of the collective behavior inside a Debye sphere of radius λ_D, it is necessary that this sphere contains enough particles. The number of particles inside a Debye sphere is $\frac{4\pi}{3} n_e \lambda_D^3$. The term $n_e \lambda_D^3$ is often called the *plasma parameter*, Λ, and the second criterion for a plasma reads

$$\Lambda = n_e \lambda_D^3 \gg 1 \tag{1.5}$$

By substituting λ_D by the expression given in Eq. (1.3) and raising each side of Eq. (1.5) to the 2/3 power, it becomes apparent that the second criterion quantifies what is meant by free particles. The mean potential energy of a particle due to its nearest neighbor, which is inversely proportional to the mean interparticle distance and thus proportional to $n_e^{1/3}$, must be much smaller than its mean energy, $k_B T_e$.

Plasma Frequency

The typical oscillation frequency in a fully ionized plasma is the electron *plasma frequency*, ω_{pe}. If the quasineutrality of the plasma is disturbed by some external force, the electrons, being more mobile than the much heavier ions, are accelerated in an attempt to restore the charge neutrality. Due to their inertia they will move back and forth around the equilibrium position, resulting in fast collective oscillations around the more massive ions. In Sec. 9.1 it will be shown that the plasma frequency depends on the

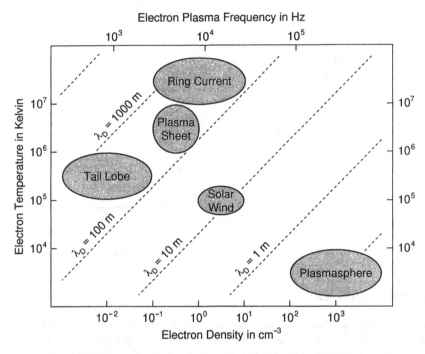

Fig. 1.2. Ranges of typical parameters for several geophysical plasmas.

square root of the plasma density. With m_e as electron mass, ω_{pe} can be written as

$$\omega_{pe} = \left(\frac{n_e e^2}{m_e \epsilon_0}\right)^{1/2} \tag{1.6}$$

Some plasmas, like the Earth's ionosphere, are not fully ionized. Here we have a substantial number of neutral particles and if the charged particles collide too often with neutrals, the electrons will be forced into equilibrium with the neutrals, and the medium does not behave as a plasma anymore, but simply like a neutral gas. For the electrons to remain unaffected by collisions with neutrals, the average time between two electron-neutral collisions, τ_n, must be larger than the reciprocal of the plasma frequency

$$\omega_{pe}\tau_n \gg 1 \tag{1.7}$$

This is the third criterion for an ionized medium to behave as a plasma.

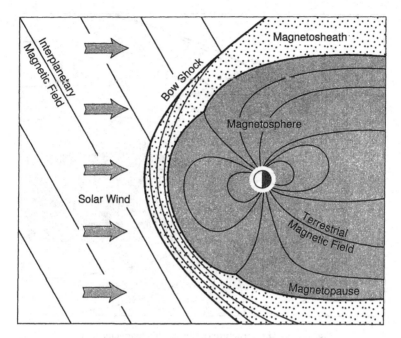

Fig. 1.3. Topography of the solar-terrestrial environment.

1.2. Geophysical Plasmas

Plasmas are not only abundant in the universe, but also in our solar system. Even in the immediate neighborhood of the Earth, all matter above about 100 km altitude, within and above the ionosphere, has to be treated using plasmaphysical methods. There are quite a number of different geophysical plasmas, with a wide spread in their characteristic parameters like density and temperature (see Fig. 1.2).

Solar Wind

The sun emits a highly conducting plasma at supersonic speeds of about 500 km/s into the interplanetary space as a result of the supersonic expansion of the solar corona. This plasma is called the *solar wind* and consists mainly of electrons and protons, with an admixture of 5% Helium ions. Because of the high conductivity, the solar magnetic field is frozen in the plasma (like in a superconductor, see Sec. 5.1) and drawn outward by the expanding solar wind. Typical values for the electron density and temperature in the solar wind near the Earth are $n_e \approx 5\,\text{cm}^{-3}$ and $T_e \approx 10^5\,\text{K}$ (1 eV = 11,600 K; see App. A.2). The *interplanetary magnetic field* is of the order of 5 nT.

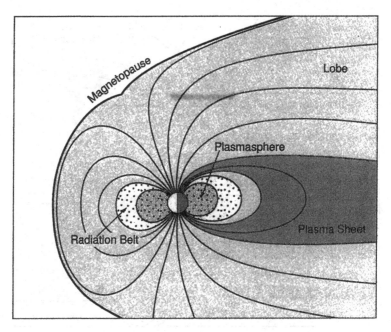

Fig. 1.4. Plasma structure of the Earth's magnetosphere.

When the solar wind hits on the Earth's dipolar magnetic field, it cannot simply penetrate it but rather is slowed down and, to a large extent, deflected around it. Since the solar wind hits the obstacle with supersonic speed, a *bow shock* wave is generated (see Fig. 1.3), where the plasma is slowed down and a substantial fraction of the particles' kinetic energy is converted into thermal energy. The region of thermalized subsonic plasma behind the bow shock is called the *magnetosheath* (see Fig. 1.3). Its plasma is denser and hotter than the solar wind plasma and the magnetic field strength has higher values in this region.

Magnetosphere

The shocked solar wind plasma in the magnetosheath cannot easily penetrate the terrestrial magnetic field but is mostly deflected around it. This is a consequence of the fact that the interplanetary magnetic field lines cannot penetrate the terrestrial field lines and that the solar wind particles cannot leave the interplanetary field lines due to the aforementioned frozen-in characteristic of a highly conducting plasma.

The boundary separating the two different regions is called *magnetopause* and the cavity generated by the terrestrial field has been named *magnetosphere* (see Figs. 1.3 and

1.4). The kinetic pressure of the solar wind plasma distorts the outer part of the terrestrial dipolar field. At the frontside it compresses the field, while the nightside magnetic field is stretched out into a long *magnetotail* which reaches far beyond lunar orbit.

The plasma in the magnetosphere consists mainly of electrons and protons. The sources of these particles are the solar wind and the terrestrial ionosphere. In addition there are small fractions of He^+ and O^+ ions of ionospheric origin and some He^{++} ions originating from the solar wind. However, the plasma inside the magnetosphere is not evenly distributed, but is grouped into different regions with quite different densities and temperatures. Figure 1.4 depicts the topography of some of these regions.

The *radiation belt* lies on dipolar field lines between about 2 and 6 R_E (1 Earth radius = 6371 km). It consists of energetic electrons and ions which move along the field lines and oscillate back and forth between the two hemispheres (see Sec. 3.2). Typical electron densities and temperatures in the radiation belt are $n_e \approx 1\,cm^{-3}$ and $T_e \approx 5 \cdot 10^7$ K. The magnetic field strength ranges between about 100 and 1000 nT.

Most of the magnetotail plasma is concentrated around the tail midplane in an about 10 R_E thick *plasma sheet*. Near the Earth, it reaches down to the high-latitude *auroral ionosphere* along the field lines. Average electron densities and temperatures in the plasma sheet are $n_e \approx 0.5\,cm^{-3}$ and $T_e \approx 5 \cdot 10^6$ K, with $B \approx 10\,nT$.

The outer part of the magnetotail is called the *magnetotail lobe*. It contains a highly rarified plasma with typical values for the electron density and temperature and the magnetic field strength of $n_e \approx 10^{-2}\,cm^{-3}$, $T_e \approx 5 \cdot 10^5$ K, and $B \approx 30\,nT$, respectively.

Ionosphere

The solar ultraviolet light impinging on the Earth's atmosphere ionizes a fraction of the neutral atmosphere. At altitudes above 80 km collisions are too infrequent to result in rapid recombination and a permanent ionized population called the *ionosphere* is formed. Typical electron densities and temperatures in the mid-latitude ionosphere are $n_e \approx 10^5\,cm^{-3}$ and $T_e \approx 10^3$ K. The magnetic field strength is of the order of $10^4\,nT$.

The ionosphere extends to rather high altitudes and, at low- and mid-latitudes, gradually merges into the *plasmasphere*. As depicted in Fig. 1.4, the plasmasphere is a torus-shaped volume inside the radiation belt. It contains a cool but dense plasma of ionospheric origin ($n_e \approx 5 \cdot 10^2\,cm^{-3}$, $T_e \approx 5 \cdot 10^3$ K), which corotates with the Earth. In the equatorial plane, the plasmasphere extends out to about 4 R_E, where the density drops down sharply to about 1 cm^{-3}. This boundary is called the *plasmapause*.

At high latitudes plasma sheet electrons can precipitate along magnetic field lines down to ionospheric altitudes, where they collide with and ionize neutral atmosphere particles. As a by-product, photons emitted by this process create the polar light, the *aurora*. These auroras are typically observed inside the *auroral oval* (see Fig. 1.5), which contains the footprints of those field lines which thread the plasma sheet. Inside of the auroral oval lies the *polar cap*, which is threaded by field lines connected to the tail lobe.

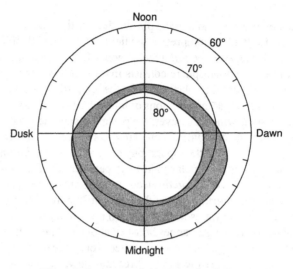

Fig. 1.5. Average auroral oval and polar cap.

1.3. Magnetospheric Currents

The plasmas discussed in the last section are usually not stationary but move under the influence of external forces. Sometimes ions and electrons move together, like in the solar wind. But at other occasions and in other plasma regions, ions and electrons move into different directions, creating an electric current. Such currents are very important for the dynamics of the Earth's plasma environment. They transport charge, mass, momentum and energy. Moreover, the currents create magnetic fields, which may severely alter or distort any pre-existing fields.

Actually, the distortion of the terrestrial dipole field into the typical shape of the magnetosphere is accompanied by electrical currents. As schematically shown in Fig. 1.6, the compression of the terrestrial magnetic field on the dayside is associated with current flow across the magnetopause surface, the *magnetopause current*. The tail-like field of the nightside magnetosphere is accompanied by the *tail current* flowing on the tail surface and the *neutral sheet current* in the central plasma sheet, both of which are connected and form a Θ-like current system, if seen from along the Earth-Sun line.

Another large-scale current system, which influences the configuration of the inner magnetosphere, is the *ring current*. The ring current flows around the Earth in a westward direction at radial distances of several Earth radii and is carried by the radiation belt particles mentioned above. In addition to their bounce motion, these particles drift slowly around the Earth. Since the protons drift westward while the electrons move in the eastward direction, this constitutes a net charge transport.

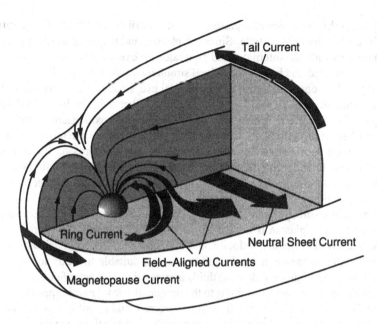

Fig. 1.6. Synopsis of magnetospheric currents.

A number of current systems exist in the conducting layers of the Earth's iono-sphere, at altitudes of 100–150 km. Most notable are the *auroral electrojets* inside the auroral oval, the *Sq currents* in the dayside mid-latitude ionosphere, and the *equatorial electrojet* near the magnetic equator.

In addition to these perpendicular currents, currents also flow along magnetic field lines. As shown in Fig. 1.6, the *field-aligned currents* connect the magnetospheric cur-rent systems in the magnetosphere to those flowing in the polar ionosphere. The field-aligned currents are mainly carried by electrons and are essential for the exchange of energy and momentum between these regions.

1.4. Theoretical Approaches

The dynamics of a plasma is governed by the interaction of the charge carriers with the electric and magnetic fields. If all the fields were of external origin, the physics would be relatively simple. However, as the particles move around, they may create local space charge concentrations and thus electric fields. Moreover, their motion can also generate electric currents and thus magnetic fields. These internal fields and their feedback onto the motion of the plasma particles make plasma physics difficult.

In general the dynamics of a plasma can be described by solving the equation of motion for each individual particle. Since the electric and magnetic fields appearing in each equation include the internal fields generated by every other moving particle, all equations are coupled and have to be solved simultaneously. Such a full solution is not only too difficult to obtain, but also of no practical use, since one is interested in knowing average quantities like density and temperature rather than the individual velocity of each particle. Therefore, one usually makes certain approximations suitable to the problem studied. It has turned out that four different approaches are most useful.

The simplest approach is the *single particle motion* description. It describes the motion of a particle under the influence of external electric and magnetic fields. This approach neglects the collective behavior of a plasma, but is useful when studying a very low density plasma, like found in the ring current.

The *magnetohydrodynamic* approach is the other extreme and neglects all single particle aspects. The plasma is treated as a single conducting fluid with macroscopic variables, like average density, velocity, and temperature. The approach assumes that the plasma is able to maintain local equilibria and is suitable to study low-frequency wave phenomena in highly conducting fluids immersed in magnetic fields.

The *multi-fluid* approach is similar to the magnetohydrodynamic approach, but accounts for different particle species (electrons, protons, and possibly heavier ions) and assumes that each species behaves like a separate fluid. It has the advantage that differences in the fluid behavior of the light electrons and the heavier ions can be taken into account. This can lead to charge separation fields and high-frequency wave propagation.

The *kinetic theory* is the most developed plasma theory. It adopts a statistical approach. Instead of solving the equation of motion for each individual particle, it looks at the development of the distribution function for the system of particles under consideration in phase space. Yet even in kinetic theory certain simplifying assumptions have to be made and there are different flavors of kinetic theory, depending on the kind of simplification made.

In the present book we will describe all these approaches and apply them to suitable geophysical plasma phenomena. We will start with the single particle approach. Subsequently, we will derive the basic equations of the kinetic theory, but then first turn to the fluid theories and their applications, before we finally go into the details of the kinetic approach.

2. Single Particle Motion

Plasmas are collections of very large numbers of electrically charged particles. It is the charged state of the particles that distinguishes plasmas from other particle collections such as normal gases or fluids. The electric charge couples the particles to the electromagnetic field, which affects their motions.

In a situation where the charged particles do not directly interact with each other and where they do not affect the external magnetic field significantly, the motion of each individual particle can be treated independently. This *single particle approach* is only valid in very rarified plasmas where collective effects are negligible. Furthermore, the external magnetic field must be rather strong, much greater than the magnetic field produced by the electric current due to the charged particle motion.

We will see later that the single particle approach can be used only in some plasmas of geophysical interest. However, in order to understand the collective behavior of the plasma, i.e., the motion of the charge carriers under the influence of electric and magnetic fields generated by the motion itself, it is very instructive to study first the motion of charged particles in prescribed electric and magnetic fields.

2.1. Field Equations

Before describing the particle motion in external electric and magnetic fields, we introduce the electromagnetic field equations. There is a twofold coupling between electric charges and electromagnetic fields. Charged particles at rest are the sources of the electrostatic field, $\mathbf{E}$, which is the origin of the *Coulomb force*

$$\mathbf{F}_C = q\,\mathbf{E} \tag{2.1}$$

they feel in the combined electrostatic field of all the other particles. On the other hand, charged particles moving with a velocity, $\mathbf{v}$, are current elements generating a magnetic field, $\mathbf{B}$, which is the origin of the *Lorentz force*

$$\mathbf{F}_L = q\,(\mathbf{v} \times \mathbf{B}) \tag{2.2}$$

11

The motion of charged particles is strongly influenced by the presence of the electromagnetic field, while at the same time it is also the source of the fields. The relation between fields and particles is described by *Maxwell's equations* (see App. A.5)

$$\nabla \times \mathbf{B} = \mu_0 \mathbf{j} + \epsilon_0 \mu_0 \frac{\partial \mathbf{E}}{\partial t} \tag{2.3}$$

$$\nabla \times \mathbf{E} = -\frac{\partial \mathbf{B}}{\partial t} \tag{2.4}$$

where $\mathbf{j}$ is the electric current density in the plasma, and ϵ_0 and μ_0 are the vacuum permittivity and susceptibility, respectively.

These equations show that the electric and magnetic fields are not independent, but are coupled by their spatial and temporal variations. Moreover, the electric current density turns out to be the source of the magnetic field and of fast fluctuations of the electric field. Since $\epsilon_0 \mu_0 = c^{-2}$ is equal to the inverse square of the light velocity, the latter will be negligible in a plasma as long as we do not consider propagation of electromagnetic waves. Hence, the second term on the right-hand side of Eq. (2.3) is small as long as no fast oscillations appear in the electric field.

In order to close the system, the first two equations have to be supplemented by two more equations, namely the conditions

$$\nabla \cdot \mathbf{B} = 0 \tag{2.5}$$

$$\nabla \cdot \mathbf{E} = \rho/\epsilon_0 \tag{2.6}$$

The first of these expressions indicates that there are no sources of the magnetic field and thus the magnetic field lines are always closed. The second condition shows that the source of the electric field is the electric space charge density, $\rho = e\,(n_i - n_e)$, which is the difference between the charge densities of the ion and the electrons. Similarly, the electric current is defined as the difference between the electron and ion fluxes as $\mathbf{j} = e\,(n_i \mathbf{v}_i - n_e \mathbf{v}_e)$ where for simplicity we have assumed that the ions are singly charged.

2.2. Gyration

The equation of motion for a particle of charge q under the action of the Coulomb and Lorentz forces given in Eqs. (2.1) and (2.2) can be written as

$$m \frac{d\mathbf{v}}{dt} = q\,(\mathbf{E} + \mathbf{v} \times \mathbf{B}) \tag{2.7}$$

where m represents the particle mass and $\mathbf{v}$ the particle velocity. Under the absence of an electric field this equation reduces to

$$m \frac{d\mathbf{v}}{dt} = q\,(\mathbf{v} \times \mathbf{B}) \tag{2.8}$$

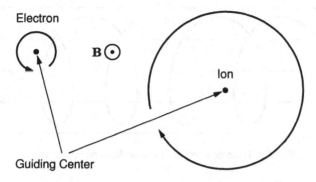

Fig. 2.1. Gyration of charged particles around a guiding center.

Taking the dot product of Eq. (2.8) with **v** and noting that $\mathbf{v} \cdot (\mathbf{v} \times \mathbf{B}) = 0$ (useful vector relations are found in App. A.4), we obtain

$$m\frac{d\mathbf{v}}{dt} \cdot \mathbf{v} = \frac{d}{dt}\left(\frac{mv^2}{2}\right) = 0 \tag{2.9}$$

which shows that the particle kinetic energy as well as the magnitude of its velocity are constants. A static magnetic field, whatever its spatial variance is, does not change the particle kinetic energy.

In a uniform magnetostatic field along the z axis, $\mathbf{B} = B\,\hat{\mathbf{e}}_z$, we get the components

$$\begin{aligned}
m\dot{v}_x &= qBv_y \\
m\dot{v}_y &= -qBv_x \\
m\dot{v}_z &= 0
\end{aligned} \tag{2.10}$$

The velocity component parallel to the magnetic field, $v_\parallel = v_z$, is constant. Taking the second derivative, we get

$$\begin{aligned}
\ddot{v}_x &= -\omega_g^2 v_x \\
\ddot{v}_y &= -\omega_g^2 v_y
\end{aligned} \tag{2.11}$$

where ω_g is the *gyrofrequency* or *cyclotron frequency*, which has opposite signs for positive and negative charges (note that ω_g is often defined as a positive number, independent of the sign of the charge; we will also use it this way in later chapters, starting on p. 227)

$$\boxed{\omega_g = \frac{qB}{m}} \tag{2.12}$$

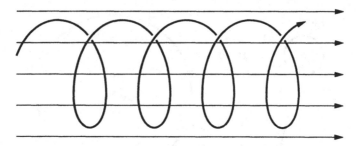

Fig. 2.2. Helicoidal ion orbit in a uniform magnetic field.

Equation (2.11) is a harmonic oscillator equation with solutions of the form

$$x - x_0 = r_g \sin \omega_g t$$
$$y - y_0 = r_g \cos \omega_g t \tag{2.13}$$

Since ω_g carries the sign of the charge, the x component has opposite signs for electrons and ions. r_g is the *gyroradius* defined as

$$r_g = \frac{v_\perp}{|\omega_g|} = \frac{m v_\perp}{|q| B} \tag{2.14}$$

where $v_\perp = (v_x^2 + v_y^2)^{1/2}$ is the constant speed in the plane perpendicular to **B**.

The components of Eq. (2.13) describe a circular orbit of the particle around the magnetic field, with the sense of rotation depending of the sign of the charge (see Fig. 2.1). The center of the orbit (x_0, y_0) is called the *guiding center*. The circular orbit of the charged particle represents a circular current and the direction of the gyration is such that the magnetic field generated by the circular current is opposite to the externally imposed field. This behavior is called *diamagnetic effect*.

In Fig. 2.1 we have neglected a possible constant velocity of the particle parallel to the magnetic field, $v_{\parallel}$. Whenever $v_{\parallel} \neq 0$, the actual trajectory of the particle is three-dimensional and looks like a helix. Such a helicoidal trajectory is shown in Fig. 2.2. The *pitch angle*, α, of the helix is defined as

$$\alpha = \tan^{-1}\left(\frac{v_\perp}{v_{\parallel}}\right) \tag{2.15}$$

and depends on the ratio between the perpendicular and parallel velocity components.

2.3. Electric Drifts

Taking the electric field into consideration will result in a drift of the particle superimposed onto its gyratory motion. The exact nature of this *electric drift* depends on whether the field is electrostatic or time-varying and whether it is spatially uniform or not.

E × B Drift

Let us now assume that an electrostatic field, E, is present. Looking for solutions of Eq. (2.7), we can again treat the perpendicular components and the component parallel to **B** separately. The parallel component

$$m\dot{v}_\parallel = qE_\parallel \tag{2.16}$$

describes a straightforward acceleration along the magnetic field. However, in geophysical plasmas most parallel electric fields cannot be sustained, since they are immediately canceled out by electrons, which are under most circumstances extremely mobile along the magnetic field lines.

Assuming that the perpendicular electric field component is parallel to the x axis, $\mathbf{E}_\perp = E_x \hat{e}_x$, the perpendicular components of Eq. (2.7) are

$$\begin{aligned}
\dot{v}_x &= \omega_g v_y + \frac{q}{m} E_x \\
\dot{v}_y &= -\omega_g v_x
\end{aligned} \tag{2.17}$$

Taking the second derivative, we obtain

$$\begin{aligned}
\ddot{v}_x &= -\omega_g^2 v_x \\
\ddot{v}_y &= -\omega_g^2 \left(v_y + \frac{E_x}{B} \right)
\end{aligned} \tag{2.18}$$

If we substitute $v_y' = v_y + E_x/B$, we get back to Eq. (2.11), where the particle is gyrating about the guiding center. Thus Eq. (2.18) describes a gyration, but with a superimposed drift of the guiding center in the $-y$ direction. This drift of the guiding center is usually called $E{\times}B$ *drift* and has the general form

$$\boxed{\mathbf{v}_E = \frac{\mathbf{E} \times \mathbf{B}}{B^2}} \tag{2.19}$$

The E×B drift is independent of the sign of the charge and thus electrons and ions move into the same direction.

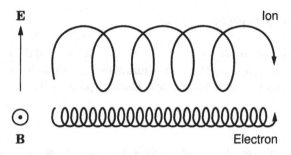

Fig. 2.3. Particle drifts in crossed electric and magnetic fields.

Figure 2.3 shows the acceleration and deceleration effect of a perpendicular electric field. An ion is accelerated into the direction of the electric field, thereby increasing its gyroradius. But it is decelerated during the second half of its gyratory orbit, now with decreasing gyroradius. The different gyroradii shift the position of the guiding center in the $\mathbf{E} \times \mathbf{B}$ direction. The electrons are accelerated when moving antiparallel to the electric field and decelerated when moving parallel. But since their sense of gyration is opposite, too, their guiding centers drift into the same direction.

It is instructive to note that the $\mathbf{E} \times \mathbf{B}$ drift has a fundamental physical root in the *Lorentz transformation* of the electric field into the moving system of the particle. In this system the electric field is given by

$$\mathbf{E}' = \mathbf{E} + \mathbf{v} \times \mathbf{B} \tag{2.20}$$

For a free particle this field must vanish, $\mathbf{E}' = 0$, which yields for the electric field

$$\mathbf{E} = -\mathbf{v} \times \mathbf{B} \tag{2.21}$$

Solving for the velocity immediately yields the expression for the electric drift in Eq. (2.19). Because the Lorentz transformation does not depend on the charge of the particles, the $\mathbf{E} \times \mathbf{B}$ drift is also independent of the sign of the charge.

Polarization Drift

We could have derived Eq. (2.19) directly from Eq. (2.7). Taking the cross-product of both sides of Eq. (2.7) with $\mathbf{B}/B^2$, we obtain

$$\mathbf{v} - \frac{\mathbf{B}\,(\mathbf{v} \cdot \mathbf{B})}{B^2} = \frac{\mathbf{E} \times \mathbf{B}}{B^2} - \frac{m}{q}\frac{d\mathbf{v}}{dt} \times \frac{\mathbf{B}}{B^2} \tag{2.22}$$

We can recognize the left-hand side as a perpendicular velocity vector and the first term on the right-hand side as the $\mathbf{E} \times \mathbf{B}$ drift. Averaging over the gyroperiod and thus neglecting temporal changes of the order of the gyroperiod or faster allows us to take the perpendicular velocity as the perpendicular drift velocity, v_d. Remembering that the magnetic field is assumed time independent, we rewrite

$$\mathbf{v}_d = \mathbf{v}_E - \frac{m}{qB^2}\frac{d}{dt}(\mathbf{v} \times \mathbf{B}) \tag{2.23}$$

which yields with Eqs. (2.12) and (2.21)

$$\mathbf{v}_d = \mathbf{v}_E + \frac{1}{\omega_g B}\frac{d\mathbf{E}_\perp}{dt} \tag{2.24}$$

Equation (2.24) describes the drift of a charged particle in crossed homogeneous magnetic and electric fields, where the electric field is allowed to vary slowly. The last term in this equation is called *polarization drift*.

$$\boxed{\mathbf{v}_P = \frac{1}{\omega_g B}\frac{d\mathbf{E}_\perp}{dt}} \tag{2.25}$$

There is an important qualitative difference between the polarization drift and the $E \times B$ drift. The $E \times B$ drift does neither depend on the charge nor on the mass of the particle, since it can be viewed as a result of the Lorentz transformation. Thus electrons, protons, and heavier ions all move into the same direction perpendicular to $\mathbf{B}$ and $\mathbf{E}$ with the same velocity. The polarization drift, on the other hand, increases proportional to the mass of the particle. It is directed along the electric field, but oppositely for electrons and ions. Accordingly, it creates a current

$$\mathbf{j}_P = n_e e\,(\mathbf{v}_{Pi} - \mathbf{v}_{Pe}) = \frac{n_e(m_i + m_e)}{B^2}\frac{d\mathbf{E}_\perp}{dt} \tag{2.26}$$

which carries electrons and ions into opposite directions and polarizes the plasma. Since $m_i \gg m_e$, the *polarization current* is mainly carried by the ions.

Electric Drift Corrections

Equation (2.24) can also describe the drift due to inhomogeneities of the electric field if the total time derivative is taken as the sum of the temporal and the convective derivative $d/dt = \partial/\partial t + \mathbf{v}\cdot\nabla$, where the velocity vector can, to a good approximation, be replaced by the $E \times B$ velocity. The convective term becomes proportional to E^2. It is a nonlinear contribution and is usually much smaller than the time derivative and therefore neglected in most cases.

The convective derivative takes into account spatial variations of the electric field in the direction of the $E \times B$ drift under the assumption that the electric field changes only gradually. When this is not the case and the electric field changes considerably over one gyroradius, there is a further correction on the electric field drift, known as *finite Larmor radius effect*. This correction is a second order effect in r_g and leads to the following more complete expression for the electric field drift

$$\mathbf{v}_E = \left(1 + \tfrac{1}{4}r_g^2\nabla^2\right)\frac{\mathbf{E}\times\mathbf{B}}{B^2} \tag{2.27}$$

The second spatial derivative takes account of the spatial variation of the electric field, averaged over the gyration orbit. Finite Larmor radius effects are normally neglected in macroscopic applications of particle motion but may become important in the vicinity of plasma boundaries, plasma transitions and small scale structures in a plasma.

2.4. Magnetic Drifts

When analyzing Eq. (2.8), we have assumed that the magnetic field is homogeneous. This is often not the case. A typical magnetic field has gradients and often field lines are curved. This inhomogeneity of the magnetic field leads to a *magnetic drift* of charged particles. Time variations of the magnetic field itself cannot impart energy to a particle, since the Lorentz force is always perpendicular to the velocity of the particle. However, since $\partial\mathbf{B}/\partial t = -\nabla\times\mathbf{E}$, the associated inhomogeneous electric field may accelerate the particles in the way described in the previous section.

Gradient Drift

Let us now assume that the magnetic field is weakly inhomogeneous, for example increasing in the upward direction. As visualized in Fig. 2.4, the gyroradius of a particle decreases in the upward direction and thus the gyroradius of a particle will be larger at the bottom of the orbit than at the top half. As a result, ions and electrons drift into opposite directions, perpendicular to both $\mathbf{B}$ and ∇B.

Since we assume that the typical scale length of a magnetic field gradient is much larger than the particle gyroradius, we can Taylor expand the magnetic field vector about the guiding center of the particle

$$\mathbf{B} = \mathbf{B}_0 + (\mathbf{r}\cdot\nabla)\mathbf{B}_0 \tag{2.28}$$

where $\mathbf{B}_0$ is measured at the guiding center and $\mathbf{r}$ is the distance from the guiding center. Inserting this relation into Eq. (2.8) we obtain

$$m\frac{d\mathbf{v}}{dt} = q\,(\mathbf{v}\times\mathbf{B}_0) + q\,[\mathbf{v}\times(\mathbf{r}\cdot\nabla)\mathbf{B}_0] \tag{2.29}$$

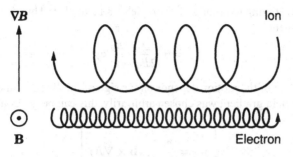

Fig. 2.4. Particle drifts due to a magnetic field gradient.

Expanding the velocity term into a gyration and a drift motion, $\mathbf{v} = \mathbf{v}_g + \mathbf{v}_\nabla$, and noting $v_\nabla \ll v_g$ yields

$$m\frac{d\mathbf{v}_\nabla}{dt} = q\,(\mathbf{v}_\nabla \times \mathbf{B}_0) + q\,(\mathbf{v}_g \times [\mathbf{r} \cdot \nabla)\mathbf{B}_0] \tag{2.30}$$

where we have omitted the terms describing gyration in a homogeneous field and ne-glected $\mathbf{v}_\nabla \times (\mathbf{r} \cdot \nabla)\mathbf{B}_0$ as a small quantity.

Since we are interested in time scales much larger than the gyroperiod, we can av-erage over one gyration. Upon this the left side vanishes since any acceleration a particle experiences when moving into the weak field region is balanced by a deceleration when moving into the strong field region in the other half of its gyratory orbit. Since we know that $\mathbf{v}_\nabla$ lies in the plane perpendicular to the magnetic field, we can follow the same line as on p. 16 and take the cross-product with $\mathbf{B}_0/B_0^2$. Then we find

$$\mathbf{v}_\nabla = \frac{1}{B_0^2}\left\langle(\mathbf{v}_g \times (\mathbf{r} \cdot \nabla)\mathbf{B}_0) \times \mathbf{B}_0\right\rangle \tag{2.31}$$

where the angle brackets denote averaging over one gyroperiod. Assuming $\mathbf{B}$ to vary only in the x direction, $\mathbf{B}_0 = B_0(x)\,\hat{\mathbf{e}}_z$, we obtain

$$\mathbf{v}_\nabla = -\frac{1}{B_0}\left\langle \mathbf{v}_g x \frac{dB_0}{dx}\right\rangle \tag{2.32}$$

Replacing $\mathbf{v}_g$ and x by the expressions given in Eq. (2.13), we get

$$\begin{aligned} v_{\nabla x} &= -\frac{v_\perp r_g}{B_0}\left\langle \sin \omega_g t \cos \omega_g t \frac{dB_0}{dx}\right\rangle \\ v_{\nabla y} &= \frac{v_\perp r_g}{B_0}\left\langle \sin^2 \omega_g t \frac{dB_0}{dx}\right\rangle \end{aligned} \tag{2.33}$$

Taking the gyroperiod average, $v_{\nabla x}$ will vanish since it contains the product of sine and cosine terms. Averaging over the $\sin^2$ term yields a factor $1/2$. Thus the drift will have only a y component

$$\mathbf{v}_\nabla = \pm \frac{v_\perp r_g}{2B_0} \frac{\partial B_0}{\partial x} \hat{\mathbf{e}}_y \tag{2.34}$$

where the direction of the motion depends on the sign of the charge. Since the direction of the magnetic field gradient was chosen arbitrarily, this can be generalized

$$\boxed{\mathbf{v}_\nabla = \frac{m v_\perp^2}{2q B^3} (\mathbf{B} \times \nabla B)} \tag{2.35}$$

showing that a magnetic field gradient leads to a *gradient drift* perpendicular to both the magnetic field and its gradient, as sketched in Fig. 2.4.

Equation (2.35) shows that ions and electrons drift into opposite directions and that, furthermore, the gradient drift velocity is proportional to the perpendicular energy of the particle, $W_\perp = \frac{1}{2} m v_\perp^2$. More energetic particles drift faster, since they have a larger gyroradius and experience more of the inhomogeneity of the field.

As in the case of the polarization drift, the opposite drift directions of electrons and ions lead to a transverse current. This *gradient drift current* has the form

$$\mathbf{j}_\nabla = n_e e \left(\mathbf{v}_{\nabla i} - \mathbf{v}_{\nabla e} \right) = \frac{n_e (\mu_i + \mu_e)}{B^2} \mathbf{B} \times \nabla B \tag{2.36}$$

when using the *magnetic moment*

$$\boxed{\mu = \frac{m v_\perp^2}{2B} = \frac{W_\perp}{B}} \tag{2.37}$$

to describe the ratio between perpendicular particle energy and magnetic field.

General Force Drift

By using Eq. (2.1) to replace $\mathbf{E}$ in Eq. (2.19) by $\mathbf{F}/q$, we get a more general form of guiding center drift, which is valid not only for the Coulomb force, but for any force acting on a charged particle in a magnetic field

$$\boxed{\mathbf{v}_F = \frac{1}{\omega_g} \left(\frac{\mathbf{F}}{m} \times \frac{\mathbf{B}}{B} \right)} \tag{2.38}$$

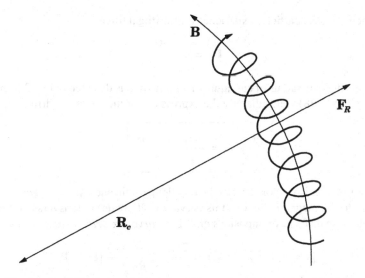

Fig. 2.5. Centrifugal force felt by a particle moving along a curved field line.

All particle drifts can be described this way by using the appropriate force terms, whenever the drift velocity of the particle is much smaller than its gyrovelocity. For the gradient, the polarization, and the gravitational drift, these forces can be written as

$$\mathbf{F}_\nabla = -\mu \nabla B \tag{2.39}$$

$$\mathbf{F}_P = -m \frac{d\mathbf{E}}{dt} \tag{2.40}$$

$$\mathbf{F}_G = -m \, \mathbf{g} \tag{2.41}$$

where $\mathbf{F}_\nabla$ denotes the gradient and $\mathbf{F}_P$ the polarization force. $\mathbf{F}_G$ is the gravitational force, which is typically much weaker than the other forces. Except for processes near the solar surface, it can usually be neglected.

Equation (2.38) shows that all drifts generated by a force other than the Coulomb force depend on the sign of the charge, since ω_g carries this sign. Hence, for these drifts ions and electrons run into opposite directions, creating a transverse current. Moreover, these drifts depend on the mass of the charge carriers and thus the drift velocities are typically quite different for ions and electrons.

Curvature Drift

The gradient drift is only one component of the particle drift in an inhomogeneous magnetic field. When the field lines are curved, a *curvature drift* appears. Due to their parallel velocity, $v_\parallel$, the particles experience a centrifugal force

$$\mathbf{F}_R = m v_\parallel^2 \frac{\mathbf{R}_c}{R_c^2} \tag{2.42}$$

where $\mathbf{R}_c$ is the local radius of curvature. This situation is depicted in Fig. 2.5. Inserting Eq. (2.42) into (2.38) yields directly the expression for the curvature drift

$$\mathbf{v}_R = \frac{m v_\parallel^2}{q} \frac{\mathbf{R}_c \times \mathbf{B}}{R_c^2 B^2} \tag{2.43}$$

The curvature drift is proportional to the parallel particle energy, $W_\parallel = \frac{1}{2} m v_\parallel^2$, and perpendicular to the magnetic field and its curvature. It creates a transverse current since ion and electron drifts have opposite signs. The *curvature drift current* has the form

$$\mathbf{j}_R = n_e e \, (\mathbf{v}_{Ri} - \mathbf{v}_{Re}) = \frac{2 n_e (W_{i\parallel} + W_{e\parallel})}{R_c^2 B^2} \, (\mathbf{R}_c \times \mathbf{B}) \tag{2.44}$$

As the associated drift, the curvature current flows perpendicular to both the curvature of the magnetic field and the magnetic field itself.

In a cylindrically symmetric field, it turns out that $-\nabla B = (B/R_c^2) \, \mathbf{R}_c$. Thus we may add the gradient to the curvature drift to obtain the total magnetic drift

$$\mathbf{v}_B = \mathbf{v}_R + \mathbf{v}_\nabla = (v_\parallel^2 + \tfrac{1}{2} v_\perp^2) \frac{\mathbf{B} \times \nabla B}{\omega_g B^2} \tag{2.45}$$

It is the transverse current associated with this full magnetic drift which creates the magnetospheric ring current mentioned in Sec. 1.3 and further detailed in Secs. 3.2–3.5.

2.5. Adiabatic Invariants

In the preceding section we encountered the magnetic moment $\mu = W_\perp / B$ of a particle. This quantity has been treated as a characteristic constant of the particle. Such quantities are called *adiabatic invariants*. Adiabatic invariants are not absolute constants like total energy or total momentum, but may change both in space and time. There essence is, however, that they change very slowly compared with some typical periodicities of the particle motion.

For particles in electromagnetic fields, adiabatic invariants are associated with each type of motion the particle can perform. The magnetic moment, μ, is associated with the gyration about the magnetic field, the longitudinal invariant, J, with the longitudinal motion along the magnetic field, and the third invariant, Φ, with the perpendicular drift.

Whenever these motions are periodic and changes in the system have angular frequencies much smaller than the oscillation frequency of the particle corresponding to one of the above motions, the action integral

$$J_i = \oint p_i dq_i \tag{2.46}$$

is a constant of the motion and describes an adiabatic invariant. The pair of variables (p_i, q_i) are the generalized momentum and coordinate of Hamiltonian mechanics, and the integration has do be done over one full cycle of q_i.

Magnetic Moment

To demonstrate that the magnetic moment of a particle does not change when the particle moves into stronger or weaker magnetic fields, we consider the energy conservation equation

$$W = W_{\parallel} + W_{\perp} \tag{2.47}$$

Since W is a constant in the absence of electric fields, its time derivative vanishes

$$\frac{dW_{\parallel}}{dt} + \frac{dW_{\perp}}{dt} = 0 \tag{2.48}$$

For the transverse energy, we can use Eq. (2.37) and obtain

$$\frac{dW_{\perp}}{dt} = \mu \frac{dB}{dt} + B \frac{d\mu}{dt} \tag{2.49}$$

Here $dB/dt = v_{\parallel} dB/ds$ is the variation of the magnetic field as seen by the particle along its guiding center trajectory. The magnetic field itself is assumed to be constant. The parallel particle energy can be derived from the parallel component of the gradient force in Eq. (2.39) which gives the parallel equation of motion

$$m \frac{dv_{\parallel}}{dt} = -\mu \nabla_{\parallel} B = -\mu \frac{dB}{ds} \tag{2.50}$$

Multiplying the left-hand side of Eq. (2.50) with $v_{\parallel}$ and the right-hand side with its equivalent, ds/dt, one finds for the time derivative of the parallel energy

$$\frac{dW_{\parallel}}{dt} = -\mu \frac{dB}{dt} \tag{2.51}$$

Adding Eqs. (2.49) and (2.51) and observing Eq. (2.48), we obtain

$$\frac{dW_{\parallel}}{dt} + \frac{dW_{\perp}}{dt} = B \frac{d\mu}{dt} = 0 \tag{2.52}$$

which yields immediately that the magnetic moment is an invariant of the particle motion. The magnetic moment is not affected by small changes in the cyclotron frequency or the gyroradius which occur when the magnetic field changes along the particle path.

Up to now we have neglected electric fields, which can accelerate particles. Thus we have neglected temporal changes of the magnetic field, since $\partial \mathbf{B}/\partial t = -\nabla \times \mathbf{E}$. However, when the magnetic field fluctuations are slow enough, the magnetic moment is still conserved. This can be demonstrated by considering the change in perpendicular particle energy caused by an electric field. This change is calculated by taking the dot product of Eq. (2.7) with $\mathbf{v}_\perp$

$$\frac{dW_\perp}{dt} = q \, (\mathbf{E} \cdot \mathbf{v}_\perp) \tag{2.53}$$

The gain in energy over one gyration is obtained by integrating over the gyroperiod

$$\Delta W_\perp = q \int_0^{2\pi/\omega_g} (\mathbf{E} \cdot \mathbf{v}_\perp) \, dt \tag{2.54}$$

If the field changes slowly, the particle orbit is closed and we can replace the time integral by a line integral over the unperturbed orbit. Using Stokes' theorem (see App. A.4) and Maxwell's equations, we obtain

$$\Delta W_\perp = q \oint_C \mathbf{E} \cdot d\mathbf{s} = q \int_A (\nabla \times \mathbf{E}) \cdot d\mathbf{A} = -q \int_A \frac{\partial \mathbf{B}}{\partial t} \cdot d\mathbf{A} \tag{2.55}$$

where $d\mathbf{s} = \mathbf{t}\,ds$ is the product of a line element, ds, of the closed gyratory orbit, C, with the line element's tangent vector, $\mathbf{t}$, and $d\mathbf{A} = \mathbf{n}\,dA$ is the product of a surface element, dA of the plane, A, enclosed by the orbit and the surface element's normal vector, $\mathbf{n}$. For changes in the field much slower than the gyroperiod, $\partial \mathbf{B}/\partial t$ can be replaced by $\omega_g \Delta B / 2\pi$, with ΔB the average change during one gyroperiod

$$\Delta W_\perp = \tfrac{1}{2} q \omega_g r_g^2 \Delta B = \mu \Delta B \tag{2.56}$$

Here we have inserted the expression for ω_g and r_g from Eqs. (2.12) and (2.14) and used the definition of μ in Eq. (2.37). On the other hand, in Eq. (2.49) the change in perpendicular energy is given by

$$\Delta W_\perp = \mu \Delta B + B \Delta \mu \tag{2.57}$$

Comparing the last two equations, we find again that

$$\Delta \mu = 0 \tag{2.58}$$

demonstrating that in the approximation of slowly variable fields the magnetic moment is invariant even when the particles are accelerated in induction electric fields.

Similarly, slow temporal variations of the electric field do not violate the invariance of the magnetic moment, since they will produce only second order time variations of the magnetic field, $\partial^2 B/\partial t^2$, on the right-hand side of Eq. (2.55), which can be neglected. For all slow variations the magnetic moment is a constant of motion. Changes in the fields merely lead to the different types of particle drifts but conserve μ.

From the adiabatic invariance of the magnetic moment, it follows that also the magnetic flux Φ_μ through the surface encircled by the gyrating particle does not change. This flux is given by $\Phi_\mu = B\pi r_g^2$ (see App. A.5), or inserting the expression for r_g from Eq. (2.14) and using the definition of μ in Eq. (2.37)

$$\Phi_\mu = \frac{2\pi m}{q^2}\mu = \text{const} \tag{2.59}$$

Hence, as a particle moves into a region of stronger magnetic field, the gyroradius of the particle will get increasingly smaller, so that the magnetic flux encircled by the orbit remains constant.

Magnetic Mirror

Let us follow the guiding center of a particle moving along an inhomogeneous magnetic field by considering its magnetic moment

$$\mu = \frac{mv^2 \sin^2\alpha}{2B} \tag{2.60}$$

where we have replaced $v_\perp$ by $v\sin\alpha$, using the pitch angle defined in Eq. (2.15). Since the magnetic moment is invariant and the total energy is a constant of motion, only the pitch angle can change when the magnetic field increases or decreases along the guiding center trajectory. The above equation also shows that the pitch angles of a particle at different locations are directly related to the magnetic field strengths at those locations according to

$$\frac{\sin^2\alpha_2}{\sin^2\alpha_1} = \frac{B_2}{B_1} \tag{2.61}$$

Thus knowing the pitch angle of a particle at one location, we can calculate this quantity at all other locations.

In a converging magnetic field geometry, a particle moving into regions of stronger fields will have its pitch angle increase and, therefore, have its transverse energy $W_\perp$ increase at the expense of its parallel energy $W_\parallel$. If B_m is a point along the field line where the pitch angle reaches $\alpha = 90°$, the particle is reflected from this *mirror point*. Here, all of the particle energy is in $W_\perp$ and the particle cannot penetrate any further,

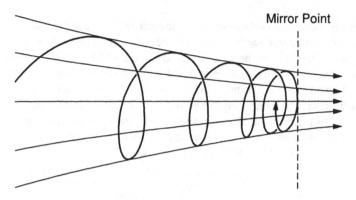

Fig. 2.6. Ion orbit and reflection in a converging magnetic field.

but is pushed back by the parallel component of the gradient force given in Eq. (2.39), the so-called *mirror force*, $-\mu \nabla_{\parallel} B$. This mirroring of a particle is visualized in Fig. 2.6.

In a symmetric magnetic field geometry with a minimum field in the middle and converging magnetic field lines on both sides, like in a dipole field, a particle may bounce back and forth between its two mirror points and become *trapped*. In this case we can describe the particle's pitch angle at a specific location by

$$\sin \alpha = \left(\frac{B}{B_m} \right)^{1/2} \tag{2.62}$$

i.e., the ratio of the field strengths at that location and at its mirror point.

Adiabatic Heating

The mirror effect described above is a consequence of the invariance of the magnetic moment when particles move along magnetic field lines. Conservation of the magnetic moment has also an important effect when particles drift across field lines. Consider a particle moving along its drift path from a region with a magnetic field strength B_1 into a region of increasing field strength B_2. Since the magnetic moment is conserved, we obtain

$$\frac{W_{\perp 2}}{W_{\perp 1}} = \frac{B_2}{B_1} \tag{2.63}$$

with $W_{\perp 2} > W_{\perp 1}$. However, in contrast to the mirror case, the particle will not bounce back and the perpendicular energy increase will remain.

The convective transport of particles into stronger magnetic fields therefore ends up with a gain of energy in the transverse direction. The work required for this process is of course taken from the drift motion which transports the particle into the stronger field. This type of particle energization is called *adiabatic heating* and is some kind of *betatron acceleration*. It increases the perpendicular energy of the particle without affecting its parallel energy. This results in an anisotropy of the particle energy. However, since magnetic fields may vary also in the parallel direction, we will discuss the generation of such an anisotropy in a more general way after introducing the second adiabatic invariant.

Longitudinal Invariant

If the field has a mirror symmetry where the field lines converge on both sides as in a dipole field, there is the possibility for a second adiabatic invariant, J. A particle moving in such a converging field will be reflected from the region of strong magnetic field and can oscillate in the field at a certain *bounce frequency*, ω_b. The *longitudinal invariant* is defined by

$$J = \oint mv_\parallel ds \tag{2.64}$$

where $v_\parallel$ is the parallel particle velocity, ds is an element of the guiding center path and the integral is taken over a full oscillation between the mirror points.

For electromagnetic variations with frequencies $\omega \ll \omega_b$, the longitudinal invariant is a constant, irrespective of weak changes in the path of the particle and its mirror points due to slow changes in the fields.

Energy Anisotropy

Invariance of the magnetic moment in magnetic fields which change in the perpendicular direction may lead to an increase in the perpendicular energy of the particle. When the magnetic field also varies in the direction parallel to the field lines, conservation of the longitudinal invariant implies that the parallel energy of the particle will also change during the combined drift and bounce motion of the particle along and across the field.

Let us define the total length of the field line between the two mirror points of the particle as ℓ, and the average parallel velocity along the field line as $\langle v_\parallel \rangle$. In terms of these quantities the longitudinal invariant can be expressed as

$$J = \oint mv_\parallel ds = 2m\ell\langle v_\parallel \rangle \tag{2.65}$$

During its drift motion from weaker into stronger fields, the particle necessarily moves from one field line of length ℓ_1 onto another field line of length ℓ_2. At the same time its average parallel velocity changes from $\langle v_\parallel \rangle_1$ to $\langle v_\parallel \rangle_2$. Conservation of the longitudinal invariant then implies that the averaged parallel energy, $\langle W_\parallel \rangle$, changes according to

$$\frac{\langle W_\parallel \rangle_2}{\langle W_\parallel \rangle_1} = \frac{\ell_1^2}{\ell_2^2}$$ (2.66)

If the length of the bounce path decreases, the parallel energy of the particle increases. This is the basic element of *Fermi acceleration*.

This result can be combined with the simultaneous increase in the perpendicular energy from Eq. (2.63) to determine the anisotropy in the energy attained by the particle. Let us define this anisotropy as

$$A_W = \frac{\langle W_\perp \rangle}{\langle W_\parallel \rangle}$$ (2.67)

Then the *energy anisotropy* of the particle changes according to

$$\boxed{\frac{A_{W2}}{A_{W1}} = \frac{B_2}{B_1}\frac{\ell_2^2}{\ell_1^2}}$$ (2.68)

which shows that the anisotropy increases if only the square of the field line length decreases less than the magnetic field increase.

Drift Invariant

The third invariant, Φ, is simply the conserved magnetic flux encircled by the periodic orbit of a particle trapped in an axisymmetric mirror magnetic field configuration when it performs closed *drift shell* orbits around the magnetic field axis. This *drift invariant* can be written as

$$\Phi = \oint v_d r d\psi$$ (2.69)

where v_d is the sum of all perpendicular drift velocities, ψ is the azimuthal angle, and the integration must be taken over a full circular drift path of the particle.

Whenever the typical frequency of the electromagnetic fields is much smaller than the drift frequency, $\omega \ll \omega_d$, Φ is invariant and essentially equal to the magnetic flux enclosed by the orbit. This can be written like in Eq. (2.59) as

$$\boxed{\Phi = \frac{2\pi m}{q^2} M = \text{const}}$$ (2.70)

where M is the magnetic moment of the axisymmetric field.

Violation of Invariance

So far we have considered variations and motions which conserve the magnetic moment, the longitudinal invariant, and the drift invariant. In nature, however, all the fields may vary in such a way that the adiabatic invariance of one or the other invariant is violated.

Time variations faster than the gyrofrequency of the particle with frequencies $\omega > \omega_g$ violate the first adiabatic invariant μ, the magnetic moment of the particle. These are high-frequency variations in either the magnetic or electric field. In this case, the concept of a gyratory orbit about a guiding center becomes obsolete and the full particle motion must be considered.

On the other hand, for frequencies $\omega_g > \omega > \omega_b$ the magnetic moment μ is conserved and the guiding center approximation is useful for the drift motion. However, under these conditions the longitudinal invariant is not conserved but violated, and the particle motion cannot be described anymore as a simple oscillation along the magnetic field between mirror points.

Finally, for frequencies $(\omega_g, \omega_b) > \omega > \omega_d$ the first two invariants will be conserved while the drift invariant becomes violated. The particles gyrate and bounce but diffuse across drift shells under the influence of the variation in the magnetic field. All these cases are realized in nature. Some of them we will encounter later.

Adiabatic invariants are not only violated when the fields vary in time but also when they abruptly change over a length scale $L < r_a$ shorter than the characteristic radius r_a of the periodic motion related to the adiabatic invariant. This can be proven, for instance, for the gyration of a particle across a magnetic field gradient which is shorter than the particle gyroradius r_g. In such a case, $r_g/L > 1$. Using Eq. (2.14) to write the perpendicular velocity of the particle as

$$v_\perp = \omega_g r_g \tag{2.71}$$

and dividing both sides of this expression by L, the gradient length of the spatial change, one finds

$$\omega = \frac{v_\perp}{L} = \omega_g \frac{r_g}{L} > \omega_g \tag{2.72}$$

Hence, for the gyrating particle the effective frequency of change in the field, $\omega > \omega_g$, is higher than its gyrofrequency, and the magnetic moment of a particle gyrating in a magnetic field which varies strongly over a length of the order $L < r_g$ is not an adiabatic invariant. Similar arguments can be applied also to the two remaining invariants.

Concluding Remarks

In concluding this chapter, a word of caution is mandatory. The guiding center theory assumes that the electromagnetic fields are prescribed. It can thus be used in geophysical plasmas where the external field is strong and will not be changed much by the motion of the particles themselves. It should not be used in weak field regions where the field will be substantially changed by the particle motion and where thus the concept of a guiding center looses its meaning.

Summary of Guiding Center Drifts

For quick reference, we summarize the expressions of the guiding center drifts and the associated transverse currents.

E×B Drift:
$$\mathbf{v}_E = \frac{\mathbf{E} \times \mathbf{B}}{B^2}$$

Polarization Drift:
$$\mathbf{v}_P = \frac{1}{\omega_g B} \frac{d\mathbf{E}_\perp}{dt} \qquad \mathbf{j}_P = \frac{n_e(m_i + m_e)}{B^2} \frac{d\mathbf{E}_\perp}{dt}$$

Gradient Drift:
$$\mathbf{v}_\nabla = \frac{m v_\perp^2}{2q B^3} (\mathbf{B} \times \nabla B) \qquad \mathbf{j}_\nabla = \frac{n_e(\mu_i + \mu_e)}{B^2} (\mathbf{B} \times \nabla B)$$

Curvature Drift:
$$\mathbf{v}_R = \frac{m v_\parallel^2}{q R_c^2 B^2} (\mathbf{R}_c \times \mathbf{B}) \qquad \mathbf{j}_R = \frac{2n_e(W_{i\parallel} + W_{e\parallel})}{R_c^2 B^2} (\mathbf{R}_c \times \mathbf{B})$$

Further Reading

For those readers who want to know more about the guiding center approach and to see the actual proof of the second and third invariant, we recommend reading one of the two monographs listed below.

[1] H. Alfvén and C. G. Fälthammar, *Cosmical Electrodynamics, Fundamental Principles* (Clarendon Press, Oxford, 1963).

[2] T. G. Northrop, *The Adiabatic Motion of Charged Particles* (Interscience Publishers, New York, 1963).

3. Trapped Particles

A *dipole magnetic field* has a field strength minimum at the equator and converging field lines in both hemispheres. As we have learned in Sec. 2.5, in such a configuration particles will be trapped and bounce back and forth between their mirror points in the northern and southern hemispheres (see Fig. 3.1). In the case of the terrestrial magnetic field, which can be approximated by a dipole field inside of about 6 R_E, these trapped populations are the energetic particles in the *radiation belts* (see Fig. 1.4). Typical energies of the ions in this region range between 3 and 300 keV, while the electrons have energies about an order of magnitude lower.

The particles do not only gyrate and bounce, but undergo a slow azimuthal drift (see Fig. 3.1). This drift is an effect of the gradient and curvature of the dipole magnetic field as described in Eq. (2.45) and is oppositely directed for ions and electrons. The ions drift westward while the electrons move eastward around the Earth. It is the current associated with this drift that constitutes the *ring current*.

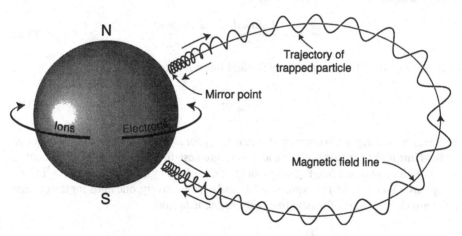

Fig. 3.1. Trajectories of particles trapped on closed field lines.

3.1. Dipole Field

At distances not too far from the Earth's surface, the geomagnetic field can be approximated by a dipole field. Introducing the *Earth's dipole moment*, $M_E = 8.05 \cdot 10^{22}$ Am^2, and choosing a spherical coordinate system with radius, r, and magnetic latitude, λ, we can write

$$\mathbf{B} = \frac{\mu_0}{4\pi} \frac{M_E}{r^3} (-2 \sin \lambda \hat{\mathbf{e}}_r + \cos \lambda \hat{\mathbf{e}}_\lambda) \tag{3.1}$$

since a dipole field is symmetric about the azimuth. Here $\hat{\mathbf{e}}_r$ and $\hat{\mathbf{e}}_\lambda$ are unit vectors in the r and λ directions. The strength of the dipole field at a specific location can easily be obtained as

$$\boxed{B = \frac{\mu_0}{4\pi} \frac{M_E}{r^3} (1 + 3 \sin^2 \lambda)^{1/2}} \tag{3.2}$$

Typical dipolar field lines are shown in Fig. 3.2. In order to construct the field lines in this figure, and for many other applications, one needs to know the *field line equation*, $r = f(\lambda)$. If $d\mathbf{s}$ is an arc element, the lines of force are defined by the differential equation

$$d\mathbf{s} \times \mathbf{B} = 0 \tag{3.3}$$

since the magnetic field vector is always tangent to the lines of force. For an axisymmetric field, this reduces to

$$\frac{dr}{B_r} = \frac{r d\lambda}{B_\lambda} \tag{3.4}$$

Using the dipole field Eq. (3.1) we obtain

$$\frac{dr}{r} = -\frac{2 \sin \lambda \, d\lambda}{\cos \lambda} = \frac{2 \, d(\cos \lambda)}{\cos \lambda} \tag{3.5}$$

Integration of Eq. (3.5) yields the dipole field line equation

$$\boxed{r = r_{eq} \cos^2 \lambda} \tag{3.6}$$

where r_{eq} is the integration constant. Since $r = r_{eq}$ for $\lambda = 0$, r_{eq} is the radial distance to the field line in the equatorial plane and thus its greatest distance from the Earth's center.

The element of arc length along a field line is given by $(ds)^2 = (dr)^2 + r^2(d\lambda)^2$. Using Eqs. (3.4) and (3.6) to express $dr/d\lambda$ and r, respectively, one finds for the change of the arc element along the field line with magnetic latitude

$$\frac{ds}{d\lambda} = r_{eq} \cos \lambda (1 + 3 \sin^2 \lambda)^{1/2} \tag{3.7}$$

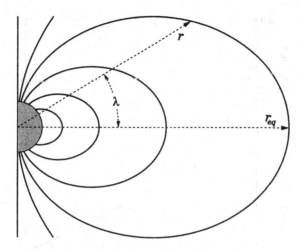

Fig. 3.2. Dipolar magnetic field lines.

By integrating this equation one can calculate the length of a field line with a given equatorial distance.

Often it is convenient to use the radius of the Earth, R_E, as the unit of distance and to introduce the *L-shell parameter* or *L-value*, $L = r_{eq}/R_E$. Using the equatorial magnetic field on the Earth's surface, $B_E = \mu_0 M_E/(4\pi R_E^3) = 3.11 \cdot 10^{-5}$ T, and inserting the field line equation (3.6), we can rewrite Eq. (3.2)

$$B(\lambda, L) = \frac{B_E}{L^3}\frac{(1 + 3\sin^2 \lambda)^{1/2}}{\cos^6 \lambda} \tag{3.8}$$

Inserting L into Eq. (3.6) yields another useful result

$$\cos^2 \lambda_E = L^{-1} \tag{3.9}$$

namely the latitude, λ_E, where a field line with a given L-value or equatorial plane distance intersects the Earth's surface.

3.2. Bounce Motion

The most prominent motion of trapped particles is their bounce motion between the mirror points. The actual trajectory of a bouncing particle is characterized by its pitch angle as defined in Eq. (2.15).

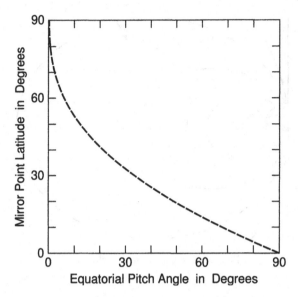

Fig. 3.3. Dipole latitudes of mirror points.

Equatorial Pitch Angle

From Eq. (2.62) we know that we can determine the pitch angle of a particle in a mirror field geometry anywhere along the field line from the ratio between the magnetic field at that location and the magnetic field at the particle's mirror point. A particular point along a field line is its intersection with the equatorial plane, where the field strength is minimum with $B_{eq} = B_E/L^3$. Inserting Eq. (3.8) into Eq. (2.62), we obtain for the *equatorial pitch angle*, α_{eq}

$$\sin^2 \alpha_{eq} = \frac{B_{eq}}{B_m} = \frac{\cos^6 \lambda_m}{(1 + 3 \sin^2 \lambda_m)^{1/2}} \qquad (3.10)$$

where λ_m is the magnetic latitude of the particle's mirror point. Equation (3.10) shows that the equatorial pitch angle of a particle depends only on the latitude of its mirror point and not on the equatorial distance of its field line or, equivalently, its L-value. Turning the argument around, the latitude of a particle's mirror point depends only on its equatorial pitch angle and is independent of the L-value.

Figure 3.3 shows the dependence of the magnetic latitude of a particle's mirror point on the equatorial pitch angle of that particle. Particles with small equatorial pitch angles have large parallel velocities and their mirror points are at high latitudes, close

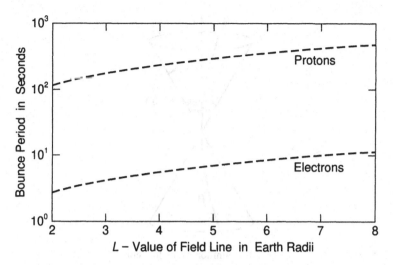

Fig. 3.4. Bounce period for 1-keV electrons and protons with $\alpha_{eq} = 30°$.

to the Earth. With increasing equatorial pitch angles, the mirror points move to more equatorial latitudes and the particles mirror close to the equatorial plane.

Bounce Period

The *bounce period*, τ_b, is the time it takes a particle to move from the equatorial plane to one mirror point, then to the other and back to the equatorial plane. It can be calculated by integrating $ds/v_{\|}$ over a full bounce path along the field line

$$\tau_b = 4 \int_0^{\lambda_m} \frac{ds}{v_{\|}} = 4 \int_0^{\lambda_m} \frac{ds}{d\lambda} \frac{d\lambda}{v_{\|}} \qquad (3.11)$$

We can make use of Eqs. (2.61) and (3.10) to replace $v_{\|} = v [1 - (B/B_{eq}) \sin^2 \alpha_{eq}]^{1/2}$. Then Eqs. (3.7) and (3.8) can be used to replace $ds/d\lambda$ and the magnetic field ratio

$$\tau_b = 4 \frac{r_{eq}}{v} \int_0^{\lambda_m} \cos \lambda \, (1 + 3 \sin^2 \lambda)^{1/2} \left[1 - \sin^2 \alpha_{eq} \frac{(1 + 3 \sin^2 \lambda)^{1/2}}{\cos^6 \lambda} \right]^{-1/2} d\lambda \qquad (3.12)$$

The integral can be solved numerically, but is usually approximated by

$$\Gamma_\alpha \approx 1.30 - 0.56 \sin \alpha_{eq} \qquad (3.13)$$

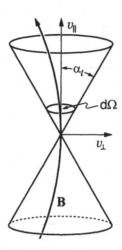

Fig. 3.5. Definition of the loss cone.

Expressing r_{eq} and v in terms of L and particle energy, W, we obtain

$$\tau_b \approx \frac{L R_E}{(W/m)^{1/2}} (3.7 - 1.6 \sin \alpha_{eq})$$

(3.14)

The bounce period depends only weakly on the equatorial pitch angle, since particles with a smaller pitch angle have a larger velocity along the field line, yet have a longer way to their mirror points (see Fig. 3.3). Naturally, it is longer for longer field lines and shorter for particles with higher energies.

Figure 3.4 gives the bounce periods for 1-keV electrons and protons with $\alpha_{eq} = 30°$. The electrons, being much lighter, have typical bounce periods of some seconds, while the heavier protons take a few minutes to complete a full bounce cycle. As explained in Sec. 2.5, the bounce period must be much shorter than any fluctuation of the magnetic field for the integral $J = \oint m v_{\parallel} ds$ to be an adiabatic invariant. For keV electrons this is usually true, but for keV ions the second adiabatic invariant may be violated by, for example, geomagnetic pulsations which have periods of the order of the ion bounce period.

Loss Cone

Even when the longitudinal invariant is conserved, not all particles are actually trapped. If a particle's mirror point lies deep in the atmosphere, it will collide too often with neutral particles (see Secs. 4.1 and 5.2) and, hence, will be absorbed by the atmosphere.

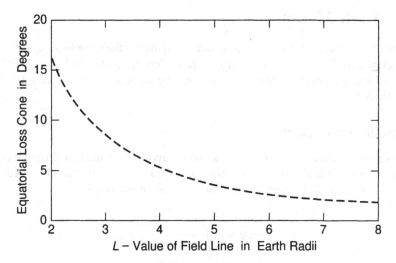

Fig. 3.6. Equatorial loss cone for different L-values.

The mirror point altitudes where particles are lost by collisions are those below about 100 km. For simplicity, we will use zero altitude since the magnetic field strength and mirror point latitude differ only by a few percent between the Earth's surface and the lower ionosphere. Under this assumption we can use Eq. (3.10) to define an *equatorial loss cone*

$$\sin^2 \alpha_\ell = \frac{B_{eq}}{B_E} = \frac{\cos^6 \lambda_E}{(1 + 3 \sin^2 \lambda_E)^{1/2}} \qquad (3.15)$$

Figure 3.5 shows the geometry of such an equatorial loss cone. All particles with equatorial pitch angles $\alpha < \alpha_\ell$ within the solid angle $d\Omega$ will be lost in the atmosphere. Since particles with $\alpha > 180° - \alpha_\ell$ will be lost in the other hemisphere, we get a double cone structure. Using Eq. (3.9) to express λ_E in terms of the L-value, we obtain

$$\boxed{\sin^2 \alpha_\ell = \left(4L^6 - 3L^5\right)^{-1/2}} \qquad (3.16)$$

The width of the loss cone is independent from the charge, the mass, or the energy of the particles, but is purely a function of the field line radius. As can be seen in Fig. 3.6, the equatorial loss cone is typically rather small for equatorial distances of more than 3 R_E. At geostationary orbit (6.6 R_E), the loss cone is less than 3° wide.

3.3.   Drift Motion

A particle in a dipole field will gyrate, bounce, and drift at the same time. Thus one has to integrate over the former two motions if one is interested in the much slower drift motion. As long as we neglect electric fields, a particle will experience a purely azimuthal magnetic drift, v_d.

Magnetic Drift Velocity

The equatorial angular drift velocity is found by dividing the angular drift that occurs during one bounce cycle, $\Delta\psi$, by the bounce period. $\Delta\psi$ is computed by integrating $v_d/r \cos\lambda$ over one full bounce cycle. With $dt = ds/v_\parallel$ we obtain

$$\Delta\psi = 4\int\limits_0^{\lambda_m} \frac{v_d}{r\cos\lambda}\frac{ds}{v_\parallel} \tag{3.17}$$

We can replace the magnetic drift velocity by Eq. (2.45) and follow a similar procedure as used for the derivation of Eq. (3.12) to obtain the angular drift velocity, $2\pi\langle\Omega_d\rangle = \Delta\psi/\tau_b$, averaged over one full bounce cycle

$$\langle\Omega_d\rangle = \frac{3LW}{\pi q B_E R_E^2 \Gamma_\alpha} \int\limits_0^{\lambda_m} \Psi(\alpha_{eq},\lambda)\,d\lambda \tag{3.18}$$

where W is the particle energy and Γ_α is the integral introduced in Eqs. (3.12) and (3.13). The function Ψ is a rather complicated expression and the integral has to be solved numerically, but an approximate solution for the ratio of the integral over Ψ and Γ_α is

$$\Gamma_\alpha^{-1}\int\limits_0^{\lambda_m} \Psi(\alpha_{eq},\lambda)\,d\lambda \approx 0.35 + 0.15\sin\alpha_{eq} \tag{3.19}$$

Hence, we get an approximation for the average *drift period*, $\langle\tau_d\rangle = \langle\Omega_d\rangle^{-1}$

$$\langle\tau_d\rangle \approx \frac{\pi q B_E R_E^2}{3LW}(0.35 + 0.15\sin\alpha_{eq})^{-1} \tag{3.20}$$

and for the average drift velocity, $\langle v_d\rangle = 2\pi L R_E/\tau_b$

$$\boxed{\langle v_d\rangle \approx \frac{6L^2 W}{q B_E R_E}(0.35 + 0.15\sin\alpha_{eq})} \tag{3.21}$$

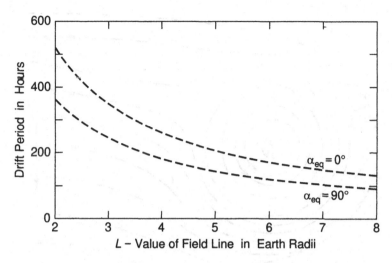

Fig. 3.7. Drift period for 1-keV particles with different equatorial pitch angles.

Both, drift period and drift velocity depend on particle charge, particle energy, and L-value, but not explicitly on the mass of the particle. Hence, electrons and protons with equal energies drift around the Earth with the same velocity, only in opposite directions. Since the drift velocity scales with L^2, the drift period is actually shorter on more distant L-shells. This is in contrast to the normal Keplerian motion, where more distant particles have lower azimuthal velocities. Figure 3.7, which gives a graphic representation of Eq. (3.20), clearly shows this effect. It also shows that the drift period (and the drift velocity) depend only weakly on the equatorial pitch angle, as was the case for the bounce period.

As can be seen in Fig. 3.7, the drift period for a 1-keV particle is of the order of several days. Since the magnetospheric field changes more frequently, it is very unlikely that the third adiabatic invariant, Φ, is conserved for typical ring current particles in the 10 keV energy range. Many ring current particles undergo *radial diffusion* across L-shells or do not complete a full revolution around the Earth. Only the most energetic ring current particles in the MeV range have a drift period short enough to perform closed orbits.

Electric Drift

However, even if the third invariant were not violated, particles in the outer ring current would not perform closed orbits. As we will see in the next section, the solar wind generates an electric field inside the magnetosphere, which is directed from dawn to dusk in the equatorial plane. Thus the particles will experience a sunward $E \times B$ drift. We can use Eqs. (2.19) and (3.8) to get an expression for the equatorial electric drift velocity,

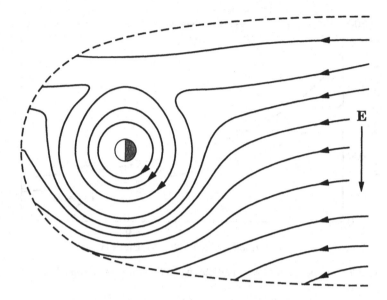

Fig. 3.8. Energetic ion drift paths in the equatorial plane.

v_E, due to a uniform equatorial transverse electric field, E_{eq}, obtaining

$$v_E = \frac{E_{eq}}{B_{eq}} = \frac{E_{eq} L^3}{B_E} \tag{3.22}$$

Electrons and ions with mirror points close to the equator will drift sunward with this velocity, in addition to their magnetic drift.

Because the magnetic drift of positive ions is directed westward, the two drifts are oppositely directed on the dawn side. For electrons, this holds on the dusk side. Since the magnetic drift velocity scales with L^2 while the E×B velocity scales with L^3, the electric drift will typically overcome the magnetic drift outside some radial distance. Hence, for the combined electric and magnetic drift we have the situation sketched in Fig. 3.8. Close to the Earth, the magnetic drift forces prevail and we have a *symmetric ring current*. Far out, the particle trajectories are dominated by the E×B drift. In the intermediate region we get a *partial ring current*, caused by the deflection of sunward drifting particles around the Earth due to the gradient and curvature forces.

As we will see in the next section, the actual magnetospheric electric field is not quite uniform, but the energetic particle trajectories in a realistic electric field are similar to those shown in Fig. 3.8.

3.4. Sources and Sinks

Figure 3.8 also indicates that the source of the ring current is the tail plasma sheet (see Fig. 1.4). The particles are brought in from the tail by the electric drift. When they reach the stronger dipolar field in the ring current region, they start to experience the gradient and curvature forces.

Adiabatic Heating

Since the particles encounter stronger magnetic fields on their inward drift while keeping their magnetic moment, they are heated adiabatically by betatron acceleration as described in Sec. 2.5. The transverse energy gain of a particle in the equatorial plane can easily be calculated by inserting Eq. (3.8) into Eq. (2.63)

$$\frac{W_\perp}{W_{\perp 0}} = \left(\frac{L_0}{L}\right)^3 \tag{3.23}$$

where $W_{\perp 0}$ denotes the transverse energy of a particle when it starts to encounter the stronger dipolar fields at L_0. Figure 3.9 visualizes this effect. A 1-keV particle starting at $L_0 = 8$ with an equatorial pitch angle $\alpha_{eq} = 90°$ and thus $W = W_\perp$ doubles its energy already after drifting inward less than $2\,R_E$ and reaches energies of some $10\,keV$ inside $L = 3$.

Since the field lines and thus the bounce paths get shorter the more the particle moves inward, electrons and ions with equatorial pitch angles $\alpha_{eq} < 90°$ also undergo Fermi acceleration as described in Eq. (2.66) to keep their longitudinal invariant constant. Calculating this energy gain is more involved, since it is somewhat difficult to calculate the length of the field line between the mirror points, but an approximate solution is

$$\frac{W_\parallel}{W_{\parallel 0}} = \left(\frac{L_0}{L}\right)^\kappa \tag{3.24}$$

where the exponent ranges between $\kappa = 2$ for particles with $\alpha_{eq} = 0°$ and $\kappa = 2.5$ for $\alpha_{eq} \to 90°$. Hence, as shown in Fig. 3.9, a 1-keV particle starting at $L_0 = 8$ with an equatorial pitch angle of $\alpha_{eq} \approx 0°$ and thus $W \approx W_\parallel$ doubles its energy after drifting inward by about $2.5\,R_E$ and reaches $10\,keV$ near $L = 2$.

It is the adiabatic heating which creates the $10–100\,keV$ ring current ions from the $1–10\,keV$ plasma sheet ions. But on their inward drift the particles do not only gain in energy but also their energy anisotropies as defined in Eq. (2.67) are increasing. Using an average $\kappa = 2.25$ we obtain

$$\frac{A_W}{A_{W0}} = \left(\frac{L_0}{L}\right)^{0.75} \tag{3.25}$$

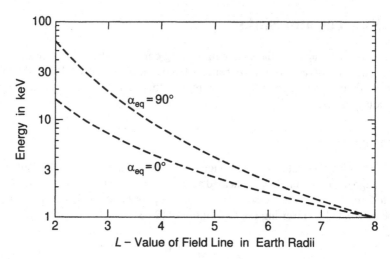

Fig. 3.9. Adiabatic heating of 1-keV particles with different equatorial pitch angles.

Accordingly, the further one gets into the inner magnetosphere, the more one will find energetic particles with large pitch angles.

Loss Processes

On average, the inward transport of particles from the tail must be balanced by loss processes in the inner magnetosphere. Ions are lost from the symmetric ring current mainly by a process called *charge exchange*. The outermost part of the neutral atmosphere has densities of some $100\,cm^{-3}$ at ring current altitudes. This density is too low for direct collisions between the ring current ions and the neutral atoms (see Sec. 4.1). However, when a cold atom from the neutral atmosphere comes close to a hot ring current ion, it looses its electron to the ion by some resonance effect. Hence, the hot ring current ion is turned into a neutral particle which either escapes from the magnetosphere, since its kinetic energy exceeds the gravitational energy, or heats the lower neutral atmosphere. The cold ion does not contribute to the ring current anymore, but mixes with the plasmaspheric particles. Typical life times of ring current ions before charge exchange are a couple of hours to some days.

Other loss processes are caused by electromagnetic variations with frequencies far above the bounce and cyclotron frequencies. In the presence of such high-frequency waves, the first and second adiabatic invariant of a particle may be violated and its pitch angle may be altered in such a way that it falls into the loss cone. The *pitch angle scattering* or *pitch angle diffusion*, may widen the otherwise narrow loss cone (see Fig. 3.5) and lead to an enhanced loss of particles in the lower neutral atmosphere. This process is discussed in our companion book, *Advanced Space Plasma Physics*. It works most effectively for the ring current electrons, since they are much lighter than the ions. In the outer ring current the mean life time of ring current electrons before they are scattered into the loss cone is of the order of hours, while it takes several days before an ion is lost by precipitation. In the inner part of the ring current the loss cone is wider and the mean precipitation life times are shorter.

3.5. Ring Current

Using the equatorial drift velocity given in Eq. (3.21) and assuming $\alpha_{eq} = 90°$, we obtain for the current density caused by ring current particles with a particular energy, W, and density, n, circulating on a given L-shell

$$j_d = \frac{3L^2 n W}{B_E R_E} \tag{3.26}$$

where j_d is an azimuthal current flowing in the westward direction. Each ring current particle on its drift around the Earth constitutes a tiny ring current. The magnetic field induced by each of these particles is negligible, but the magnetic disturbance due to the total current is noticable even on the Earth's surface.

Magnetic Disturbance

The total current, I_L, caused by all ring current particles on a particular L-shell is related to the current density by $I_L dl = j_d dV$. When integrating over the circumference element dl and the volume element dV, we may note that the total energy of all ring current electrons and ions at that particular radial distance is given by $U_L = \int n W \, dV$, and that the total circumference is simply $\int dl = 2\pi L R_E$

$$I_L = \frac{3 U_L L}{2\pi B_E R_E^2} \tag{3.27}$$

From Biot-Savart's law we can evaluate the magnetic field disturbance caused by such a circular current loop at the Earth's center (see App. A.5)

$$\delta B_d = -\frac{\mu_0 I_L}{2L R_E} = -\frac{\mu_0}{4\pi} \frac{3 U_L}{B_E R_E^3} \tag{3.28}$$

where we have introduced the minus sign to account for the fact that the disturbance field of the westward ring current is directed opposite to the terrestrial dipole magnetic field. Since the magnetic disturbance does not depend on radial distance, we can replace U_L

by the total energy, U_R, and obtain for the disturbance field generated by the drift of all ring current particles

$$\Delta B_d = -\frac{\mu_0}{4\pi} \frac{3U_R}{B_E R_E^3} \tag{3.29}$$

The total magnetic field perturbation caused by the ring current must also include the dia-magnetic contribution due to the cyclotron motion of the ring current particles. Again as-suming $\alpha_{eq} = 90°$, we get the diamagnetic field at the Earth's center caused by a charged particle orbiting on a particular L-shell by replacing M_E in Eq. (3.1) by the particle mag-netic moment

$$\delta B_\mu = \frac{\mu_0}{4\pi} \frac{\mu}{L^3 R_E^3} \tag{3.30}$$

Using the definition of the magnetic moment given in Eq. (2.37) and replacing the field strength by the dipole field value given in Eq. (3.8) we find

$$\delta B_\mu = \frac{\mu_0}{4\pi} \frac{W}{B_E R_E^3} \tag{3.31}$$

As for the drift current, the magnetic disturbance does not dependent on radial distance and we can replace W by the total energy of all ring current particles, U_R, and obtain for the diamagnetic field generated by all ring current particles

$$\Delta B_\mu = \frac{\mu_0}{4\pi} \frac{U_R}{B_E R_E^3} \tag{3.32}$$

This disturbance adds to the terrestrial dipole field, since the Earth's dipole moment and the magnetic moments of the ring current particles are co-aligned. The total magnetic field depression caused by the ring current $\Delta B_R = \Delta B_d + \Delta B_\mu$ at the Earth's center is

$$\boxed{\Delta B_R = -\frac{\mu_0}{2\pi} \frac{U_R}{B_E R_E^3}} \tag{3.33}$$

Magnetic Storms

At certain times more particles than usual are injected from the tail into the ring current, mainly by an enhanced duskward electric field. This way the total energy of the ring cur-rent is increased and the additional depression of the surface magnetic field can clearly be seen in near-equatorial magnetograms like shown in Fig. 3.10. For about one day, the equatorial terrestrial field was depressed by more than 300 nT, i.e., by more than 1% of its total value. Such strong depressions of the terrestrial field have been noticed in mag-netograms long before one knew about the ring current and have been called *magnetic storms*.

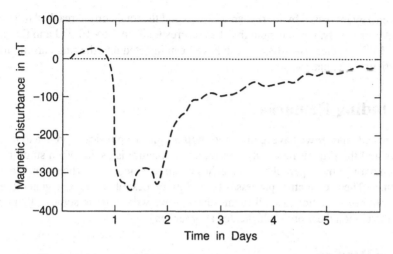

Fig. 3.10. Magnetic field depression during a major magnetic storm.

A magnetic storm has two distinct phases. For some hours or days, an enhanced electric field injects more and more particles into the inner magnetosphere, building up the strong storm-time ring current and the associated magnetic disturbance field. After a day or two, the electric field amplitude and the rate of injection get back to the normal level. Now the disturbance field starts to recover, since the ring current looses more and more storm-time particles due to charge exchange and pitch angle scattering. As can be seen in Fig. 3.10, this recovery phase typically lasts several days.

The depression of the terrestrial dipole field given in Eq. (3.33) can be used to estimate the amount of additional energy deposited in the ring current during a magnetic storm, if one approximates ΔB_R by the average magnetic field depression on the Earth's surface near the equator, usually taken from the *Dst index*. The *Dst* index represents the average disturbance field at the Earth's equator and is calculated on the basis of recordings from four low-latitude magnetic observatories (see App. B.3). Since *Dst* is also influenced by sources other than the ring current (induction effects, changes in the magnetopause topology, etc.), which may contribute more than 30% of its value, one may use half the *Dst* value to get a lower limit for

$$\Delta U_R \geq \pi \mu_0^{-1} B_E R_E^3 \, Dst \qquad (3.34)$$

Putting numbers into Eq. (3.34) one finds that a depression of 1 nT is caused by an increase in ring current energy of more than $2 \cdot 10^{13}$ J. Assuming all current to be concentrated at $L = 5$ and using Eq. (3.27) we find that a ring current of about $4 \cdot 10^4$ A produces a 1 nT depression. During the big magnetic storm shown in Fig. 3.10 the ring current energy increased by more than $5 \cdot 10^{15}$ J and the total current reached more than 10^7 A. Alternatively, we may note that 1 eV corresponds to $1.6 \cdot 10^{-19}$ J and that more than $3 \cdot 10^{30}$ particles with 10 keV each have been injected into the ring current during that particular magnetic storm.

Concluding Remarks

In the present chapter we have applied the simplest plasma physics approach to trapped particles and the ring current. The single particle picture is useful when studying the basic behavior of these particles, but reaches its limits where the particles interact with each other. These collective processes have only be described in more general terms or have not been mentioned at all in this chapter. We will return to some of them after having developed the more sophisticated approaches.

Further Reading

More about the physics of the radiation belt and the ring current is found in the relevant chapters of the monographs and articles listed below.

[1] L. R. Lyons and D. J. Williams, *Quantitative Aspects of Magnetospheric Physics* (D. Reidel Publ. Co., Dordrecht, 1984).

[2] A. Nishida, *Geomagnetic Diagnosis of the Magnetosphere* (Springer Verlag, Heidelberg, 1978).

[3] J. G. Roederer, *Dynamics of Geomagnetically Trapped Radiation* (Springer Verlag, Heidelberg, 1970).

[4] M. Schulz and L. J. Lanzerotti, *Particle Diffusion in Radiation Belts* (Springer Verlag, Heidelberg, 1974).

[5] R. A. Wolf, in *Solar-Terrestrial Physics*, eds. R. L. Carovillano and J. M. Forbes (D. Reidel Publ. Co., Dordrecht, 1983), p. 303.

4. Collisions and Conductivity

So far we have considered only the motion of single particles in external and slowly variable electromagnetic fields, but have neglected any interaction between the particles. Interaction in plasmas is, however, unavoidable. The simplest kind of interaction between single particles is a direct collision. This is not yet a collective plasma effect, since it involves interactions between individual particles and not between large groups of particles in the plasma or the plasma as a whole, but under the presence of collisions the particles already behave quite differently from what would be expected when using the single particle picture.

4.1. Collisions

One distinguishes two types of plasmas: *collisional* and *collisionless*. In the latter collisions are so infrequent compared with any relevant variation in the fields or particle dynamics that they can be safely neglected. Most space plasmas belong to this type. In the former collisions are sufficiently frequent to influence or even dominate the behavior of the plasma.

Collisional plasmas can again be divided into two classes: *partially ionized* plasmas and *fully ionized* plasmas. Partially ionized plasmas contain a large amount of residual neutral atoms or molecules, while fully ionized plasmas consist of electrons and ions only. It is clear that the types of collisions in both cases must be very different because neutrals do not respond to Coulomb fields. In partially ionized gases direct collisions between the charge carriers and neutrals dominate, while in fully ionized plasmas direct collisions are replaced by Coulomb collisions.

Partially Ionized Plasmas

In a partially ionized plasma most collisions occur between charged and neutral particles. Neutral particles affect the motion of charged particles by their mere presence as heavy compact obstacles. Hence, collisions between a charge and a neutral can be treated as head-on. Such a collision occurs only when a charged particle directly hits a neutral atom or molecule along its orbit. The *neutral collision frequency*, v_n, that is the number of collisions per second, is therefore proportional to the number of neutral particles in a column of the cross-section of an atom or molecule, $n_n \sigma_n$, where n_n is the neutral particle density and $\sigma_n = \pi d_0^2$ the molecular cross-section, and to the average velocity, $\langle v \rangle$, of the charged particles

$$\boxed{v_n = n_n \sigma_n \langle v \rangle} \tag{4.1}$$

The molecular cross-section can be approximated by $\sigma_n \approx 10^{-19}\,\mathrm{m}^2$. Similarly, it is possible to define the *mean free path length* a charged particle can propagate between two collisions with neutrals as

$$\boxed{\lambda_n = \frac{\langle v \rangle}{v_n} = (n_n \sigma_n)^{-1}} \tag{4.2}$$

The reason for using the average velocity instead of the actual velocities of the particles themselves in these definitions is that collisions are unpredictable and that it is meaningful only to define average collision frequencies and mean free path lengths.

Collisions between charged and neutral particles become important in the denser and partially ionized regions of space plasmas as the terrestrial ionosphere, where they contribute to recombination, electrical resistivity, and current flow, and cause charged particle diffusion across the magnetic field.

Fully Ionized Plasmas

In fully ionized plasmas the charged particles interact via their electric Coulomb fields. The existence of these fields implies that the particles are deflected at interparticle distances much larger than the atomic radius. The Coulomb potential therefore enhances the cross-section of the colliding particles, but also leads to a preference for small angle deflections. Both these facts considerably complicate the calculation of a collision frequency in a fully ionized plasma.

A further complication arises from the fact that in a plasma with many particles in a Debye sphere, the Coulomb potential is screened and the electric field is approximately confined to the Debye sphere (see Fig. 1.1). One could think that the effective radius of the cross-section would become equal to the Debye length. But this is not the case because the Debye sphere is transparent for particles of sufficiently high energy. Since the potential increases steeply when approaching the center of the sphere, deflections will occur predominantly inside the Debye radius, but large angle deflections will still be rare. Formally, the *Coulomb collision frequency* in a fully ionized plasma has the same functional dependence as Eq. (4.1). The problem lies in determining the Coulomb

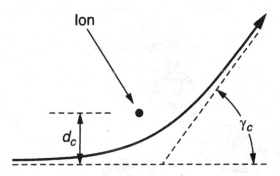

Fig. 4.1. Electron orbit during a Coulomb collision with an ion.

collisional cross-section, σ_c. In the following we present a simplified derivation of the Coulomb collision frequency between electrons and ions

$$\nu_{ei} = n_e \sigma_c \langle v_e \rangle \tag{4.3}$$

in a fully ionized plasma and later include the modification introduced by the predominance of small angle deflections.

Consider the collision between a single heavy ion and an electron. Because of the much larger mass, the ion can be considered at rest. When the electron approaches the ion it will be deflected in the Coulomb field of the ion as shown in Fig. 4.1 due to its attraction toward the ion. In a fully ionized plasma the temperature and consequently the energy of the electron is so high that the ion cannot trap the electron. The electron will turn around the ion and escape. Its orbit is a hyperbola which at large distances from the ion can be approximated by straight lines and close to the ion by a section of a circle of radius d_c.

The distance d_c is called *collision parameter* or *impact parameter*. The simplest method to determine this quantity is to consider the Coulomb force an ion is exerting on an electron of mass m_e, charge $q = -e$, and velocity v_e

$$F_C = -\frac{e^2}{4\pi \epsilon_0 d_c^2} \tag{4.4}$$

This force is felt by the electron only during an approximate average time $\tau \approx d_c/v_e$ when it passes the ion. The change in momentum $|\Delta(m_e v_e)|$ it experiences during this time is approximately given by the product $\tau|F_C|$ or

$$|\Delta(m_e v_e)| \approx \frac{e^2}{4\pi \epsilon_0 v_e d_c} \tag{4.5}$$

For large deflection angles, $\gamma_c \approx 90°$, the change in the particle momentum is of the

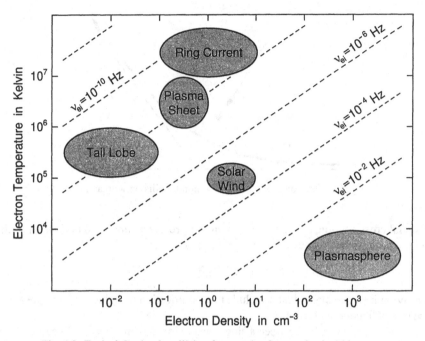

Fig. 4.2. Typical Coulomb collision frequencies for geophysical plasmas.

same order as the momentum itself, $\Delta(m_e v_e) \approx m_e v_e$. Inserting this crude approximation in the above equation enables us to determine d_c for a given velocity

$$d_c \approx \frac{e^2}{4\pi \epsilon_0 m_e v_e^2} \tag{4.6}$$

yielding the maximum cross-section as

$$\sigma_c = \pi d_c^2 \approx \frac{e^4}{16\pi \epsilon_0^2 m_e^2 \langle v_e \rangle^4} \tag{4.7}$$

where we have replaced v_e in the denominator by the average electron velocity, $\langle v_e \rangle$, since the bulk of the electrons move at the average velocity. Multiplying this equation by the electron plasma density, n_e, and the average electron velocity, one obtains the collision frequency between electrons and ions defined in Eq. (4.3)

$$\nu_{ei} = n_e \sigma_c \langle v_e \rangle \approx \frac{n_e e^4}{16\pi \epsilon_0^2 m_e^2 \langle v_e \rangle^3} \tag{4.8}$$

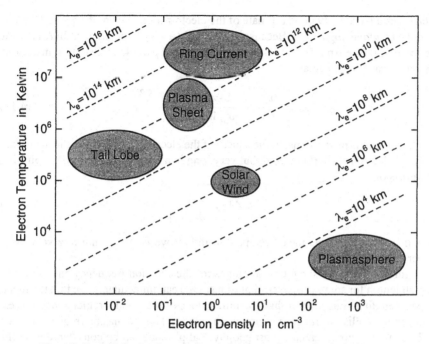

Fig. 4.3. Typical Coulomb mean free path lengths for geophysical plasmas.

Moreover, one can use the average thermal electron energy given by $k_B T_e = \frac{1}{2} m_e \langle v_e \rangle^2$ and apply the formula for the plasma frequency given in Eq. (1.6) to obtain the simplified expression

$$\nu_{ei} \approx \frac{\sqrt{2}\, \omega_{pe}^4}{64 \pi n_e} \left(\frac{k_B T_e}{m_e} \right)^{-3/2} \qquad (4.9)$$

The collision frequency turns out to be proportional to the plasma density and inversely proportional to the 3/2 power of the electron temperature. It increases with increasing density and decreases with increasing electron temperature.

This formula is not exact, insofar as we have to include a correction factor $\ln \Lambda$ to correct for the predominance of weak deflection angles as well as for the different velocities electrons assume in thermal equilibrium in the plasma (see App. B.1). Λ is within a factor 4π equal to the plasma parameter introduced in Eq. (1.5) which is proportional to the number of particles in a Debye sphere. Multiplying Eq. (4.9) with $\ln \Lambda$ and using the definition of Λ in Eq. (1.5) to simplify the expression, we obtain

$$\boxed{\nu_{ei} \approx \frac{\omega_{pe}}{64\pi} \frac{\ln \Lambda}{\Lambda}} \qquad (4.10)$$

which gives a reasonably good estimate of the electron-ion collision frequency. $\ln \Lambda$ is called the *Coulomb logarithm*. Because Λ is usually a very large number, $\ln \Lambda$ is of the order of 10–30 (see App. B.2. As in the case of neutral particle collisions, a mean free path length can be defined as

$$\lambda_e = \frac{\langle v_e \rangle}{\nu_{ei}} \approx \frac{64\pi n_e}{\omega_{pe}^4 \ln \Lambda} \left(\frac{k_B T_e}{m_e} \right)^2 \qquad (4.11)$$

Since this length is proportional to the square of the electron temperature, the mean free path of an electron is short in a cold, but very long in a hot plasma. Again, simplifying the expression

$$\boxed{\lambda_e \approx 64\pi \lambda_D \frac{\Lambda}{\ln \Lambda}} \qquad (4.12)$$

shows that the ratio of the mean free path to the Debye length is indeed a very large number in a plasma.

Figures 4.2 and 4.3 show typical ranges for the collision frequency and the mean free path length for some geophysical plasmas. One can immediately see that the mean free path length is much larger than the dimensions of the plasma regions themselves. Moreover, the collision frequencies are much smaller than the plasma frequencies (see Fig. 1.2) in all regions. Hence, most geophysical plasmas can be considered as collisionless.

Collisions occur also between electrons and electrons or ions and ions. These particles have equal masses and the collision affects the motion of both particles. Collisions between particles of equal masses quickly equilibrize their velocities via momentum exchange. Particles of one species will readily assume a well defined average temperature, while the temperatures between particles of unlike masses may be different.

4.2. Plasma Conductivity

In the presence of collisions we have to add a collisional term to the equation of motion Eq. (2.7) for a charged particle under the action of the Coulomb and Lorentz forces. Assuming all collision partners to move with the velocity $\mathbf{u}$, we obtain for a charged particle moving with the velocity $\mathbf{v}$

$$m \frac{d\mathbf{v}}{dt} = q \, (\mathbf{E} + \mathbf{v} \times \mathbf{B}) - m\nu_c(\mathbf{v} - \mathbf{u}) \qquad (4.13)$$

The collisional term on the right-hand side describes the momentum lost through collisions occurring at a frequency ν_c. It is often called frictional term since it impedes motion. Equation (4.13) holds both for Coulomb and neutral collisions.

Unmagnetized Plasma

Let us assume a steady state in an unmagnetized plasma with $\mathbf{B} = 0$, where all electrons move with the velocity $\mathbf{v}_e$ and all collision partners (ions in the case of a fully ionized or neutrals in a partially ionized plasma) are at rest. Then we get

$$\mathbf{E} = -\frac{m_e v_c}{e}\mathbf{v}_e \tag{4.14}$$

Since the electrons move with respect to the ions, they carry a current

$$\mathbf{j} = -en_e\mathbf{v}_e \tag{4.15}$$

Combining these two equations yields for the electric field

$$\mathbf{E} = \frac{m_e v_c}{n_e e^2}\mathbf{j} \tag{4.16}$$

which is the familiar Ohm's law with

$$\boxed{\eta = \frac{m_e v_c}{n_e e^2}} \tag{4.17}$$

where η is the *plasma resistivity*. It has the same form for fully and partially ionized plasmas and differs only in the collision frequency used.

For a fully ionized plasma, we may introduce the Coulomb collision frequency from Eq. (4.10) into Eq. (4.17) to get for the *Spitzer resistivity* in a fully ionized plasma

$$\boxed{\eta_s = \frac{1}{64\pi\epsilon_0\omega_{pe}}\frac{\ln\Lambda}{\Lambda}} \tag{4.18}$$

The Spitzer resistivity is actually independent from the plasma density, since ω_{pe} is directly and Λ inversely proportional to the square of the electron density. This has its roots in the fact that if one tries to increase the current by adding more charge carriers one also increases the collision frequency and the frictional drag and, by this, decreases the velocity of the charge carriers and, hence, the current.

Magnetized Plasma

In the magnetized case, the plasma may move with velocity $\mathbf{v}$ across a magnetic field and we have to add the $\mathbf{v} \times \mathbf{B}$ electric field resulting from the Lorentz transformation to Eq. (4.16), yielding

$$\mathbf{j} = \sigma_0(\mathbf{E} + \mathbf{v} \times \mathbf{B}) \tag{4.19}$$

where we have replaced the resistivity by its inverse, the *plasma conductivity*

$$\sigma_0 = \frac{n_e e^2}{m_e \nu_c} \tag{4.20}$$

Equation (4.19) is a simple form of the *generalized Ohm's law*, which is valid in all fully ionized geophysical plasmas where the typical collision frequencies are extremely low (see Fig. 4.2) and the plasma conductivity can be taken as near-infinite.

While treating the plasma conductivity as a scalar is warranted in the dilute, fully ionized magnetospheric and solar wind plasmas with their near-infinite conductivity, there is one place where we have to take the anisotropy introduced by the presence of the magnetic field into account. This is the lower part of the partially ionized terrestrial ionosphere where abundant collisions between the ionized and the neutral part of the upper atmosphere in the presence of a strong magnetic field lead to a finite anisotropic conductivity tensor.

Starting again from Eq. (4.13) and assuming a steady state, where all electrons move with the velocity $\mathbf{v}_e$ and all collision partners are at rest, but now in a magnetized plasma, we obtain

$$\mathbf{E} + \mathbf{v}_e \times \mathbf{B} = -\frac{m_e \nu_c}{e} \mathbf{v}_e \tag{4.21}$$

Using the definition of σ_0 in Eq. (4.20) and Eq. (4.15) to express $\mathbf{v}_e$ by the current yields another form of Ohm's law

$$\mathbf{j} = \sigma_0 \mathbf{E} - \frac{\sigma_0}{n_e e} \mathbf{j} \times \mathbf{B} \tag{4.22}$$

Let us now assume that the magnetic field is aligned with the z axis, $\mathbf{B} = B\,\hat{\mathbf{e}}_z$. Taking into account the definition of the electron cyclotron frequency given in Eq. (2.12) and remembering that the cyclotron frequency carries the sign of the charge, we obtain

$$\begin{aligned}
j_x &= \sigma_0 E_x + \frac{\omega_{ge}}{\nu_c} j_y \\
j_y &= \sigma_0 E_y - \frac{\omega_{ge}}{\nu_c} j_x \\
j_z &= \sigma_0 E_z
\end{aligned} \tag{4.23}$$

Combining the first two equations to eliminate j_y from the first and j_x from the second equation yields

$$\begin{aligned}
j_x &= \frac{\nu_c^2}{\nu_c^2 + \omega_{ge}^2} \sigma_0 E_x + \frac{\omega_{ge}\nu_c}{\nu_c^2 + \omega_{ge}^2} \sigma_0 E_y \\
j_y &= \frac{\nu_c^2}{\nu_c^2 + \omega_{ge}^2} \sigma_0 E_y - \frac{\omega_{ge}\nu_c}{\nu_c^2 + \omega_{ge}^2} \sigma_0 E_x \\
j_z &= \sigma_0 E_z
\end{aligned} \tag{4.24}$$

This set of component equations can be written in dyadic notation (see App. A.4)

$$\mathbf{j} = \sigma \cdot \mathbf{E} \tag{4.25}$$

For a magnetic field aligned with the z direction, the conductivity tensor reads

$$\sigma = \begin{pmatrix} \sigma_P & -\sigma_H & 0 \\ \sigma_H & \sigma_P & 0 \\ 0 & 0 & \sigma_\parallel \end{pmatrix} \tag{4.26}$$

and the tensor elements are given by

$$\boxed{\begin{aligned} \sigma_P &= \frac{v_c^2}{v_c^2 + \omega_{ge}^2} \sigma_0 \\ \sigma_H &= -\frac{\omega_{ge} v_c}{v_c^2 + \omega_{ge}^2} \sigma_0 \\ \sigma_\parallel &= \sigma_0 = \frac{n_e e^2}{m_e v_c} \end{aligned}} \tag{4.27}$$

The tensor element σ_P is called *Pedersen conductivity* and governs the *Pedersen current* in the direction of that part of the electric field, $\mathbf{E}_\perp$, which is transverse to the magnetic field. The *Hall conductivity*, σ_H, determines the *Hall current* in the direction perpendicular to both the electric and magnetic field, in the $-\mathbf{E} \times \mathbf{B}$ direction (remember that ω_{ge} is a negative number). The element $\sigma_\parallel$ is called *parallel conductivity* since it governs the magnetic *field-aligned current* driven by the parallel electric field component, $E_\parallel$. The parallel conductivity is equal to the plasma conductivity in the unmagnetized case.

When the magnetic field has an arbitrary angle to the axes of the chosen coordinate system, one can rewrite Eq. (4.25) into the form

$$\mathbf{j} = \sigma_\parallel \mathbf{E}_\parallel + \sigma_P \mathbf{E}_\perp - \sigma_H (\mathbf{E}_\perp \times \mathbf{B})/B \tag{4.28}$$

This expression can be derived directly from Eq. (4.22) by taking the cross-product of Eq. (4.22) with $\mathbf{B}$ and using the result to eliminate the $\mathbf{j} \times \mathbf{B}$ term from Eq. (4.22).

The dependence of the conductivity tensor elements on the ratio of the cyclotron frequency to the collision frequency is shown in Fig. 4.4. In a highly collisional plasma containing a weak magnetic field we have $|\omega_{ge}| \ll v_c$. The set of Eqs. (4.27) then shows that $\sigma_P = \sigma_\parallel = \sigma_0$ and $\sigma_H = 0$, and the conductivity tensor becomes isotropic and reduces to a scalar. For a dilute, nearly collisionless plasma with a strong magnetic field we are in the opposite regime $|\omega_{ge}| \gg v_c$, where $\sigma_\parallel = \sigma_0$ and $\sigma_P \approx \sigma_H \approx 0$. Hence, in such a plasma the current flows essentially along the field lines.

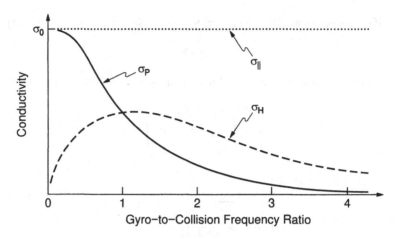

Fig. 4.4. Dependence of the conductivities on the frequency ratio ω_g / ν_c.

The conductivity is most anisotropic for plasmas with $|\omega_{ge}| \approx \nu_c$. For $|\omega_{ge}| < \nu_c$ the Pedersen conductivity dominates, since in such a domain the electrons are scattered in the direction of the electric field before they can start to gyrate about the magnetic field. For $|\omega_{ge}| > \nu_c$ the electrons experience the $E \times B$ drift for many gyrocycles, before a collision occurs, and the Hall conductivity dominates. For $|\omega_{ge}| \approx \nu_c$ the electrons are scattered about once per gyration. Hence, both $E \times B$ drift and motion along the transverse electric field are equally important and the Pedersen and Hall conductivities are of the same order. In this latter case, the electrons will, on average, move at an angle of $45°$ with both the direction of the transverse electric field and the $\mathbf{E} \times \mathbf{B}$ direction.

4.3.   Ionosphere Formation

The *ionosphere* forms the base of the magnetospheric plasma environment of the Earth. It is the transition region from the fully ionized magnetospheric plasma to the neutral atmosphere. This implies that it consists of a mixture of plasma and neutral particles and will therefore have an electrical conductivity to which Coulomb and especially neutral collisions may contribute.

Before doing so, we need to know how the plasma density and the collision frequency varies in dependence of height, latitude, and time of day, and possibly also during times of magnetospheric disturbance. What interests us first is, how the plasma density depends on height and how the ionization of the ionosphere is created. Two main sources of ionization can be identified: the solar ultraviolet radiation and energetic particle precipitation from the magnetosphere into the atmosphere.

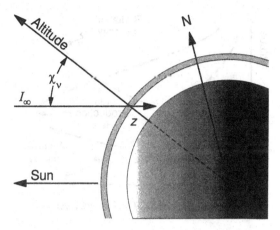

Fig. 4.5. Solar UV absorption in the ionosphere.

Solar Ultraviolet Ionization

In order to produce ionization, the solar photons must have energies higher than the ionization energy of the atmospheric atoms. Thus the photons should come from the ultraviolet spectral range or higher, but at higher frequencies (or photon energies) the solar radiation intensity becomes very weak and sporadic and is therefore unimportant when considering the average state of the ionosphere.

The ionosphere is horizontally structured. Its dominant variation occurs with altitude, z, and is prescribed by the variation of the neutral atmosphere density, $n_n(z)$. In a one-component isothermal atmosphere the density changes with height according to the *barometric law*

$$n_n(z) = n_0 \exp(-z/H) \tag{4.29}$$

H is the *scale height* for an isothermal atmosphere with atoms of mass m_n and temperature T_n. It is defined as

$$\boxed{H = k_B T_n / m_n g} \tag{4.30}$$

where g is the gravitational acceleration and n_0 is the atmospheric density at $z = 0$.

Solar ultraviolet radiation impinging onto the atmosphere at height z under an angle χ_v (see Fig. 4.5) hits atmospheric atoms of density n_n and looses its energy due to ionization. At this height the radiation is partially absorbed. The interaction of radiation with the atmospheric atoms takes place along the oblique ray path, i.e., along $z/\cos \chi_v$. The diminution of radiation intensity, I, with altitude z along the ray path element $dz/\cos \chi_v$ is given by

$$dI = \sigma_v n_n \frac{dz}{\cos \chi_v} I \tag{4.31}$$

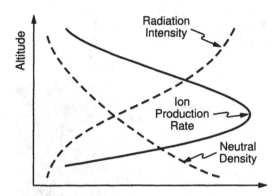

Fig. 4.6. Formation of an ionized layer.

where σ_ν is the radiation absorption cross-section. The equation shows that the differential decrease in radiation intensity is proportional to the incident intensity, to the number density of absorbing neutral gas particles, to the absorption cross-section, and to the path length of the radiation in the atmosphere. Using the barometric law (4.29), one may integrate Eq. (4.31) to find the height variation of the intensity

$$\int_{I_\infty}^{I(z)} \frac{dI}{I} = \int_\infty^z \exp(-z/H)\frac{\sigma_\nu n_0}{\cos \chi_\nu}dz \qquad (4.32)$$

where I_∞ is the solar flux at the top of the atmosphere. Solving for $I(z)$ yields

$$I(z) = I_\infty \exp\left[-\frac{\sigma_\nu n_0 H}{\cos \chi_\nu}\exp(-z/H)\right] \qquad (4.33)$$

which shows the exponential increase of the radiation intensity with height schematically plotted in Fig. 4.6.

The number of electron-ion pairs locally produced by the solar ultraviolet radiation, the *photoionization rate* per unit volume at a particular height, $q_\nu(z)$, is proportional to the absorbed fraction of radiation in the altitude interval dz and to the photoionization efficiency, κ_ν, the fraction of the absorbed radiation that goes into ionization

$$q_\nu(z) = \kappa_\nu \cos \chi_\nu dI/dz \qquad (4.34)$$

Using Eq. (4.31) to replace the dI/dz by the intensity itself, one obtains

$$q_\nu(z) = \kappa_\nu \sigma_\nu n_n I \qquad (4.35)$$

Equations (4.29) and (4.33) can be used to introduce the explicit height dependence of neutral density and ray intensity

$$q_\nu(z) - \kappa_\nu \sigma_\nu n_0 I_\infty \exp\left[\frac{z}{H} - \frac{\sigma_\nu n_0 H}{\cos \chi_\nu} \exp(-z/H)\right] \tag{4.36}$$

This *Chapman production function* can be written in a simpler form

$$q_\nu(\zeta) = q_{\nu 0} \exp\left[1 - \zeta - \frac{\exp(-\zeta)}{\cos \chi_\nu}\right] \tag{4.37}$$

To get this expression one considers the variation of the ionization with altitude. As shown in Fig. 4.6, the density decreases with height while the solar intensity increases. Thus it is clear that the ionization will have a pronounced maximum at a particular height z_m. The value of z_m can be calculated by setting the derivative of Eq. (4.36) to zero

$$\begin{aligned} z_m &= z_0 + H \ln\left(1/\cos \chi_\nu\right) \\ z_0 &= H \ln\left(\sigma_\nu n_0 H\right) \end{aligned} \tag{4.38}$$

Here z_0 is the height of the maximum ionization rate for vertical incidence of the solar radiation ($\chi_\nu = 0$). It is a constant which depends only on gravity, ion mass, scale height, and ground level atmospheric density. The maximum value of the ionization rate, $q_{\nu m}$, is then given by

$$\begin{aligned} q_{\nu m} &= q_{\nu 0} \cos \chi_\nu \\ q_{\nu 0} &= \kappa_\nu I_\infty (eH)^{-1} \end{aligned} \tag{4.39}$$

where $q_{\nu 0}$ is the maximum ionization rate at vertical incidence. One now introduces the new variable $\zeta = (z - z_0)H^{-1}$, inserts it into Eq. (4.36) and obtains Eq. (4.37).

Equation (4.37) must be evaluated numerically. One finds that the height of maximum ionization, z_m, is restricted to a narrow range of altitudes. Its position depends on χ_ν in such a way that for smaller χ_ν the maximum is found at lower altitudes. Moreover, the maximum weakens with increasing χ_ν. Since χ_ν is a function of geographic latitude and longitude, the photoionization layer in the ionosphere exhibits a strong dependence on geographic latitude, time of day, and season.

Ionization by Energetic Particles

In addition to photoionization, ionospheric ionization is produced in those regions where sufficiently energetic particles, in the first place electrons, impinge onto the atmosphere. Because such particles must follow the magnetic field lines, one naturally expects that this type of ionization will dominate at high magnetic latitudes in the auroral zone (see Sec. 1.2), where photoionization becomes less important. Also during nighttime, when photoionization ceases, ionization due to particle impact can maintain the ionosphere.

Ionization by electrons precipitating into the atmosphere along magnetic field lines from the magnetosphere is collisional. It requires electron energies $W_e > W_{ion}$, where W_{ion} is the ionization energy needed to extract an electron out of an atom or molecule. For oxygen atoms the ionization energy is about 35 eV. The *collisional ionization rate* per unit volume at a particular height, $q_e(z)$, is proportional to the energy loss, $dW_e(z)$, a precipitating electron will experience at this altitude. Hence, it is proportional to the product of ionization energy, W_{ion}, collisional ionization efficiency, κ_e, and the number of collisions per unit height, v_c/v_z, at this altitude. How the two latter are distributed over altitude requires precise knowledge of the altitude profile of atmospheric density and composition. Moreover, the number of collisions per unit height depends on the pitch angle of the precipitating electron, which determines how much time a particle spends at a given altitude.

However, one can get a fairly good idea on the altitude variation of the ionization rate changes if one considers a field-aligned electron beam impinging vertically onto a neutral atmosphere governed by a simple barometric law and assumes the ionization efficiency to be constant. For an electron moving purely vertical, the number of collisions per unit height is given by the inverse mean free path length defined in Eq. (4.2), yielding for the energy loss

$$dW_e(z) = \kappa_e W_{ion} \sigma_n n_n dz \tag{4.40}$$

The variation of the energy loss with altitude does not depend on the original energy of the precipitating particle. The collisional ionization rate in the height interval dz produced by a flux of precipitating electrons, F_e, is

$$q_e(z) = F_e dW_e/dz \tag{4.41}$$

Replacing the energy loss by Eq. (4.40) and the neutral density profile by the barometric law (4.29), we obtain

$$q_e(z) = \kappa_e F_e W_{ion} \sigma_n n_0 \exp(-z/H) \tag{4.42}$$

Accordingly, the height profile of the collisional ionization rate is simply determined by the altitude variation of the neutral density. Independent of particle energy, the ion production rate increases exponentially with decreasing altitude.

However, the energy of the precipitating electron enters when considering the lowest altitude reached. An electron with energy W_e can penetrate only down to a stopping height, z_s, where it will have lost all its energy by collisions. Naturally, more energetic electrons penetrate deeper into the atmosphere and produce more electron-ion pairs by

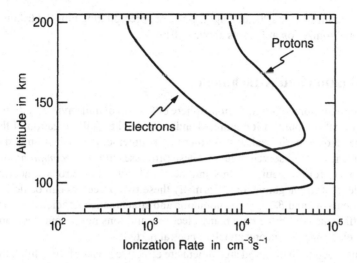

Fig. 4.7. Ion production due to precipitating electrons and protons.

collisions because more energy can be distributed. The stopping height can be calculated by integrating the energy loss

$$W_e = \int_0^{W_e} dW_e = \int_\infty^{z_s} \kappa_e W_{ion} \sigma_n n_0 \exp(-z/H) dz \qquad (4.43)$$

Solving for z_s, one obtains

$$z_s = H \ln \left(\kappa_e \sigma_n n_0 H W_{ion} / W_e \right) \qquad (4.44)$$

showing that more energetic particles reach lower altitudes. At this stopping altitude they deposit the largest fraction of their energy and thus produce the largest fraction of electron-ion pairs. In a realistic atmosphere electrons of 300 keV energy penetrate down to about 70 km, while electrons of 1 keV energy are stopped at an altitude of about 150 km. Due to the exponential ionization rate profile, the ionization maximum is more pronounced for more energetic electrons.

Our rather crude model predicts the shape of real ionization rate profiles (see Fig. 4.7) quite well. Here, the ionization rate profile labelled 'electrons' was computed from rocket observations of precipitating electrons with energies of about 10 keV, using realistic profiles of ionization efficiency, atmospheric density, and composition. As also indicated in Fig. 4.7, ions are stopped at greater heights than electrons of the same energy, because their ionization efficiency is lower. The electrons are responsible for the ionization measured near 100 km height, while the precipitating ions contribute most to the collisional ionization at heights above 130 km.

Recombination and Attachment

The production of ionization in the ionosphere either by solar ultraviolet radiation or by energetic particles would, if it continued endlessly, lead to full ionization of the upper atmosphere. However, in reality two processes counteract the ionization and in equilibrium limit it to its observed values. These processes are the *recombination* of ions and electrons to reform neutral atoms and the *attachment* of electrons at neutral atoms or molecules to form negative ions. Formally these two processes can be described by two coefficients, α_r and β_r, the recombination and attachment coefficients, respectively. These coefficients determine how many electrons and ions per second recombine and how many electrons per second attach to neutral particles.

Because recombination and attachment are effective losses of ionization, they contribute negatively to the ionization. Recombination is proportional to both the number density of electrons and of ions which can recombine, while attachment is proportional only to the number of electrons available to attach to a neutral particle. Observing that in equilibrium the ionospheric plasma should be quasineutral, $n_i \approx n_e$, the continuity equation for the electron density becomes

$$\frac{dn_e}{dt} = q_{v,e} - \alpha_r n_e^2 - \beta_r n_e \tag{4.45}$$

The first term on the right-hand side is the ionization due to solar ultraviolet radiation and collisions with precipitating particles. It acts as the source of the electron density, while the two other terms are sinks of ionization. The coefficients α_r and β_r contain a number of complicated photochemical processes which are responsible for the differences of the ionospheric composition at different heights.

In equilibrium the time dependence of the density is zero. Hence, setting the left-hand side of Eq. (4.45) to zero, one finds the equilibrium electron density of the ionosphere. At lower altitudes, recombination is more important than attachment and we obtain by setting $\beta_r = 0$

$$\boxed{n_e = \left(\frac{q_{v,e}}{\alpha_r} \right)^{1/2}} \tag{4.46}$$

Hence, in the lower ionosphere the equilibrium electron density is proportional to the square root of the ratio between the ion production rate and the recombination coefficient. At greater altitudes attachment is the dominant loss process. Here we can neglect

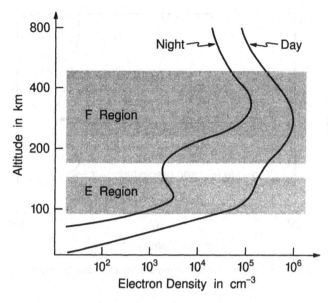

Fig. 4.8. Vertical profiles of mid-latitude electron density.

the recombination term and find

$$n_e = \frac{q_{v,e}}{\beta_r} \tag{4.47}$$

Thus at higher altitudes the electron density is proportional to the ion production rate.

Ionospheric Layers

The real structure of the Earth's ionosphere is not as simple as given by Eq. (4.37) and sketched in Fig. 4.6. The actual electron density profile is determined by the specific absorption properties of the gaseous constituents of the atmosphere, which due to the different barometric laws for different molecular components vary drastically. In addition, the altitude variation of the recombination and attachment coefficients has to be taken into account. As a result of these processes, the electron density in the Earth's ionosphere exhibits three different layers.

The lower ionosphere below a height of about 90 km is called *D-region*. It is very weakly ionized and due to high collision frequencies mostly dominated by neutral gas dynamics and chemistry and cannot be considered a plasma (see Sec. 1.1). The upper ionosphere is the region above 90 km. It is highly but still partially ionized, containing

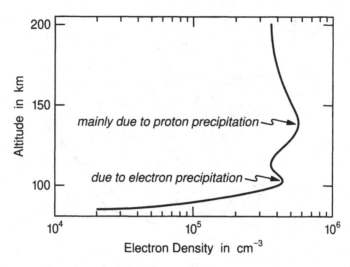

Fig. 4.9. Vertical profile of electron density over diffuse aurora.

a substantial contribution of neutral gas up to about 500 km height. The upper ionosphere consists of two well separated layers of ionized matter, the *E-region*, which has
its ionization peak at about 110 km, and the *F-region* around 300 km altitude. Figure
4.8 shows height profiles of the ionospheric electron density during day and night hours
in mid-latitudes. The distinction between the two layers is obvious from the nightside
profiles. During daytime the gap between the E- and F-region is partially filled.

The E-region is formed by the absorption of longer wavelength ultraviolet radiation (approximately 90 nm) which passes the higher altitudes until the density of molecular oxygen becomes high below about 150 km height. Thus oxygen ions dominate the
E-region. At higher latitudes ionization due to precipitating energetic electrons and protons contributes significantly to the formation of the E-region (see Fig. 4.9).

The F-region splits into two layers, the *F1-region* at around 200 km, and the *F2-
region* around 300 km height. The former is a dayside feature created in the same way as
the E-region, but the absorbed ultraviolet wavelengths are shorter (20–80 nm) because
of different absorbing molecules. The more important layer is the F2-region. Its formation is basically determined by the height variation of the neutral densities and the
recombination and attachment rates for the different atmospheric constituents. In the
lower F2-region ionization of atomic oxygen and recombination play the key roles. At
greater altitudes the decreasing neutral density and attachment limit the increase in electron density. In this way the competition between ionization and attachment leads to the
F2-region peak at roughly 300 km seen in Fig. 4.8. The F2-region peak contains the
densest plasma in the Earth's environment, with electron densities up to 10^6 cm^{-3}.

4.4. Ionospheric Conductivity

In Sec. 4.2 we derived the conductivity tensor due to collisions between moving electrons and unspecified scatter centers at rest. For a partially ionized ionosphere, the collision partners are the neutral atmosphere particles and the general collision frequency v_c in Eq. (4.27) is replaced by the *electron-neutral collision frequency*, v_{en}.

Conductivity Tensor

In the terrestrial ionosphere, not only the electrons are scattered by the neutrals but also the ions. Since the current caused by the finite *ion-neutral collision frequency*, v_{in}, is governed by the same equation as the current carried by the electrons, we can retain the generalized Ohm's law first given in Eq. (4.28)

$$\mathbf{j} = \sigma_\parallel \mathbf{E}_\parallel + \sigma_P \mathbf{E}_\perp - \sigma_H (\mathbf{E} \times \mathbf{B})/B \tag{4.48}$$

if we add the ion contribution to the electron conductivity tensor elements given in Eq. (4.27). The ion conductivities are simply found by replacing ω_{ge} and v_{en} by ω_{gi} and v_{en}. Since we have defined the cyclotron frequency as carrying the sign of the charge, the latter is automatically taken care of and we get

$$
\begin{aligned}
\sigma_P &= \left(\frac{v_{en}}{v_{en}^2 + \omega_{ge}^2} + \frac{m_e}{m_i} \frac{v_{in}}{v_{in}^2 + \omega_{gi}^2} \right) \frac{n_e e^2}{m_e} \\
\sigma_H &= -\left(\frac{\omega_{ge}}{v_{en}^2 + \omega_{ge}^2} + \frac{m_e}{m_i} \frac{\omega_{gi}}{v_{in}^2 + \omega_{gi}^2} \right) \frac{n_e e^2}{m_e} \\
\sigma_\parallel &= \left(\frac{1}{v_{en}} + \frac{m_e}{m_i} \frac{1}{v_{in}} \right) \frac{n_e e^2}{m_e}
\end{aligned}
\tag{4.49}
$$

where we have used the simplified assumption that there is only one type of ions in the terrestrial ionosphere.

Conductivity Profile

Figure 4.10 shows typical altitude profiles of ion and electron cyclotron and collision frequencies in the E-region ionosphere at mid latitudes. In the narrow altitude range shown the value of the geomagnetic dipole field is about constant, and correspondingly the cyclotron frequencies are constant as indicated by the dashed vertical lines in Fig. 4.10 (note that the ion cyclotron frequency is governed by the heavy oxygen ions which dominate the E-region). The collision frequencies decrease across the E-region. Below altitudes of about 75 km the electron collision frequency exceeds the electron cyclotron frequency. Inside the shaded region in Fig. 4.10, the electron collision frequency is lower

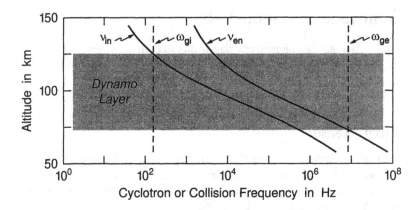

than the electron cyclotron frequency, but the ion collision frequency is still larger than the ion cyclotron frequency. Ions are therefore coupled to the neutral gas while electrons are partially decoupled.

Figure 4.10 explains how the Hall and Pedersen currents are generated and which are the primary charge carriers for these currents. Except for the lower bottom of the shaded region, where the density of the ionized component is too low to allow any appreciable current, the lower two thirds of the *dynamo layer* are governed by $\nu_{in} \gg \omega_{gi}$ and $\nu_{en} \ll \omega_{ge}$. Around 100 km the electrons can already do about a hundred gyrocycles before they collide with neutrals and experience a somewhat impeded $E \times B$ drift. The ions, on the other hand, still collide with neutrals about a hundred times per gyrocycle and thus move with the neutrals. Hence, a Hall current is carried by the electron motion transverse to the electric and magnetic fields. At the top of the dynamo layer, around 125 km altitude, the ion cyclotron and collision frequencies become comparable and the ions will not be perfectly coupled to the neutrals anymore. Instead they will move in the direction of the electric field and carry a Pedersen current, while the electrons still move at right angles to the fields and carry a Hall current. Above the E-region, both electrons and ions will undergo the same $E \times B$ drift and no current flows.

Using the cyclotron and collision frequencies in Fig. 4.10 and electron density profiles like those in Figs. 4.8 and 4.9, one can calculate height profiles of the E-region conductivities. Figure 4.11 shows such profiles of Pedersen, Hall, and parallel conductivities. They have been normalized to the maximum Hall conductivity, σ_{peak}, since the actual value of σ_{peak} scales about linearly with the electron density at the center of the E-region, while the height dependence is similar for the day and nightside ionosphere at all latitudes. The Pedersen conductivity peaks around 130 km and the Hall conductivity has its peak around 100 km altitude, in accordance with our above reasoning. The peak

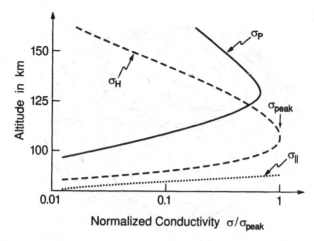

Fig. 4.11. Height profiles of normalized conductivities.

Hall conductivity is always larger than the peak Pedersen conductivity.

The actual peak value of the conductivity differs widely between the dayside low-latitude ionosphere, where $n_e \approx 2 \cdot 10^5$ cm^{-3} and $\sigma_{\mathrm{peak}} \approx 10^{-3}$ S/m, and the nightside ionosphere, where $n_e \approx 2 \cdot 10^3$ cm^{-3} and $\sigma_{\mathrm{peak}} \approx 10^{-5}$ S/m. In the auroral zone, the electron density caused by energetic electrons precipitating from the magnetosphere may be even higher, up to $10^6 - 10^7$ cm^{-3}, and the peak Hall conductivity can reach values of more than 10^{-2} S/m. The peak Pedersen conductivity is typically half the peak Hall conductivity. The parallel conductivity is much higher than the other two. In the E-region it may reach 10^2 S/m and in the F-region and above it approaches the conductivity of a fully ionized plasma and can usually be taken as infinite.

4.5. Ionospheric Currents

We have already mentioned in the previous section that the ions and, to a lesser degree, also the electrons in the E-region are coupled to the neutral components of the atmosphere and follow their dynamics. Atmospheric winds and tidal oscillations of the atmosphere force the E-region ion component to move across the magnetic field lines, while the electrons move much slower at right angles to both the field and the neutral wind. The relative movement constitutes an electric current and the separation of charge produces an electric field, which in turn affects the current. Because of this the E-region bears the name dynamo layer, the generator of which is the atmospheric wind motion.

To see the relation between current, conductivity, electric field, and neutral wind velocity, we have to write down Ohm's law. As in Sec. 4.2, we have to add a $\mathbf{v}_n \times \mathbf{B}$

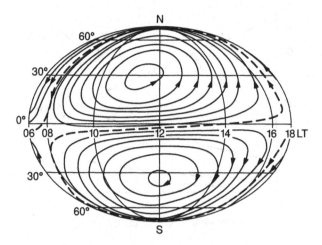

Fig. 4.12. Global view of the average Sq current system.

term to the Ohm's law given in Eq. (4.25)

$$\mathbf{j} = \sigma \cdot (\mathbf{E} + \mathbf{v}_n \times \mathbf{B}) \tag{4.50}$$

This relation is valid throughout the Earth's ionosphere. For mid- and low-latitude dynamo currents, the dominant driving force for the current is the $v \times B$ field induced by the ion motion across the magnetic field. For auroral oval current systems at higher latitudes, on the other hand, the neutral wind term is usually much smaller than the electric field term and can safely be neglected. In the present section we will study two current systems confined to the mid and low-latitude ionosphere. The auroral oval current systems will be described later in Secs. 5.4 through 5.7, after we have discussed their external driving force, the convection electric field.

Sq Current

The most important dynamo effect at mid-latitudes is the daily variation of the atmospheric motion caused by the tides of the atmosphere. The tides with the lowest and the largest amplitudes are the diurnal and semi-diurnal oscillations which are excited by the solar radiation heating of the atmosphere. The current system created by this tidal motion of the atmosphere is called solar quiet or Sq current.

The Sq currents create a magnetic field disturbance, which can be measured on the ground by magnetometers and permits to determine the extent of the currents. Records of these daily magnetic variations obtained at many different stations distributed across the globe can be used to construct the Sq current system, using Biot-Savart's law and

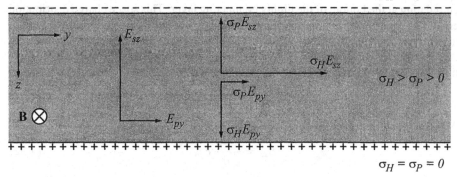

Fig. 4.13. Enhancement of the effective conductivity at the magnetic equator.

methods of potential theory. More sophisticated methods use measured wind patterns, conductivities, and disturbance magnetic fields and calculate electric fields and currents based on Eq. (4.50) and Biot-Savart's law (see App. A.5).

Figure 4.12 presents a global view of the average Sq current system as seen from above the terrestrial ionosphere. In this figure, the lines give the direction of the current while the distance between the lines is inversely proportional to the height-integrated current density. The Sq currents form two vortices, one in the northern and the other in the southern hemisphere, which touch each other at the geomagnetic equator. In accordance with the day-night contrast in the low and mid-latitude E-region electron densities (see Fig. 4.8), the Sq currents are concentrated in the dayside region.

Equatorial Electrojet

At the geomagnetic equator, the Sq current systems of the southern and northern hemispheres touch each other and form an extended nearly jet-like current in the ionosphere, the *equatorial electrojet*. However, the electrojet would not be so strong as it is if were formed only by the concentration of the Sq current. The special geometry of the magnetic field at the equator together with the nearly perpendicular incidence of solar radiation cause an equatorial enhancement in the effective conductivity which leads to an amplification of the jet current.

To see this combined action consider a situation where the magnetic field is about horizontal to the Earth's surface as is the case at the equator. The direction of the magnetic field is from south to north, along the x axis. The primary Sq Pedersen current flows eastward in the y direction, parallel to the primary ionospheric electric field, E_{py}. As sketched in Fig. 4.13, this primary electric field drives a Hall current which flows vertically downward in the z direction, causing a charge separation in the equatorial ionosphere with negative charges accumulating on the top boundary and positive charges accumulating at the bottom of the highly conducting layer. This space charge distribution creates a secondary *polarization electric field*, E_{sz}, vertically directed from the bottom to the top of the conducting ionosphere. The polarization electric field drives a vertical Pedersen current opposing the Hall current until it compensates it. The resulting equilibrium condition in which no vertical current flows is

$$j_z = \sigma_H E_{py} + \sigma_P E_{sz} = 0 \tag{4.51}$$

which yields for the secondary vertical electric field

$$E_{sz} = -\frac{\sigma_H}{\sigma_P} E_{py} \tag{4.52}$$

In addition, the secondary polarization electric field component generates a secondary Hall current component flowing into the y direction

$$j_{sy} = -\sigma_H E_{sz} = \frac{\sigma_H^2}{\sigma_P} E_{py} \tag{4.53}$$

The total current into the eastward direction consists of the sum of the primary Pedersen and the secondary Hall current

$$j_y = j_{py} + j_{sy} = \left(\sigma_P + \frac{\sigma_H^2}{\sigma_P} \right) E_{py} \tag{4.54}$$

The conductivity term appearing on the right-hand side is called *Cowling conductivity*

$$\boxed{\sigma_C = \sigma_P + \frac{\sigma_H^2}{\sigma_P}} \tag{4.55}$$

For typical Hall-to-Pedersen conductivity ratios of 3–4 the Cowling conductivity is an order of magnitude higher than the Pedersen conductivity, explaining the amplification and concentration of the equatorial electrojet current above the equator. The strong horizontal jet current causes a magnetic field disturbance which weakens the horizontal terrestrial magnetic field at the Earth's surface over a distance of about 600 km across the equator (similar to the effect of the ring current field; see Sec. 3.5). Typical disturbance fields near the noon magnetic equator are of the order of 50–100 nT.

4.6. Auroral Emissions

Precipitating electrons do not only produce ionization (see Sec. 4.3). At higher latitudes, where the geomagnetic field lines map to the plasma sheet, collisions between precipitating energetic electrons and neutral atmosphere ions also produce radiation in the visible

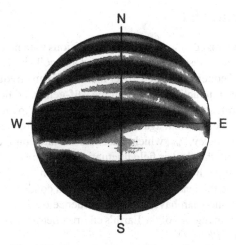

Fig. 4.14. Auroral arcs over Kiruna, Sweden.

wavelength range. This light has been named *aurora* and can be seen from the ground with the naked eye. Intense aurora has an *emission rate* of several million Rayleigh (1 R $= 10^6$ photons/cm^{2}s). The aurora is visible at latitudes of about 70° all around the globe inside a belt called *auroral oval* (see Sec. 1.2 and Fig. 1.5). However, the aurora is not a kind of continuous glow inside this belt, but organized into distinctive structures or auroral forms. Most often it appears as shown in the 180° all-sky camera picture displayed in Fig. 4.14, namely as a couple of thin band-like structures which are aligned with the east-west direction and called *auroral arcs*.

The auroral emissions are the result of processes where neutral atoms or molecules are excited by collisions with precipitating electrons and fall back into their ground state by emitting photons. Often the excitation is accompanied by ionization. The commonly observed green color in the aurora is due to the *auroral green line* of atomic oxygen at 557.7 nm, typically observed at altitudes between 100 and 200 km. At higher altitudes one may observe the *auroral red line* of atomic oxygen at 630.0 nm. Both lines represent so-called forbidden lines. The excited energy states are relatively long-lived (metastable) and under normal atmospheric pressure the excited atom would loose its energy by collisions with other particles rather than by light emission.

Molecular nitrogen produces some weaker violet or blue auroral emission lines at 391.4, 427.0 and 470.0 nm. In the ultraviolet range one finds molecular nitrogen emission lines (around 150 nm) and an atomic oxygen line at 130.4 nm. Auroras also have infrared emissions, for example the molecular oxygen lines at 1270 and 1580 nm.

Concluding Remarks

Except for the partially ionized ionosphere where collisions with neutrals play an important role, most geophysical plasmas are fully ionized and collisionless, in the sense that the Coulomb collision frequency is much lower than the plasma frequency. However, in these plasmas collective interactions, in which the self-generated fields of the particles take over a correlative role in scattering the electrons, may lead to *anomalous collisions*. The plasma may become collisional again, with the collision frequency replaced by an *anomalous collision frequency*, ν_{an}, which must be calculated on the basis of the interaction of the particles with the electric field fluctuations. A good example for anomalous collisions is the electron pitch angle scattering mentioned in Sec. 3.4. Anomalous collisions may also decrease the near-infinite normal plasma conductivity and thus lead to *anomalous resistivity*, which can have a critical influence on the interrelation between plasma and fields in certain regions of the Earth's environment (see our companion book, *Advanced Space Plasma Physics*).

Further Reading

A more complete description of the physics of collisions is found in the first and especially the last of the monographs listed below. Further material on the physics of ionospheres is contained in the second and fourth monograph. The third monograph gives a comprehensive treatment of all phenomena associated with the aurora, including collisional ionization.

[1] J. A. Bittencourt, *Fundamentals of Plasma Physics* (Pergamon Press, Oxford, 1986).

[2] J. K. Hargreaves, *The Solar-Terrestrial Environment* (Cambridge University Press, Cambridge, 1992).

[3] A. V. Jones, *Aurora* (D. Reidel Publ. Co., Dordrecht, 1974).

[4] J. A. Rishbeth and O. K. Garriot, *Introduction to Ionospheric Physics* (Academic Press, New York, 1969).

[5] L. Spitzer, *Physics of Fully Ionized Gases* (Interscience Publishers, New York, 1962).

5. Convection and Substorms

In Chap. 3 we found that energetic particles move across magnetic field lines under the influence of magnetic gradient and curvature forces. Cold plasma particles with near-zero energy, on the other hand, do not feel the magnetic forces because their energy is too low. In the absence of external electric fields, cold particles do not drift at all but stay close to that field line they gyrate about. But that does not mean that the cold plasma in the Earth's magnetosphere is stagnant. Actually, magnetospheric plasma and field lines move together and circulate under the influence of two external forces. The major energy source for the circulation of plasma and field lines in the outer magnetosphere is the kinetic energy of the solar wind. In the inner magnetosphere, the plasma motion is driven by the daily rotation of the Earth.

5.1. Diffusion and Frozen Flux

Cold plasma particles in the collisionless magnetospheric plasma are bound to a specific field line which they cannot leave unless collective effects like the anomalous collisions mentioned on p. 72 violate the adiabatic invariants. Even the more energetic particles will, in regions without strong magnetic field gradients or curvature, stay with the field line they gyrate about. This has an important consequence. Whenever a field line moves due to the action of external forces, cold plasma tied to that field line is also set into motion. The same holds for a moving plasma. Since it cannot leave the field line, the cold plasma will transport the field line along with it. Hence, the motions of the plasma and of the associated flux tube are intimately related.

In order to study the transport of field lines and plasma more quantitatively, we may use the generalized Ohm's law (4.19) to eliminate the electric field in Faraday's law (2.4), obtaining

$$\frac{\partial \mathbf{B}}{\partial t} = \nabla \times (\mathbf{v} \times \mathbf{B} - \mathbf{j}/\sigma_0) \qquad (5.1)$$

Using Ampère's law (2.3) without the $\partial \mathbf{E}/\partial t$ term and noting that $\nabla \cdot \mathbf{B} = 0$, we get a general induction equation for the magnetic field (for the differential vector algebra see

App. A.4)

$$\frac{\partial \mathbf{B}}{\partial t} = \nabla \times (\mathbf{v} \times \mathbf{B}) + \frac{1}{\mu_0 \sigma_0} \nabla^2 \mathbf{B} \qquad (5.2)$$

where σ_0 is the plasma conductivity due to Coulomb or neutral collisions defined in Eq. (4.20). The magnetic field at a point in a plasma can be changed by a motion of the plasma described in the first term on the right-hand side. It can also be changed by diffusion due to the second term on the right-hand side.

Magnetic Diffusion

Assuming the plasma to be at rest and dropping the first term on the right-hand side, Eq. (5.2) becomes a diffusion equation for the magnetic field

$$\frac{\partial \mathbf{B}}{\partial t} = D_m \nabla^2 \mathbf{B} \qquad (5.3)$$

with the *magnetic diffusion coefficient* given by

$$\boxed{D_m = (\mu_0 \sigma_0)^{-1}} \qquad (5.4)$$

Under the influence of a finite resistance in the plasma the magnetic field tends to diffuse across the plasma and to smooth out any local inhomogeneities. Consider the sketch in Fig. 5.1, where we start out with magnetic field lines confined to small regions of space at time t_1. With time the field lines expand away in a diffusive manner, moving through the plasma. When field lines from different regions come to overlap, at a later time t_2, they must be added together vectorially, thereby possibly changing the topological structure of the magnetic field.

The characteristic time of the magnetic field diffusion is found by replacing the vector derivative by the inverse of the characteristic gradient of the magnetic field, L_B. Then the local solution of the diffusion equation is

$$B = B_0 \exp(\pm t / \tau_d) \qquad (5.5)$$

where τ_d is the *magnetic diffusion time* given by

$$\boxed{\tau_d = \mu_0 \sigma_0 L_B^2} \qquad (5.6)$$

Whenever $\sigma_0 \to \infty$ or when the characteristic length, L_B, is very large, the decay or diffusion time can become extremely long, in which case the magnetic field is not able to diffuse efficiently across the plasma.

Such situations may be realized in many geophysical plasmas where the conductivities are high while at the same time the characteristic length scales are huge. Sometimes

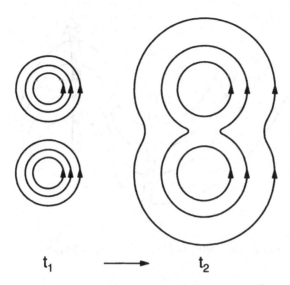

Fig. 5.1. Diffusion of magnetic field lines.

the combination of both yields diffusion times longer than the age of the object or even longer than the age of the universe. It is then clear that the magnetic field does not efficiently diffuse across the plasma.

For an example we consider the solar wind. Its density is of the order of 5 cm^{-3} and its electron temperature is about 50 eV. Since there are no neutral particles, the sole collision frequencies entering the problem are Coulomb collisions between electrons and protons. Using Eq. (4.18) and (5.6) permits us to estimate the magnetic diffusion time as $\tau_d \approx 0.3\, L_B^2$ (if τ_d is given in seconds and L_B is given in meters). The time the solar wind needs to flow with its typical velocity of 500 km/s across the Sun-Earth distance of $1.5 \cdot 10^{11}$ m is $\tau_{sw} \approx 3 \cdot 10^5$ s or 3.5 days. Setting this transit time equal to τ_d, we find that the magnetic field is permitted to diffuse across the solar wind over the very short distance of only $L_B \approx 1.9\,\sqrt{\tau_{sw}} \approx 10^3$ m during the 3.5 days it needs to travel across the Sun-Earth distance. Hence, the magnetic field is practically frozen in the solar wind and carried along with the particle stream (see Sec. 8.1).

The situation changes in the lower E-region where the collision frequency becomes high due to collisions with neutrals (see Fig. 4.10) and the conductivities are of the order of 10^{-3} S/m. Here $\tau_d \approx 10^{-9} L_B^2$, and structures with widths of the order of 10 km become diffusive in times of the order of 1 s. All smaller structures will readily become diffusive, implying that on these temporal and spatial scales the magnetic field can slip through the plasma and vice versa. Narrow structures persisting for longer than a second are not magnetized any more and may move independent of the magnetic field.

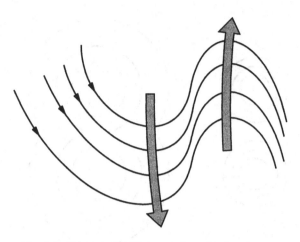

Fig. 5.2. Magnetic field lines moving with the plasma.

Hydromagnetic Theorem

In cases where magnetic diffusion is negligible one speaks about *frozen-in magnetic flux*. We will discuss the meaning of this term by considering the other limiting case where the plasma is in motion but has negligible electrical resistance. In collisionless plasmas with infinite conductivity Eq. (5.2) reduces to

$$\frac{\partial \mathbf{B}}{\partial t} = \nabla \times (\mathbf{v} \times \mathbf{B}) \tag{5.7}$$

This equation for the magnetic field is identical with the equation for the vorticity in the theory of non-viscous fluids, and is interpreted in that theory as implying that any vortex lines move with the fluid. Equivalently, Eq. (5.7) implies that any field changes are such as if the magnetic field lines are constrained to move with the plasma. For example, if patches of plasma populating different sections of a bundle of field lines move into different directions, the field lines will be deformed in the manner shown in Fig. 5.2

In fact, it can be shown that Eq. (5.7) implies that the total magnetic induction encircled by a closed loop remains unchanged even if each point on this closed loop moves with a different local velocity. The field lines are frozen to the plasma and can actually be identified by the plasma glued to it. We will call this identifiable field lines *flux tubes*, where we define a flux tube to be a volume bounded by a surface which is generated by moving a closed loop parallel to the magnetic field lines it intersects at a given time. Thus a flux tube is a kind of generalized cylinder containing a constant amount of magnetic flux. The frozen-in concept implies that all particles and all magnetic flux contained in a certain flux tube at a certain instant will stay inside the flux tube at all instants, independent from any motion of the flux tube or any change in the form of its bounding surface.

Due to the analogy with hydrodynamics, Eq. (5.7) is usually called the *hydromagnetic theorem*. One also finds the name *frozen-in flux theorem* and often the above equation is represented by its equivalent

$$E + v \times B = 0 \qquad (5.8)$$

where we used Faraday's law (2.4) to replace $\partial B/\partial t$. Equation (5.8) shows that in an infinitely conducting plasma there are no electric fields in the frame moving with the plasma. Electric fields can only result from a Lorentz transformation (see Sec. 2.3).

Moreover, Eq. (5.8) contains another important point. Since the cross-product between any velocity component parallel to the magnetic field and the field itself is zero, we can immediately see that any component of the electric field parallel to the magnetic field must vanish in an infinitely conducting plasma.

Magnetic Reynolds Number

The term frozen-in can be given a more precise definition. The induction Eq. (5.2) can be rewritten in simple dimensional form as

$$\frac{B}{\tau} = \frac{VB}{L_B} + \frac{B}{\tau_d} \qquad (5.9)$$

In this equation B is the average magnetic field strength and V represents the average plasma velocity perpendicular to the field, while τ denotes the characteristic time of magnetic field variations, and L_B is again the characteristic length over which the field varies. The second term on the right-hand side describes the diffusion of the magnetic field through the medium, while the first term has the form of a convective derivative, which describes the convective motion of the field with the plasma. The ratio of the first and second term yields the so-called *magnetic Reynolds number*

$$\boxed{R_m = \mu_0 \sigma_0 L_B V} \qquad (5.10)$$

This number is very useful in deciding if a medium is diffusion or flow dominated. In particular, when $R_m \gg 1$ the diffusion term in the induction equation can be entirely neglected. In this case the flow dominates, and the magnetic field simply moves together with the flow: it is frozen-in into the flow. For example, the solar wind magnetic Reynolds number is about $R_m \approx 7 \cdot 10^{16}$, indeed very much larger than one and justifying our previous estimate about the negligible diffusion of the magnetic field in the solar wind. Of course, only the perpendicular velocity enters the frozen-in convective term. Any flow parallel to the magnetic field has no consequences. On the other hand, when

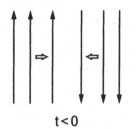 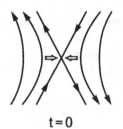 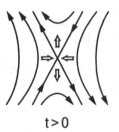

Fig. 5.3. Evolution of field line merging.

$R_m \approx 1$, diffusion becomes important and may dominate. The magnetic field is not any more frozen into the plasma and may slip across the plasma. In particular, in a diffusion dominated region, the plasma can freely stream across the magnetic field without any remarkable effect on the latter.

The magnetic Reynolds number R_m is proportional to the conductivity, the length scale, and the velocity. Increase in any of these quantities will therefore lead to the dominance of frozen-in conditions. In most large-scale and dilute plasmas of natural origin one therefore finds that the magnetic fields are frozen-in for most of the time scales of interest. Reasonable plasma flow velocities are normally restricted to low values. So the dominant quantities determining the frozen-in behaviour are the conductivities and length scales.

Magnetic Merging

There is one particular situation where both the frozen-in flux concept and its breakdown are equally important. This is the process of *magnetic merging*, where field lines are cut and reconnected to other field lines, thus changing the magnetic topology. This process is theoretically quite complicated and will be detailed in a later chapter. Here, we will deal only with the basic structure of this process as illustrated in Fig. 5.3.

Consider a magnetic topology with antiparallel field lines frozen into the plasma, like sketched in the left-hand diagram of Fig. 5.3. Such a topology exists around thin current sheets like at the magnetopause and in the tail neutral sheet (see Fig. 1.6). If the field lines on both sides of the current sheet are stagnant and do not move, such a topology may be quite stable over long times.

However, when plasma and field lines on both sides move toward the current sheet, the situation may change. Whenever the magnetic Reynolds number becomes equal or greater than one in even a small volume of space due to anomalous collisions (see p. 72), the magnetic field may vanish due to diffusion at a particular point. This results in the X-type configuration shown in the middle panel of Fig. 5.3, with the magnetic field being zero at the center of the X, the magnetic *neutral point*. The field lines forming the

X and passing through the neutral point are called *separatrix*.

As a result we will get the situation sketched on the right-hand side of Fig. 5.3. Plasma and field lines are being transported toward the neutral point from either side. At the neutral point the antiparallel field lines are cut into halves and the field line halves from one side are reconnected with those from the other side. The merged field lines are then expelled from the neutral point. The merged field lines will be populated by a mixture of plasma from both sides of the current sheet. As long as oppositely directed flux tubes are being pushed toward each other from both sides and as long as anomalous resistivity lets the magnetic field vanish inside a small volume of space, the process of field line merging continues.

5.2. Convection Electric Field

The concurrent drift of the plasma and the field lines as a whole is often called *convection*. Due to the infinite conductivity, the electric field is zero in the frame of reference moving with the plasma and the flux tubes at a velocity v_c. However, according to the Lorentz transformation (2.21), an observer in the Earth's fixed frame of reference will measure an electric field

$$\mathbf{E}_c = -\mathbf{v}_c \times \mathbf{B} \qquad (5.11)$$

It is this electric field which is referred to by the name *convection electric field*.

Merging and Reconnection

The ultimate source of magnetospheric convection is the momentum of the solar wind flow. According to Eq. (5.11) the flow of the magnetized solar wind represents an electric field in the Earth's frame of reference. But since the solar wind cannot penetrate the magnetopause (see Sec. 1.2), this electric field cannot directly penetrate into the magnetosphere either. However, when the interplanetary magnetic field has a southward component, the northward directed terrestrial field lines at the dayside magnetopause are allowed to merge with the interplanetary magnetic field (see Sec. 5.1).

When a southward directed interplanetary field line (denoted by 1 in Fig. 5.4) encounters the magnetopause, it will merge with the closed terrestrial field line 1, which has both footpoints on the Earth and is transported to the magnetopause from the inside. The merged field lines will split into the two open field lines marked by 2, each of which has one end connected to the Earth and the other stretching out into the solar wind. Subsequently, the solar wind will transport this field line down-tail across the polar cap (field lines marked 3–6) and due to the stiffness of the field line, the magnetic tension (see Sec. 7.3), the magnetospheric part of the field line (inside the shaded region) will also be transported down-tail.

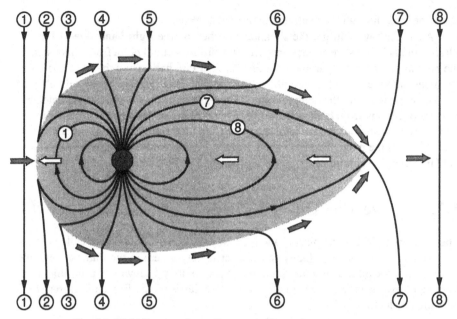

Fig. 5.4. Field line merging and reconnection at the magnetopause.

At the nightside end of the magnetosphere, around 100–200 R_E downtail, the two open field line halves will meet again and reconnect, leaving a closed but stretched terrestrial field line in the magnetotail and an open solar wind field line down-tail of the magnetosphere (denoted by 7 and 8 in Fig. 5.4). Due to magnetic tension, the stretched tail field line marked by 8 will relax and shorten in the Earthward direction. During this relaxation it transports the plasma to which it is frozen toward the Earth. This is the reason for the Earthward convective flow of plasma in the magnetotail. Moreover, under equilibrium conditions, the field line will eventually be brought back to the frontside magnetosphere and replace the terrestrial field line denoted by 1 in Fig. 5.4, since otherwise the dayside magnetosphere would soon be devoid of magnetic flux. Provided the interplanetary magnetic field still has a southward component, the same cycle can be repeated.

Frontside merging and tail *reconnection* do not occur at singular points but rather along a line. Such a line is called *X-line*, since along this line the magnetic field lines have a topology resembling this letter (see Fig. 5.3). The X-line is aligned more or less perpendicular to the plane shown in Fig. 5.4. It also bears the name *neutral line* since the magnetic field strength vanishes at the points forming this line. The latter name is typically used for the nightside X-line.

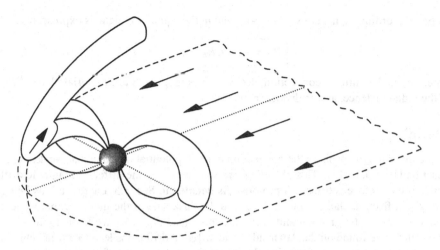

Fig. 5.5. Flux tube and plasma convection caused by magnetic merging.

Convection Electric Field

As explained above, the circulation from the frontside over the polar cap to the magnetotail and back to the dayside near the equatorial plane is not only experienced by the terrestrial field lines but also by all the particles gyrating about them. Hence, as visualized in Fig. 5.5, magnetic merging will drive a tailward plasma flow on open field lines across the polar caps and the magnetospheric lobes and complete the cycle by sunward convection in the inner magnetosphere.

Magnetic merging works most efficient for a southward orientation of the interplanetary magnetic field. In fact, there is a lot of experimental evidence that magnetospheric convection is strongly enhanced during periods of southward interplanetary magnetic field. For example, the magnetic storms discussed in Sec. 3.5, which are caused by enhanced convection, occur mainly during periods of prolonged southward interplanetary magnetic field.

The sunward transport of plasma in the inner magnetosphere caused by magnetic reconnection at the Earth's magnetopause is, for an observer on the Earth, equivalent to an electric convection field. The total potential difference between the dawn and dusk magnetopause (or, equivalently, across the polar cap) corresponds to about 50–100 kV. Taking an average cross-section of the magnetosphere of about 30 R_E, this amounts to a dawn-to-dusk directed field of some 0.2–0.5 mV/m.

The dawn-to-dusk directed convection electric field can be assumed as homogeneous in a first approximation. The same holds for the associated convection electric potential defined by

$$\mathbf{E}_c = -\nabla \phi_c \qquad (5.12)$$

In polar coordinates, the convection potential in the equatorial plane is expressed as

$$\phi_c = -E_c L R_E \sin \psi$$ (5.13)

where E_c is the uniform convection electric field strength in the equatorial plane, $L R_E$ is the radial distance, and ψ denotes azimuth.

Shielding

In the inner magnetosphere the convection electric potential is somewhat weaker than described by Eq. (5.13). This *shielding effect* is caused by the different magnetic drift paths of energetic electrons and protons. As detailed in Sec. 3.3, energetic ions being brought in from the tail tend to drift toward the dusk side of the inner magnetosphere due to the magnetic gradient and curvature force (see Fig. 3.8), while energetic electrons tend to be found on the dawn side. The different drift paths lead to a weak charge separation with a surplus of positive charges on the dusk side and of negative charges on the dawn side. The polarization electric field associated with this charge separation is directed from dusk to dawn and will shield the inner magnetosphere from the full dawn-to-dusk directed cross-tail convection electric field. Taking the shielding effect into account leads to a more realistic form of the convection potential

$$\phi_{cs} = -A_\gamma (L R_E)^\gamma \sin \psi$$ (5.14)

where γ is the shielding factor and A_γ is a constant described by

$$A_\gamma = 0.5 \Delta\phi \Delta y^{-\gamma}$$ (5.15)

with $\Delta\phi$ denoting the cross-tail potential difference and Δy half the distance between the dawn and dusk magnetopause along the $\psi = \pm 90°$ axis. Under typical conditions, the shielding factor γ ranges between 2 and 3. Calculating the electric field amplitude from Eq. (5.14) one obtains

$$E_{cs} = A_\gamma (L R_E)^{\gamma-1} \left[(\gamma^2 - 1) \sin^2 \psi + 1 \right]^{1/2}$$ (5.16)

For $\gamma = 1$ we recover a uniform electric field, while for a realistic shielding factor $\gamma \approx 2 - 3$ the electric field amplitude decreases toward the inner magnetosphere and varies with local time.

5.3.   Corotation and Plasmasphere

To obtain the full magnetospheric plasma motion, one must add yet another electric field to the convection field. This field is again caused by a movement of the field lines and the plasma tied to it. But in this case the plasma motion is not caused by the solar wind but by the Earth's rotation.

The ionospheric plasma is only partially ionized and the neutral collision frequency defined in Sec. 4.1 is rather high. Hence, the neutral atmosphere particles which corotate with the Earth will force the ionospheric plasma into *corotation* via ion and electron-neutral collisions. As we have seen in Sec. 5.1, the ionospheric conductivity is finite due to the high collision frequency and the magnetic field is not totally frozen-in. But as long as no other forces are exerted on the field lines, they will corotate with the Earth. This condition is typically valid in the low and mid-latitude ionosphere. In the high-latitude ionosphere, the solar wind-induced convection exerts a controlling influence on the field lines and causes them to slip through the collision-dominated and thus not infinitely conducting ionospheric plasma (see p. 75 and Sec. 5.4).

Corotation Electric Field

If we neglect the aforementioned slippage of the magnetic field lines at auroral and polar latitudes, the corotation of plasma and flux tubes is, for a non-rotating observer, equivalent to an electric field

$$\mathbf{E}_{cr} = -(\mathbf{\Omega}_E \times \mathbf{r}) \times \mathbf{B} \tag{5.17}$$

where $\Omega_E = 7.27 \cdot 10^{-5}$ rad s^{-1} is the angular velocity of the Earth's rotation. Using Eq. (3.8) to express the magnetic field, we obtain for the corotation electric field strength in the equatorial plane

$$E_{cr} = \Omega_E B_E R_E / L^2 \tag{5.18}$$

Here, the corotation field is directed radially inward and decreases with the square of the L-value. We can calculate the equatorial plane electric potential and obtain

$$\boxed{\phi_{cr} = -\Omega_E B_E R_E^2 / L} \tag{5.19}$$

where the term $\Omega_E B_E R_E^2$ amounts to 92 kV. In the equatorial plane the potential distribution is radially symmetric and consists of concentric circles with the inter-circle distances decreasing with L.

Plasmasphere

The radial decrease of the corotation potential leads to a weakening of the influence of corotation with distance. At larger distances the convection potential will take over and dominate the drift of the (cold) plasma. Neglecting the shielding effect, the total electric potential in the equatorial plane is given by the sum of the two potentials $\phi_c + \phi_{cr}$ and is depicted in Fig. 5.6.

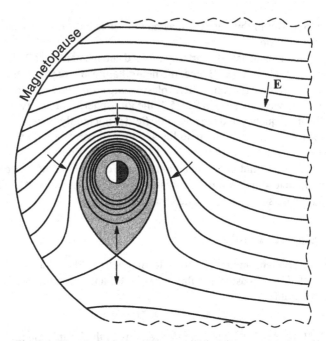

Fig. 5.6. Electric equipotential contours in the equatorial plane.

Figure 5.6 shows two topologically different regions. Close to the Earth a region of closed equipotential contours exists, marked by the shading. Here, within the *plasmasphere*, the electric field is directed inward and corotation dominates. The plasma content of a flux tube is nearly constant and, accordingly, the plasma density is rather high, several 10^3 cm^{-3}. The plasmaspheric particles originate from the ionosphere and are cold, with energies in the range of 1 eV. Outside this region, the potential contours are open, the electric field has a strong duskward component and a magnetic flux tube will, at some time, encounter the dayside magnetopause and loose its plasma to the magnetosheath. This is the reason for the sharp density gradient between the plasmasphere and the outer region, where less than one particles is found per cubiccentimeter. Figure 5.7 shows an average ion density profile in the night-side magnetosphere. The sharp density gradient at the outer boundary of the plasmasphere is called *plasmapause* and not only observed on the night side, but everywhere along the border of the shaded region in Fig. 5.6.

An interesting region is the *stagnation point* in Fig. 5.6, This is the point on the evening side, where the plasmasphere has a *bulge* and where an equipotential contour crosses itself. Here the eastward directed corotation and the westward directed convection have the same velocity and the plasma is stagnant. Since $\phi_c = \phi_{cr}$ and $\sin \psi = 1$ at this point, the distance of the stagnation point from the Earth's surface can easily be

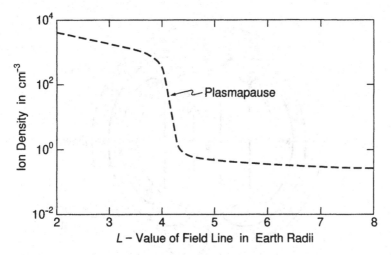

Fig. 5.7. Average ion density profile in the night-side magnetosphere.

calculated as

$$L_{sp} = \left(\frac{\Omega_E B_E R_E}{E_c} \right)^{1/2}$$ (5.20)

where $\Omega_E B_E R_E = 1.45 \cdot 10^{-2}$ Vm. Thus, for a convection electric field of 1 mV/m the stagnation point is located at a distance of 3.81 R_E. The radial distance of the stagnation point depends on the strength of the convection electric field: whenever the convection electric field increases, the radius of the plasmasphere decreases and vice-versa.

5.4. High-Latitude Electrodynamics

Until now we have considered the motion of plasma and flux tubes in the equatorial plane of the magnetosphere. But the circulation of the flux tubes is not restricted to the equatorial plane, it also affects the lower ends of the flux tubes, the foot points of the field lines in the high-latitude ionosphere. Since the ionosphere is conducting, the electric field associated with the ionospheric convection will drive ionospheric currents.

Ionospheric Convection

The motion of the flux tubes across the polar cap due to magnetic merging depicted in Fig. 5.4 also moves the ionospheric footpoint of the flux tube and the plasma tied to it across the polar cap to the nightside. Similarly, the sunward convection of magnetospheric flux tubes shown in Fig. 5.5 leads to a sunward convection of the foot points of

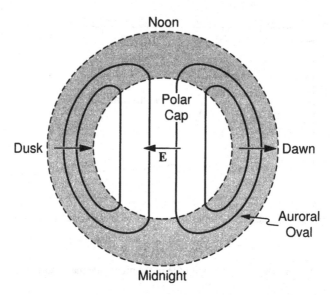

Fig. 5.8. Equipotential contours of the high-latitude electric field.

these flux tubes in the dawn and duskside high-latitude ionosphere, inside the auroral oval shown in Fig. 1.5. This leads to a two-cell convection pattern in the polar ionosphere.

The convection pattern is equivalent to an electric potential pattern. This can be recognized from the definition of the E×B drift in Eq. (2.19) when using Eq. (5.12) to substitute the electric field by the gradient of the electric potential

$$\mathbf{v}_E = -(\nabla\phi \times \mathbf{B})/B^2 \qquad (5.21)$$

Because the gradient of the electric potential, $\nabla\phi$, is perpendicular to the contours of constant potential, ϕ, and the E×B drift is perpendicular to both $\nabla\phi$ and $\mathbf{B}$, the convection streamlines are also equipotential contours. Cold particles will drift along these contours. Drawing equipotential contours and drawing E×B drift trajectories of the plasma is therefore equivalent.

Hence, we can translate the two-cell convection pattern into a two-cell pattern of equipotential contours shown in Fig. 5.8. This equipotential pattern is equivalent to an ionospheric electric field that is directed toward dusk in the northern polar cap. Inside the northern hemisphere auroral oval the electric field is directed toward the pole on the duskside, while it has a southward direction in the morning hours.

Height-Integrated Ohm's Law

Since the ionospheric conductivity is a tensor with three different components (see Sec. 4.2), three types of currents will be generated by the convection electric field. The first type are the field-aligned currents flowing parallel to the magnetic field into and out of the ionosphere. Secondly, there are the Pedersen currents which flow perpendicular to the magnetic field lines and parallel to the ionospheric convection field. Finally, Hall currents will flow perpendicular to both the magnetic and the electric field.

As discussed in Sec. 4.2, the ionospheric conductivity along the geomagnetic field lines, $\sigma_\parallel$, is always much higher than the transverse conductivities, σ_P and σ_H. For reasons of current continuity, the electric field component along the magnetic field lines, $E_\parallel$, will, therefore, generally be much smaller than the perpendicular field, $E_\perp$. Often the magnetic field lines may be considered as perfectly conducting and the parallel electric field vanishes (see p. 77). At high latitudes, where the field lines are nearly vertical, the horizontal electric field becomes almost height-independent and one may introduce the height-integrated quantities

$$\Sigma_P = \int \sigma_P dz$$
$$\Sigma_H = \int \sigma_H dz \qquad (5.22)$$
$$\mathbf{J}_\perp = \int \mathbf{j}_\perp dz$$

The height-integrated conductivities or *conductances* and the height-integrated horizontal current density are related through the height-integrated version of Ohm's law. The latter follows by integrating the transverse component of Eq. (4.28), yielding

$$\mathbf{J}_\perp = \Sigma_P \mathbf{E}_\perp - \Sigma_H (\mathbf{E}_\perp \times \mathbf{B})/B \qquad (5.23)$$

Here, we have neglected the contribution of the neutral wind, which is important at low- and mid-latitudes (see Sec. 4.5), but at high latitudes is much weaker than the motion due to the convection electric field.

The field-aligned current is not specified by Eq. (5.23). But since the total current flow must be continuous, it can be calculated from the divergence of the height-integrated horizontal current. For vertical field lines one obtains

$$j_\parallel = \nabla_\perp \cdot \mathbf{J}_\perp \qquad (5.24)$$

where $\nabla_\perp$ denotes the vector derivative in the horizontal plane. Inserting Eq. (5.23) into (5.24) we obtain after some vector algebra (see App. A.4)

$$j_\parallel = (\nabla_\perp \Sigma_P) \cdot \mathbf{E}_\perp - (\nabla_\perp \Sigma_H) \cdot (\mathbf{E}_\perp \times \mathbf{B})/B + \Sigma_P (\nabla_\perp \cdot \mathbf{E}_\perp) \qquad (5.25)$$

where we haven't taken into account that the ionospheric magnetic field is both stationary and uniform on the typical scales involved and thus $\nabla_\perp \cdot (\mathbf{E}_\perp \times \mathbf{B}) = 0$. Accordingly, field-aligned currents are generated at gradients of the Pedersen conductance along the electric field direction, at gradients of the Hall conductance which are aligned perpendicular to the horizontal electric field, and in regions where the divergence of the electric field is non-zero.

Polarization Electric Fields

The primary electric field in Eq. (5.23) is the magnetospheric convection electric field mapped down to the ionosphere as sketched in Fig. 5.8. Current continuity at conductivity gradients is usually served for by field-aligned currents. However, despite the fact that the parallel conductivity is near-infinite, the field-aligned current density cannot grow above certain levels. As we will see in more detail later, the field-aligned current density is restricted by the number of charge carriers that can be moved along the field line with a high-enough velocity. Moreover, the field-aligned currents have to be closed somewhere in the magnetosphere and the current transverse to the magnetic field may also be limited.

At strong conductance gradients the field-aligned current density due to the first two terms on the right-hand side of Eq. (5.25) may become larger than the limiting value. It is hardly possible to alter the conductivity pattern since it is imposed by the structure of the particle precipitation. But field-aligned current resulting from the $\nabla \Sigma$ terms may be balanced by altering the electric field distribution in the last term on the right-hand side of Eq. (5.25).

Maxwell's equation (2.6) tells us that changing the divergence of the electric field is equivalent to adding polarization charges. Hence, the secondary electric field that has to be added to the convection electric field in order to limit the field-aligned current is a polarization electric field introduced in Sec. 4.5.

Joule Heating

Ionospheric currents may heat the atmosphere by Ohmic dissipation or Joule heating. The *Joule heating rate* is proportional to the current flowing parallel to the electric field. The height-integrated Joule heating rate can be written as

$$Q_J = \mathbf{J}_\perp \cdot \mathbf{E}_\perp \tag{5.26}$$

Since the Hall currents flow perpendicular to the electric field, they do not contribute to Ohmic dissipation. Thus Eq. (5.26) can be rewritten as

$$Q_J = \Sigma_P E_\perp^2 \tag{5.27}$$

This expression shows that the amount of Ohmic heat produced in the auroral ionosphere depends more on the level of convection than on the number of precipitating particles.

5.5. Auroral Electrojets

Since particles precipitating into the auroral oval cause significant ionization (see Sec. 4.3), its conductivity is much higher than that of the polar cap. As a result, the high-latitude current flow is concentrated inside the auroral oval, where it forms the *auroral electrojets*. The auroral electrojets are the most prominent currents at auroral latitudes. They carry a total current of some 10^6 A. This is the same order of magnitude as the total current carried by the ring current discussed in Sec. 3.5 , but since the auroral electrojets flow only 100 km above the Earth's surface, they create the largest ground magnetic disturbance of all current systems in the Earth's environment. The disturbance fields have typical magnitudes of 100–1000 nT, but may reach 3000 nT during the largest magnetic storms. The latter is a sizable fraction (about 5%) of the terrestrial dipole field at high latitudes. The present knowledge about the conductivity structure, the electric field distribution and the current flow associated with the auroral electrojet system is summarized in Fig. 5.9.

Conductances and Electric Fields

Inside the auroral oval, which is approximately an off-center ring shifted by an average 5° from the magnetic pole toward magnetic midnight, the ionospheric conductivity is enhanced above the solar ultraviolet-induced level due to the ionization of neutral atoms and molecules by precipitating electrons and ions (see Secs. 4.3 and 4.4). The energetic particles drift toward and around the Earth (see Fig. 3.8) and precipitate, depending on their energy and pitch angle, in different local time sectors. The precipitation pattern is reflected in the conductivity structure. The weakest conductivities are found near the noon sector while the conductivity maximum lies in the midnight sector where typical values of 7–10 S and 10–20 S for Pedersen and Hall conductances, respectively, are found.

The electric field pattern in the auroral oval reflects the large-scale pattern of magnetospheric plasma convection. The transport of open and closed flux tubes results in a convection pattern with two cells sketched in Fig. 5.8. In reality, the convection pattern is slightly more complicated but still resembles a two-cell pattern. The auroral zone electric field associated with the two-cell system of plasma transport has typical values between 20 and 50 mV/m. It is poleward directed in the afternoon and early evening sector, points equatorward in the postmidnight and morning sector and rotates from north over west to south in the premidnight sector. This region of field rotation is called the *Harang discontinuity* region. The clockwise rotation of the pattern, as compared to the idealized picture in Fig. 5.8, has its origin in a weak polarization electric field (see p. 88), which is directed from midnight to noon and stems from electric charges built up at the boundary between the highly conducting auroral zone and the polar cap.

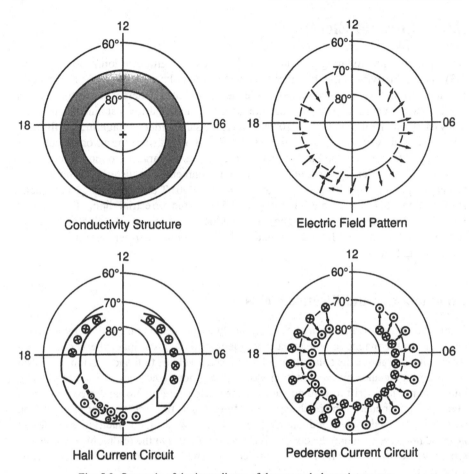

Fig. 5.9. Synopsis of the ingredients of the auroral electrojet system.

Ionospheric and Field-Aligned Currents

As summarized in the lower left panel of Fig. 5.9, the eastward and westward electrojets are primarily Hall currents which originate around noon and are fed by downward
field-aligned currents. Typical sheet current densities range between 0.5 and 1 A/m and
increase toward midnight due to the increasing Hall conductance. The eastward electrojet flows in the afternoon sector and terminates in the region of the Harang discontinuity
where it partially flows up magnetic field lines and partially rotates northward, joining
the westward electrojet. The westward electrojet flows through the morning and midnight sector and typically extends into the evening sector along the poleward border of

the auroral oval where it also diverges as upward field-aligned currents.

The Pedersen currents shown in the lower right panel of Fig. 5.9 have typical densities of 0.3–0.5 A/m. They flow northward in the eastward electrojet region and are connected to sheets of downward and upward field-aligned currents in the southern and northern half of the afternoon-evening northern hemisphere auroral oval, respectively. In the midnight-to-noon sector the Pedersen current flows equatorward and field-aligned currents provide continuity by flowing upward in the southern and downward in the poleward half of the auroral oval. The sheets of field-aligned currents in the poleward half of the auroral oval were named *Region-1 currents* and those in the equatorward half are called *Region-2 currents*. Inside the Harang discontinuity region, the evening and morning side Pedersen current circuits overlap, leading to three sheets of field-aligned currents.

As sketched in Fig. 1.6, the Region-1 and -2 field-aligned current belts extend into the magnetosphere. The Region-2 currents in the equatorward belt are closed by the westward ring current in the near-Earth equatorial plane. On the other hand, the Region-1 currents in the poleward half flow along the high-latitude boundary of the plasma sheet. Deep in the magnetotail, they merge with the neutral sheet current. The upward and downward directed field-aligned currents continuing the Hall currents in the midnight and noon sector are closed in the magnetosphere by the evening-side partial ring current mentioned in Sec. 3.3.

Joule Heating and *AE* Index

Inserting the average values for the Pedersen conductivity and the electric field given above into Eq. (5.27), one finds that typical Joule heating rates due to the auroral electrojets are in the range 5–50 mW/m^2. While this number does not seem large, the total heat input into the auroral zone is quite significant. The total Joule heat input into one hemisphere can be estimated from the *auroral electrojet index*, which was introduced as a measure of global auroral electrojet activity (see App. B.3).

The *AE* index is based on readings of the northward magnetic disturbance component from twelve auroral zone observatories located in different local time zones and calculated for any given instant of universal time as the difference between the maximum northward and southward disturbance field (which are thought to represent the maximum current in the eastward and westward electrojet). Empirically one has found that a total hemispheric Joule heat input of 0.3 GW is equivalent to 1 nT in *AE*.

Taking into account both hemispheres and adding the heating of the upper atmosphere due to particle precipitation, which empirically is known to be about half as effective as the Joule heating for the same level of *AE*, we find that the total heating of the upper atmosphere amounts to about 1 GW per nT in *AE*. During major magnetic storms, when *AE* can easily reach 1000 nT and above, the total heat input at high latitudes amounts to more than 10^{12} W.

5.6. Magnetospheric Substorms

Until now we have treated convection as a stationary process. First, we tacitly assumed that magnetic merging between interplanetary and terrestrial field lines at the frontside magnetopause occurs always at the same rate. Second, it was implicit in our scenario shown in Fig. 5.4 that dayside merging and reconnection in the distant magnetotail are in equilibrium.

Let us now look closer at what really happens on the dayside. The amount of dayside magnetic flux merged per unit time, the dayside *reconnection rate*, depends on the number of southward oriented interplanetary field lines that get into contact with the Earth's magnetopause during a given time interval. Thus it depends on the solar wind velocity and on the magnitude of the southward interplanetary magnetic field component. In particular the latter is rather variable. Moreover, for considerable periods of time the interplanetary field is directed northward, rendering any dayside merging impossible. Hence, we have intervals when the magnetosphere is very quiet and convection will cease as well as other periods where a lot of flux is merged on the dayside and the magnetospheric plasma will be active.

Eventually, all magnetic flux transported to the tail has to be reconnected and convected back to the frontside magnetosphere. But there is no need for identical instantaneous dayside and nightside reconnection rates, only the average rates must be equal. Actually, it has been found that only part of the flux transported into the tail is reconnected instantaneously and convected back to the dayside. The remaining field lines become added to the tail lobes, where they increase the magnetic flux density. After about one hour these intermediately stored field lines are suddenly reconnected in the tail and their magnetic energy is explosively released.

The sudden reconnection of previously stored flux tubes has rather dramatic effects on the magnetospheric plasma and associated phenomena like aurora and magnetospheric and ionospheric currents. These effects, which last for about one or two hours and will be detailed below, have been summarized under the term *magnetospheric substorm*, since they form the basic element of the magnetic storm described in Sec. 3.5. Whenever a couple of major substorms occur in a row, the ring current increases significantly and the low-latitude magnetograms exhibit a magnetic storm.

Substorm Growth

The substorm starts when the dayside merging rate is distinctively enhanced, typically due to a southward turning of the interplanetary magnetic field. The flux eroded on the dayside magnetopause is transported into the tail. Part of the flux is reconnected and convected back to the frontside of the magnetosphere. The enhanced convection due to the driven substorm component causes enhanced current flow in the auroral electrojets and an associated growth of the AE index.

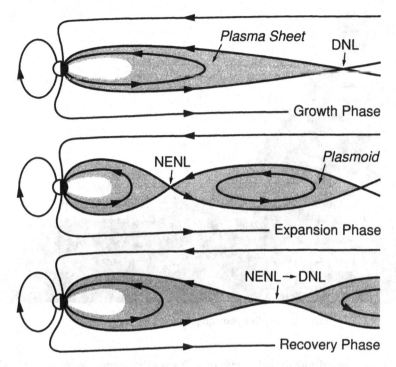

Fig. 5.10. Reconfiguration of the plasma sheet during a substorm.

At the same time the magnetic flux eroded from the dayside magnetopause which is not reconnected is added to the tail lobes. Since the magnetic field in the tail lobes and the neutral sheet current are related by Biot-Savart's law (see App. A.5), the growth of the tail lobe magnetic field must be accompanied by a growing neutral sheet current. The growth of the latter will also stretch the field lines threading the plasma sheet into a more tail-like configuration. This is sketched in the upper panel of Fig. 5.10.

The period of enhanced convection and loading of the tail with magnetic flux is called *substorm growth phase*. It typically lasts for about one hour. After that time period too much magnetic flux and thus magnetic energy has been accumulated in the tail. The tail becomes unstable and tries to get rid of the surplus energy. This is the time of *substorm onset* and the beginning of the *substorm expansion phase*.

Substorm Onset and Expansion

During the substorm expansion phase, which typically lasts about 30–60 min, rather dramatic changes are seen in the magnetosphere and auroral zone ionosphere. One of those

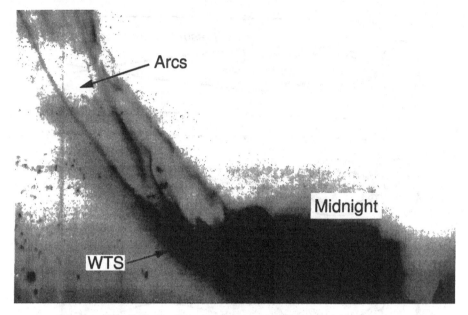

Fig. 5.11. Satellite photograph of evening-side aurora.

can even be observed with the naked eye, or with a satellite camera as in Fig. 5.11, which shows a negative image of the evening auroral oval over Scandinavia and western Russia taken by a satellite flying high above the auroral oval. During the growth phase the aurora appears in the form of *auroral arcs*. These are structures which are widely stretched azimuthally along the auroral oval, but very thin in their north-south extent (see Figs. 4.14 and 5.11). At substorm onset one of these arcs suddenly brightens and fills the whole sky. The *auroral break-up* first occurs in the midnight sector, but then expands rapidly northward and, in particular, westward. As can be seen from Fig. 5.11, the westward motion of the break-up aurora looks like that of a surge. Accordingly, this type of aurora has been named *westward traveling surge* or WTS.

Not only the aurora changes dramatically at substorm onset. The sharp increase in the *AE* index on the left-hand side of Fig. 5.12 to values of about 500 nT indicates that the ionospheric current flow is strongly enhanced. Moreover, the stretched magnetic field in the plasma sheet suddenly becomes more dipolar again, as is evident from the right-hand panel of Fig. 5.12 where the average magnetic field elevation in the plasma sheet, i.e., the angle between the magnetic field direction and the equatorial plane, rises from less than $10°$ to more than $30°$ during the first 20 min after substorm onset.

The dipolarization of the magnetotail field is the signature of a dramatic reconfiguration of the plasma sheet sketched in the middle panel of Fig. 5.10. Around $30\,R_E$

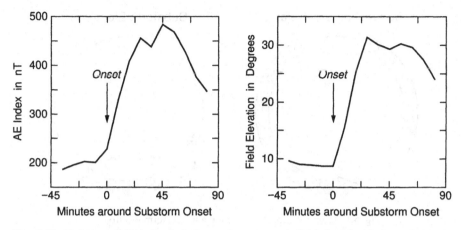

Fig. 5.12. Variation of AE index and plasma sheet magnetic field elevation during substorms.

downtail, a new neutral line is formed. To distinguish it from the distant neutral line at around 100–200 R_E, the newly formed X-line is usually called *near-Earth neutral line*. The excess magnetic flux deposited in the tail during the growth phase is reconnected along this new X-line during the expansion phase. After the stretched field lines are reconnected, they move back to their normal more dipolar shape.

The large region of the tail between the two neutral lines forms a *plasmoid*. The magnetic field inside the plasmoid has a peculiar structure. Its field lines are neither connected to the terrestrial nor to the interplanetary magnetic field, but form closed loops.

We should add a word of caution. The exact configuration of the magnetic field in the vicinity of the near-Earth neutral line is not fully known yet. It may be a single large-scale X-line like the one shown in Fig. 5.10. But there is also the possibility that reconnection proceeds along multiple neutral lines. Similarly, if the magnetic field in the plasmoid has a dawn-dusk component, the field lines may form a three-dimensional helix rather than a two-dimensional closed loop.

Substorm Recovery

Figure 5.12 shows that about 45 min after substorm onset the ionospheric current flow and the strong dipolar field orientation in the tail plasma sheet start to decrease again. At this instant also the aurora starts to fade and retreats to higher latitudes. In general, reconnection at the near-Earth neutral line ceases and substorm activity settles. This is the begin of the *substorm recovery phase*. This phase lasts for about 1–2 hours and ends when the magnetosphere has returned to a quiet state. However, during intervals when the interplanetary magnetic field has a stable southward orientation, the recovery phase

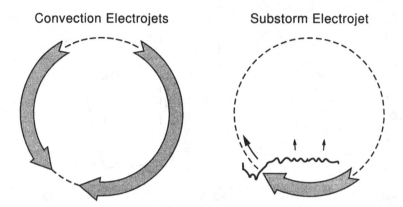

Fig. 5.13. Comparison of convection and substorm electrojets.

of one substorm may coincide with the growth phase of the next substorm.

Besides the general decrease in activity, satellite observations have identified a important process that occurs during the recovery phase. As sketched in the lower panel of Fig. 5.10, the near-Earth neutral line starts to retreat tailward at the time of maximum expansion. In doing so, it pushes the plasmoid tailward until the latter is finally ejected from the magnetotail. Its plasma is lost to the downtail solar wind and the former near-Earth neutral line has become the distant neutral line.

5.7. Substorm Currents

During magnetospheric substorms the ionospheric current flow is affected in two ways. On the one hand, the current flow in the auroral electrojets sketched on the left-hand side of Fig. 5.13 increases along with the enhanced convection. The strengthening of the convection electrojets is already seen during the growth phase and is caused mainly by the increasing convection electric field.

In addition to the overall growth of the auroral electrojet current, the unloading of magnetic flux previously stored in the magnetotail leads to the formation of a substorm electrojet with strongly enhanced westward current flow in the midnight sector. The *substorm electrojet* is concentrated in the region of active break-up aurora and expands westward during the course of the expansion phase along with the westward traveling surge. In contrast to the convection electrojets, the strength of the substorm electrojet current is mainly determined by the strong increase in ionospheric conductance due to the strong particle precipitation in the bright substorm aurora.

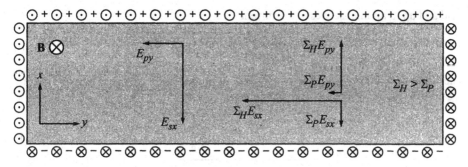

Fig. 5.14. Generation of a Cowling channel inside the westward traveling surge.

Substorm Electrojet

Since the substorm electrojet is governed by the strong increase of the conductivities inside the westward traveling surge, the situation is similar to that in the equatorial electrojet described in Sec. 4.5. However, in the present case the Cowling channel is not perfect since field-aligned currents will remove part of the space charge deposited at the boundaries of the highly conducting channel.

The generation of an imperfect Cowling channel is summarized in Fig. 5.14. Assuming that the conductivities are negligible outside the highly conducting strip, we have the following situation. The primary convection electric field possesses a substantial westward component in the midnight sector (see Fig. 5.9). This primary westward electric field component, E_{py}, drives a primary northward Hall current, $\Sigma_H E_{py}$, across the highly conducting strip. It turns out that only a fraction of this current can be closed via field-aligned currents at the strong conductivity gradients along the northern and southern boundaries (see p. 88). The excess Hall current deposits positive charges at the northern border of the high-conductivity channel while negative charges build up at its southern boundary. These charges give rise to a southward polarization electric field, E_{sx}, which drives a southward Pedersen current, $\Sigma_P E_{sx}$, to balance that part of $\Sigma_H E_{py}$ which is not continued via field-aligned currents.

Introducing the fraction, α_p, of the primary Hall current which is not closed via field-aligned current, we obtain a relation between the primary and secondary electric fields

$$\alpha_p \Sigma_H E_{py} - \Sigma_P E_{sz} = 0 \qquad (5.28)$$

which yields for the secondary southward polarization electric field

$$E_{sx} = \alpha_p \frac{\Sigma_H}{\Sigma_P} E_{py} \qquad (5.29)$$

The westward currents due to the primary convection and secondary polarization electric field add up to an intense westward current

$$J_y = \Sigma_P E_{py} + \Sigma_H E_{sz} = \left(\Sigma_P + \alpha_p \frac{\Sigma_H^2}{\Sigma_P} \right) E_{py} \qquad (5.30)$$

The sum of conductivities appearing on the right-hand side of this expression can be regarded as an imperfect Cowling conductance

$$\Sigma_C' = \Sigma_P + \alpha_p \frac{\Sigma_H^2}{\Sigma_P} \qquad (5.31)$$

For a typical Hall-to-Pedersen conductance ratio of about four and $\alpha_p = 0.5$, the Cowling current is about one order of magnitude stronger than the normal westward Pedersen current.

Also the westward Cowling current encounters conductance gradients at the western end eastern boundary of the channel. But here all current can be closed via field-aligned currents. The strong conductance gradient at the western edge leads to intense localized upward field-aligned currents near the head of the westward traveling surge. Under normal conditions, the density of these upward field-aligned currents would be much higher than what can be carried by the ambient plasma. However, the intense precipitation of 10–30 keV electrons in the same area constitutes an upward current, which is large enough to serve for current continuity. The conductance gradient at the eastern boundary is much more gradual than the gradients at the other borders of the conductance channel The downward field-aligned currents flowing here are less intense and more wide-spread than those produced at the other boundaries and can be carried by the ambient particles.

Typical values for conductances, fields, and currents inside the westward traveling surge are as follows. Within the active region the Hall conductance reaches peak values of more than 100 S. The convection electric field pattern is distorted by the superposition of a southward polarization electric field with a strength of up to 50 mV/m inside the westward traveling surge. The ionospheric current has sheet current densities of 500–1000 mA/m, comparable with typical westward electrojet values. The westward component of the ionospheric current is connected to very localized and intense (about 5–10 μA/m^2) upward field-aligned currents at the western border, near the head of the surge, and to more wide-spread downward current of lower density (1–2 μA/m^2) in the eastern part. The northward current is connected to field-aligned current sheets of 1–2 μA/m^2 at the southern and northern boundaries of the active region.

Substorm Current Wedge

The substorm electrojet is the ionospheric part of the *substorm current wedge* sketched in Fig. 5.15. The substorm current wedge diverts part of the neutral sheet current along

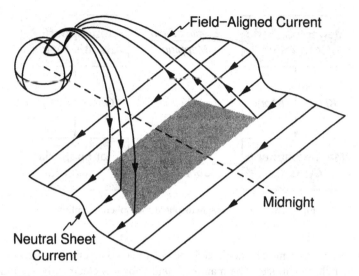

Fig. 5.15. Diversion of neutral sheet current through the ionosphere.

magnetic field lines through the ionosphere. The formation of the current wedge is naturally associated with the formation of the near-Earth neutral line. When the tail magnetic field dipolarizes in the vicinity of a neutral line, the cross-tail neutral sheet current (see Secs. 1.3 and 7.4) must be reduced in that region. Indeed the collapse of the tail-like field lines to a dipolar configuration would not occur if the cross-tail current remained at its original level.

However, the question of cause and effect is not quite settled yet. The more common view is that the near-Earth neutral line is the cause of the diversion of the cross-tail current and the formation of the substorm current wedge. Another school of thought believes that the strong enhancement of the westward ionospheric current inside the westward traveling surge and the need to close it via field-aligned currents and transverse current flow in the tail leads to a diversion of the neutral sheet current. In this model the formation of the near-Earth neutral line is a consequence of the diversion of the cross-tail current and the associated dipolarization.

Concluding Remarks

In the present chapter we have treated the ionosphere as a relatively passive medium with its dynamics governed by the processes that take place in the magnetosphere. The real situation is more complex. The magnetosphere is dominated by a collision-free plasma, while the ionosphere is the region where the effects of collisions of charged particles

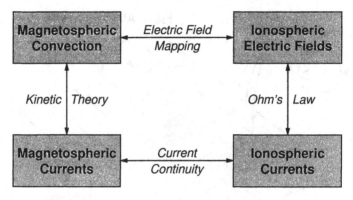

Fig. 5.16. Logic of magnetosphere-ionosphere coupling.

with neutral particles cannot be neglected and electrical conductivities transverse to the geomagnetic field maximize. The magnetic field connects electrically the ionosphere and the magnetosphere, causing an exchange or coupling of energy and momentum between the two regions. In a sense, the *magnetosphere-ionosphere coupling* is the interaction of different physical processes taking place in either of these two regions. Strong coupling occurs since the two regions are connected by magnetic field lines.

A simplified version of the logic governing the entire magnetosphere-ionosphere coupling system is shown in Fig. 5.16. One can understand the physical mechanisms fully only if one examines the coupled system in its entirety, since a change in one of the boxes will imply changes in all other boxes. For example, a variation in magnetospheric convection will modify the electric field mapped to the ionosphere. This will change the ionospheric current flow related to the electric field through Ohm's law and, by the requirement of current continuity, the field-aligned currents and the magnetospheric current distribution. Since the latter is, at least partially, caused by the magnetic drift of energetic particles which may change the magnetospheric electric field distribution due to shielding (see Sec. 5.2), the whole chain must be readjusted to reach an equilibrium situation.

The situation is further complicated by the fact that all logical connections between the boxes work in both directions. For example, the flow of ionospheric currents is governed by the ionospheric electric field, but the currents may affect the field through polarization charges set up at conductivity gradients, whenever current continuity would require field-aligned current densities higher than what can be carried by the ambient plasma. Subsequently, these ionospheric electric polarization fields will affect the magnetospheric electric field and, hence, the convection of the magnetospheric plasma.

The coupled set of equations represented by the logic diagram of Fig. 5.16 has been solved only recently. For example, the Rice Convection Model can reproduce the large-scale features of the coupled magnetosphere-ionosphere system in near-equilibrium situations. For small-scale features associated with auroral forms or the highly dynamic situations like during substorm onsets only partial solutions exist, linking at most three of the boxes in Fig. 5.16.

Actually, magnetosphere-ionosphere coupling cannot be dealt with using only the single particle approach plus collisions. Especially, the link between magnetospheric convection and magnetospheric currents involves collective plasma effects already for large-scale steady state situations if more than just the ring current is discussed. For small-scale and highly dynamical situations collective effects govern the whole chain, and we need the more sophisticated plasma physics approaches, which will be addressed in the remainder of this book and in our companion volume, *Advanced Space Plasma Physics*. Only near the end of the companion book we will be able to come back to some microphysical aspects of magnetosphere-ionosphere coupling.

Further Reading

A thorough description of the physics of solar wind-magnetosphere coupling is given in the second article. A full proof of the frozen-in flux theorem can be found in monograph [6]. A good discussion about corotation and its limits is found in reference [8]. Additional material on current systems and magnetosphere-ionosphere coupling is found in references [4] and [7]. Readers interested in the aurora should have a look at the third article. A good summary of our present knowledge about substorms is given in reference [5] while recent results on the behavior of the magnetotail during substorms are presented in the first article. The Rice Convection Model mentioned above is described in the last article.

[1] W. Baumjohann, *Space Science Rev.* **64** (1993) 141.

[2] S. W. H. Cowley, *Rev. Geophys. Space Phys.* **30** (1982) 531.

[3] T. J. Hallinen, in *Geomagnetism, Vol. 4*, ed. J. A. Jacobs (Academic Press, London, 1991), p. 741.

[4] Y. Kamide and W. Baumjohann, *Magnetosphere-Ionosphere Coupling* (Springer Verlag, Heidelberg, 1993).

[5] R. L. McPherron, in *Geomagnetism, Vol. 4*, ed. J. A. Jacobs (Academic Press, London, 1991), p. 593.

[6] D. R. Nicholson, *Introduction to Plasma Theory* (Wiley & Sons Inc., New York, 1983).

[7] J. Untiedt and W. Baumjohann, *Space Science Rev.* **63** (1993) 245.

[8] V. M. Vasyliunas, in *Solar-Terrestrial Physics*, eds. R. L. Carovillano and J. M. Forbes (D. Reidel Publ. Co., Dordrecht, 1983), p. 479.

[9] R. A. Wolf, in *Solar-Terrestrial Physics*, eds. R. L. Carovillano and J. M. Forbes (D. Reidel Publ. Co., Dordrecht, 1983), p. 303.

6. Elements of Kinetic Theory

Collective behavior leads to entirely new and otherwise unknown effects. Some of these effects have already been mentioned in Secs. 1.1 and 4.1. The collective interactions were hidden behind such terms as the Debye length, plasma frequency, plasma parameter, and Coulomb logarithm. These quantities attribute common average properties to large parts of the plasma consisting of very many particles or to the plasma as such, describing its average behavior in a global way.

The collective behavior has its roots in the many-particle character of plasmas. The reason for its appearance is the existence of long-range interparticle interactions between the charged particle components of the plasma due to the electric fields, $E(x, t)$, connected to each point charge, q, and the magnetic fields, $B(x, t)$, generated when the charges move at a given velocity, v. The other charged particles in the plasma respond to these fields in a way which leads to momentum and energy exchange between the particles as well as the fields. As a consequence, for a plasma consisting of many particles, with each of them generating its own field and reacting to the microscopic fields of other particles, the actual field configuration is the sum over all the microscopic contributions of the particles to the fields. This average field is of extremely complicated spatial structure and, in addition, varies on a variety of different time scales. At the same time, the motion of the particles in all the microscopic fields is far from the simple motion of single particles discussed before. Consequently, accounting for all the fields and the full particle dynamics is a very complex and almost untreatable task.

One way to obtain more realistic field and particle configurations than assumed in the previous chapters is to consider all the self-generated microscopic fields, to calculate the trajectories of all particles in these fields by solving their equations of motion and to self-consistently account for their self-generated fields during this motion. Subsequently, one should average over the fastest time scales which one is not interested in when describing average properties of the plasma. It is obvious that such a kind of description immediately runs into enormous computational difficulties which cannot be overcome by even the fastest computers available. Hence, one seeks for a more approximate description. Such a description is necessarily of statistical nature and is found in the so-called *kinetic plasma theory*, the basic elements of which will be developed in this chapter as a preparation for later use and further simplifications.

103

6.1. Exact Phase Space Density

In contrast to the point of view taken in the previous chapters, where the plasma consisted of single particles, we now assume that it is a strongly interacting system of very many particles, each having a time dependent position $x_i(t)$ and velocity $v_i(t)$. It is then useful to take these positions and velocities as independent coordinates in a hypothetical six-dimensional space with the coordinate axes $(\mathbf{x}, \mathbf{v})$, called the *phase space*. The particle at a certain time t_0 is then characterized as one point in this space inside a phase space volume element, $d\mathbf{x}d\mathbf{v}$ (see Fig. 6.1), and the particle path at subsequent times $t_1, t_2, \ldots, t_5, \ldots$ is a curve in this phase space as indicated in Fig. 6.2, actually a six-dimensional curve.

Exact Particle Density

For many particles, as in the case of a plasma, one may define an exact number density $\mathcal{F}_i$ of the i-th particle through

$$\mathcal{F}_i(\mathbf{x}, \mathbf{v}, t) = \delta(\mathbf{x} - \mathbf{x}_i(t))\delta(\mathbf{v} - \mathbf{v}_i(t)) \tag{6.1}$$

where $\delta(\mathbf{x} - \mathbf{x}_i) = \delta(x - x_i)\delta(y - y_i)\delta(z - z_i)$ and $\delta(\mathbf{v} - \mathbf{v}_i)$ are three-dimensional Dirac delta functions. Equation (6.1) tells that the particle density in phase space is different from zero only at the position and velocity of the i-th particle at time t which is the dynamic path of the particle it performs under the action of all the forces it experiences. For the single i-th particle this density is singular because its position in phase space, $\mathbf{x} = \mathbf{x}_i$ and $\mathbf{v} = \mathbf{v}_i$, is nothing else but the singular point in the small phase space vol-

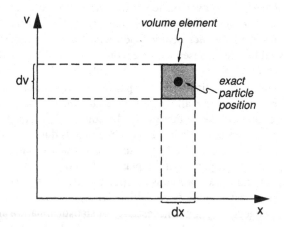

Fig. 6.1. Particle position in a phase space volume element.

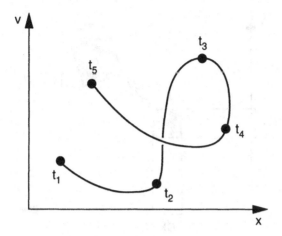

Fig. 6.2. Path of a particle in phase space.

ume of Fig. 6.1. Integrating over the full phase space gives the value one, which means that the particle can be found with certainty somewhere in phase space.

Since the history of the particle in phase space is described by the set of all points it assumed between the initial time t_0 and the actual time t, the exact particle density is a function of the phase space coordinates and time, as written explicitly in Eq. (6.1). The total exact particle density function of the plasma is then the sum over all the single exact particle densities given by Eq. (6.1)

$$\mathcal{F}(\mathbf{x}, \mathbf{v}, t) = \sum_i \delta(\mathbf{x} - \mathbf{x}_i(t))\delta(\mathbf{v} - \mathbf{v}_i(t)) \tag{6.2}$$

If the plasma consists of several components, this equation holds separately for each component. To obtain the total exact phase space density, one has to sum Eq. (6.2) over all particle species.

The geometrical content of the definition Eq. (6.2) is that the phase space volume occupied by the plasma consists of all the phase space points of the single particles or, equivalently, of all the single particle phase space volume elements of Fig. 6.1. Because the particles building up the phase space volume of the plasma are subject to the action of forces, and the forces are different for each of the particles, the phase space volume of the plasma will deform under the action of forces. Because the number of particles does not change, however, the volume remains constant, merely changing its shape. As shown in Fig. 6.3, the simplest kind of change is a mere rotation and stretching of the volume caused by a slight change in the particle velocity, $\mathbf{v}$, and position, $\mathbf{x}$, under the action of a microscopic force, e.g., microscopic electric fields, collisions between particles, anomalous collisions, etc., during the time dt.

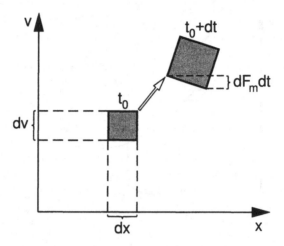

Fig. 6.3. Deformation of $dxdv$ due to a microscopic force.

Equation of Motion

It is important to remember that x and v are independent coordinates in phase space. The particle position itself is determined by its equation of motion under the action of all the microscopic electromagnetic fields, where the instantaneous particle velocity is given by $v_i(t) = dx_i(t)/dt$ and d/dt indicates the total derivative with respect to time. Denoting the microscopic fields by the index m, the equation of motion (2.7) reads

$$\frac{d}{dt}v_i(t) = \frac{q}{m}\left[E_m(x_i(t), t) + v_i(t) \times B_m(x_i(t), t)\right] \tag{6.3}$$

The microscopic electric and magnetic fields depend on the particle position, which is a function of time, and also explicitly on time. They are defined as the fields generated by all the particles in the plasma at the exact instantaneous position of the i-th particle and satisfy the microscopic Maxwell equations

$$\nabla \times B_m(x, t) = \mu_0 j_m(x, t) + \epsilon_0 \mu_0 \frac{\partial}{\partial t} E_m(x, t) \tag{6.4}$$

$$\nabla \times E_m(x, t) = -\frac{\partial}{\partial t} B_m(x, t) \tag{6.5}$$

$$\nabla \cdot E_m(x, t) = \frac{1}{\epsilon_0} \rho_m(x, t) \tag{6.6}$$

$$\nabla \cdot B_m(x, t) = 0 \tag{6.7}$$

The coupling of the i-th particle to all other plasma particles proceeds via the microscopic electric space charge, ρ_m, and current densities, j_m, of all particles, which generate the electric and magnetic fields $\mathbf{E}_m$ and $\mathbf{B}_m$, respectively. These charge and current densities are defined as

$$\rho_m(\mathbf{x}, t) = \sum_s q_s \int \mathcal{F}_s(\mathbf{x}, \mathbf{v}, t) \, d^3v \qquad (6.8)$$

$$\mathbf{j}_m(\mathbf{x}, t) = \sum_s q_s \int \mathcal{F}_s(\mathbf{x}, \mathbf{v}, t) \, \mathbf{v} \, d^3v \qquad (6.9)$$

where $d^3v = d\mathbf{v} = dv_x dv_y dv_z$ and the sum has to be taken over all particle species, electrons, protons and other ions, with exact phase space densities $\mathcal{F}_s$ and charges q_s.

The equation of motion (6.3) together with the microscopic Maxwell equations, which govern the electric and magnetic fields, form a system of equations which is exact and self-consistent. It describes all particle motions and all fields in the plasma in a given phase space. Computer simulations of plasmas can be based on it. One simply assumes an initial given exact phase space particle density in the simulation box together with an initial given field configuration and then solves the equation of motion for all the particles together with the Maxwell equations for the fields at each point for successive times. The results of such calculations are phase space evolution plots. However, due the large particle numbers in a plasma such calculations require enormous amounts of computer time.

Klimontovich-Dupree Equation

If no particle is lost from or added to the plasma, the exact phase space density (6.2) is conserved during the dynamic evolution of the plasma. Thus the total time derivative of $\mathcal{F}(\mathbf{x}, \mathbf{v}, t)$, taken along the dynamic path of all the particles in the plasma, i.e., along the path of the phase space volume occupied by the plasma, must vanish

$$\frac{d}{dt} \mathcal{F}(\mathbf{x}, \mathbf{v}, t) = 0 \qquad (6.10)$$

Since both $\mathbf{x}$ and $\mathbf{v}$ depend on time, we have to use the differential chain rule

$$\frac{d}{dt} f[g(t)] = \frac{df}{dg} \frac{dg}{dt} \qquad (6.11)$$

to write the total time derivative in six-dimensional phase space as

$$\frac{d}{dt} = \frac{\partial}{\partial t} + \mathbf{v} \cdot \nabla_{\mathbf{x}} + \frac{d\mathbf{v}}{dt} \cdot \nabla_{\mathbf{v}} \qquad (6.12)$$

where the indices of the two ∇ operators indicate differentiation with respect to particle position and velocity. Equation (6.12) is a *convective derivative* since it gives the time derivative of a quantity along a particle trajectory in phase space.

Using the equation of motion (6.3) to replace the derivative of the velocity, one obtains the following expression

$$\frac{\partial \mathcal{F}}{\partial t} + \mathbf{v} \cdot \nabla_x \mathcal{F} + \frac{q}{m}(\mathbf{E}_m + \mathbf{v} \times \mathbf{B}_m) \cdot \nabla_v \mathcal{F} = 0 \qquad (6.13)$$

Equation (6.13) is the evolution equation of the exact particle density in phase space. It is known under the name *Klimontovich-Dupree equation* and describes the plasma state in phase space at all times.

6.2. Average Distribution Function

The Klimontovich-Dupree equation still contains the exact microscopic fields and solving it is a difficult task. One therefore asks for a simpler way by averaging over a larger number of particles, considering them as being statistically correlated in time, space, and velocity by their mutual interactions.

Kinetic Equation

One can define an ensemble *averaged phase space density*, $\langle \mathcal{F}(\mathbf{x}, \mathbf{v}, t) \rangle = f(\mathbf{x}, \mathbf{v}, t)$, and express the exact phase space density as the sum of its average and a fluctuation, $\delta \mathcal{F}$, which accounts for the deviation of the exact phase space density from the average distribution as

$$\mathcal{F}(\mathbf{x}, \mathbf{v}, t) = f(\mathbf{x}, \mathbf{v}, t) + \delta \mathcal{F}(\mathbf{x}, \mathbf{v}, t) \qquad (6.14)$$

Since the fluctuations should form a statistical ensemble, the ensemble average over the fluctuation is equal to zero, $\langle \delta \mathcal{F} \rangle = 0$. The term ensemble average is in statistical mechanics not well defined; one understands under an ensemble average an average over all the particle properties over all particles in the ensemble. Clearly then, considering one property of the particles, relative to the average of this property in the ensemble there must be equal numbers of deviations of this property in both directions, and the average over these deviations must vanish. This is the content of the above decomposition. In a similar way one composes the fields as sums of averages and fluctuations

$$\begin{aligned}
\mathbf{E}_m(\mathbf{x}, \mathbf{v}, t) &= \mathbf{E}(\mathbf{x}, \mathbf{v}, t) + \delta \mathbf{E}(\mathbf{x}, \mathbf{v}, t) \\
\mathbf{B}_m(\mathbf{x}, \mathbf{v}, t) &= \mathbf{B}(\mathbf{x}, \mathbf{v}, t) + \delta \mathbf{B}(\mathbf{x}, \mathbf{v}, t)
\end{aligned} \qquad (6.15)$$

with $\langle \delta \mathbf{B} \rangle = 0$ and $\langle \delta \mathbf{E} \rangle = 0$. Inserting Eqs. (6.14) and (6.15) into Eq. (6.13) and taking the ensemble average yields the *kinetic equation* for the average phase space density of a plasma

$$\frac{\partial f}{\partial t} + \mathbf{v} \cdot \nabla_{\mathbf{x}} f + \frac{q}{m}(\mathbf{E} + \mathbf{v} \times \mathbf{B}) \cdot \nabla_{\mathbf{v}} f = -\frac{q}{m}\langle(\delta\mathbf{E} + \mathbf{v} \times \delta\mathbf{B}) \cdot \nabla_{\mathbf{v}}\delta\mathcal{F}\rangle \qquad (6.16)$$

The kinetic equation describes the evolution of the coarse-grained phase space density in time and space under the action of average fields. This coarse-grained density is called the *distribution function* of the particles in phase space and is interpreted as the probability of a particle to be found in a certain phase space volume element $d\mathbf{x}d\mathbf{v}$ (see Fig. 6.1).

The fields in Eq. (6.16) are average fields. They are governed by average Maxwell equations, which can be obtained from the microscopic equations by inserting the decomposed fields from Eq. (6.15) and similarly decomposed charge and current densities into the latter and taking the ensemble average. The form of the resulting average Maxwell equations is the same as that of the microscopic equations, but now for the average fields and densities.

The kinetic equation has the advantage that both the average distribution $f(\mathbf{x}, \mathbf{v}, t)$ and the average fields do not depend any more on the single coordinates of all the single particles of a species, but only depend on the phase space coordinates (x,v,t). The ensemble average has smeared out the exact positions of the particles over the phase space volume occupied by the particle group under consideration. Hence, the average particle distribution does not describe any more the exact position of the particles in the volume but instead accounts for the probability to find the ensemble in the interval $\{\mathbf{x}, \mathbf{x} + d\mathbf{x}\}, \{\mathbf{v}, \mathbf{v} + d\mathbf{v}\}$. The function $f(\mathbf{x}, \mathbf{v}, t)$ has thus become a probability distribution function, and the above equation is the dynamic equation for its evolution under the action of the average fields.

Boltzmann Equation

The term in angular brackets on the right-hand side of Eq. (6.16) contains all the correlations between the fields and particles. Its calculation poses a very serious problem. Various steps of evaluation can be imagined. Since $f(\mathbf{x}, \mathbf{v}, t)$ does not distinguish anymore between single particles but accounts only for their dependence on space and velocity, evaluation of the correlation term must provide this information in terms of two-point correlations, three-point correlations, and so on.

One way to simplify the kinetic equation is to neglect the correlations between the fields and to account only for correlations between the particles themselves via collisions. Then Eq. (6.16) can be written as

$$\frac{\partial f}{\partial t} + \mathbf{v} \cdot \nabla_{\mathbf{x}} f + \frac{q}{m}(\mathbf{E} + \mathbf{v} \times \mathbf{B}) \cdot \nabla_{\mathbf{v}} f = \left(\frac{\partial f}{\partial t}\right)_c \qquad (6.17)$$

where the right-hand side is the time rate of change of $f(\mathbf{x}, \mathbf{v}, t)$ due to all kinds of collisions. This equation is the generalized *Boltzmann equation*, which is well-known from statistical mechanics.

In order to solve Eq. (6.17), the exact functional form of its right-hand side has to be specified. For hard-core collisions between particles, the collision term on the right-hand side has been evaluated already by Boltzmann himself in the past century. In the simplest case of collisions in a plasma that is collisions between charged particles and neutrals in a partially ionized plasma, the collision term can be approximated by the so-called *Krook collision term*

$$\left(\frac{\partial f}{\partial t}\right)_c = \nu_n (f_n - f)$$ (6.18)

where $f_n(\mathbf{x}, \mathbf{v}, t)$ is the distribution function of the neutral atoms and ν_n is the neutral collision frequency defined in Sec. 4.1. In a fully ionized plasma the long-range Coulomb interactions between the particles complicate the situation considerably. Their inclusion replaces the collision term with the *Landau collision integral* which can be approximated by taking into account the Coulomb collision frequency (see Sec. 4.1). Since the latter depends on density and temperature, the collision term is further complicated. In the case of fully ionized collisionless dilute plasmas, the collision term vanishes, but the right-hand side of the Boltzmann equation is then dominated by correlations between the particles which are caused by their contributions to the field variations. This correlation term becomes a rather complicated function of the change of velocity of the particles. The equation including this quasi-collision term

$$\left(\frac{\partial f}{\partial t}\right)_c = \nabla_\mathbf{v} \cdot (\mathbf{D} \cdot \nabla_\mathbf{v} f)$$ (6.19)

is called *Fokker-Planck equation*. Here, the diffusion coefficient, $\mathbf{D}(\mathbf{v})$, is a function of velocity and is a tensor derived from the averages over the first- and second-order fluctuations, $\langle \Delta \mathbf{v} \rangle$ and $\langle \Delta \mathbf{v} \Delta \mathbf{v} \rangle$, of the particle velocities. The resulting collisions are not collisions in the usual sense but instead describe changes of the particle velocities and thus a diffusion in phase space. In a later chapter we will encounter such an equation when considering wave particle interactions in a plasma.

Vlasov Equation

Since space plasmas are collisionless, except for the ionosphere, one can often entirely neglect the collision term in the Boltzmann equation. This results in the simplest possible form of kinetic equation of a plasma, the *Vlasov equation*

$$\boxed{\frac{\partial f}{\partial t} + \mathbf{v} \cdot \nabla_\mathbf{x} f + \frac{q}{m}(\mathbf{E} + \mathbf{v} \times \mathbf{B}) \cdot \nabla_\mathbf{v} f = 0}$$ (6.20)

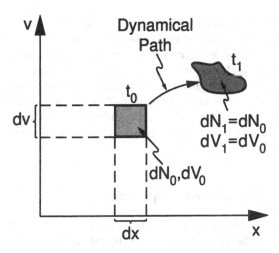

Fig. 6.4. Illustration of Liouville's theorem.

In the absence of collisions the phase space density remains constant under the interaction of the particles with the ensemble averaged self-consistent fields in the Lorentz force as it is convected with the particles. This behavior is the content of *Liouville's theorem*, which states that the phase space volume can be deformed but its density is not changed during the dynamic evolution of the plasma. The Liouville theorem holds exactly for the exact distribution function. In the case of the Vlasov equation, the Liouville theorem holds only if collisions and correlations between the particles and microscopic fields can be neglected. For most applications this approximation is valid.

The Liouville theorem states that a phase space volume element, dV_0, moves under the action of the Lorentz force like an incompressible fluid in phase space, because $\nabla \cdot \mathbf{v} = 0$ holds for the phase space coordinates. This behavior is visualized in Fig. 6.4. Imagine a phase space element $dx dv$ with density $f(\mathbf{x}_0, \mathbf{v}_0, t_0)$. At time t_0 all particles in this volume element have nearly the same position and velocity. At later time t_1 the particles will have moved to different positions, with their slightly different initial velocities and under the action of the Lorentz forces acting on them. This leads to deformation of the phase space volume element. However, because the number of the particles in the element is conserved, the volume of the phase space element, dV_1, i.e., the total number of points occupied by the dN_0 particles, is conserved. It is constant along the dynamical trajectories of all the particles it contains. Hence, the phase space density, f, is constant along such an orbit such that $f(\mathbf{x}_1, \mathbf{y}_1, t_1) = f(\mathbf{x}_0, \mathbf{v}_0, t_0)$.

The Vlasov equation forms the basis of all kinetic theory in collisionless plasmas like those in the magnetosphere and solar wind. Its mathematical structure is that of a partial differential equation which, though only of first order, is coupled to the full set of Maxwell's equations (for the ensemble averaged fields, charges and currents) through the last term on the left-hand side of Eq. (6.20). In addition, the electromagnetic fields are determined by the charge and current densities which, as will be proved below, are themselves given as integrals over the distribution function. One therefore realizes that the Vlasov equation is in fact a highly nonlinear equation in six-dimensional phase space which is very difficult to solve in full generality. This difficulty forces one to seek for further approximative methods to find solutions under special conditions and in special regimes of the plasma. The Vlasov equation forms the basis for the discussion of plasma processes in this book.

Approximations of the Vlasov Equation

The first reasonable approximation to the Vlasov equation is the assumption of an un-magnetized plasmas in the absence of currents. In this case there is no magnetic field, and the electric field is a pure potential field $\mathbf{E} = -\nabla_{\mathbf{x}}\phi$ with potential ϕ. This simplifies the third term in Eq. (6.20) and frees one from three of the Maxwell equations. One is left with only two scalar equations, the *electrostatic Vlasov equation*

$$\frac{\partial f}{\partial t} + \mathbf{v} \cdot \nabla_{\mathbf{x}} f - \frac{q}{m}(\nabla_{\mathbf{x}}\phi) \cdot \nabla_{\mathbf{v}} f = 0 \qquad (6.21)$$

and the Poisson equation

$$\epsilon_0 \nabla_{\mathbf{x}}^2 \phi = \sum_s q_s \int f_s \, d^3 v \qquad (6.22)$$

where the sum is performed over the different particle species. Since these equations are scalar they describe the most simple kinetic form of a plasma. However, even in this simple case one observes that the system is highly nonlinear with f appearing on the right-hand side of Poisson's equation, and $\nabla_{\mathbf{x}}\phi$ being multiplied with the velocity derivative of f. This system will later serve us as starting point for investigating the microscopic theory of waves in a plasma.

The full electromagnetic Vlasov equation (6.20) is exact in the sense of neglecting particle correlations and using ensemble averaged fields. It represents a total differential along the dynamical orbit of the phase space element. This property can be used to easily transform the Vlasov equation to other, sometimes more convenient coordinates. One particular set of coordinates are guiding center coordinates $(\mathbf{X}, v_{\parallel}, \mathbf{v}_{\perp}, \psi, t)$, where $\mathbf{X}$ is the guiding center position and ψ the particle gyration phase-angle. Introducing such coordinates becomes useful if one considers the behavior of the distribution function averaged over the gyratory motion, $\langle f(\mathbf{X}, v_{\parallel}, \mathbf{v}_{\perp}, \psi, t)\rangle$. The relation between the guiding center and the exact particle coordinates is

$$\mathbf{x} = \mathbf{X} - \frac{\mathbf{v} \times \mathbf{B}}{\omega_g B}$$

$$\mathbf{v}_\perp = v_\perp (\hat{\mathbf{e}}_x \cos \psi - \hat{\mathbf{e}}_y \sin \psi)$$

(6.23)

where the magnetic field direction is thought to be aligned with the z axis and $\hat{\mathbf{e}}_x$ and $\hat{\mathbf{e}}_y$ are unit vectors along the two other axes. Let us for convenience restrict to the electrostatic limit when no magnetic field changes occur ($\partial \mathbf{B}/\partial t \approx 0$). Introducing these expressions into Eq. (6.20), performing the various first-order differentiations and rearranging, one obtains the Vlasov equation expressed in the new coordinates as

$$\frac{\partial f}{\partial t} + \left(\frac{v_\parallel \mathbf{B}}{B} + \mathbf{v}_E \right) \cdot \nabla_\mathbf{X} f + \frac{q E_\parallel}{m} \frac{\partial f}{\partial v_\parallel} + \frac{q \mathbf{E}_\perp \cdot \mathbf{v}_\perp}{m v_\perp} \frac{\partial f}{\partial v_\perp} + \omega_g \left(1 - \frac{\mathbf{v}_E \cdot \mathbf{v}_\perp}{v_\perp^2} \right) \frac{\partial f}{\partial \psi} = 0$$

(6.24)

where $\nabla_\mathbf{X}$ denotes the vector derivative with respect to guiding center position. In this representation only the electric field drift, $\mathbf{v}_E$, is included while gradient and curvature drifts have been neglected. One must now average over the ensemble gyratory motion of the particles to obtain the evolution of $\langle f \rangle$ in phase space. This average assumes that the distribution is to first-order constant over the gyro-orbit, i.e., the gyroradius is much smaller than the typical scale length over which the density varies. Then $\partial \langle f \rangle / \partial \psi = 0$ and $\langle \mathbf{E}_\perp \cdot \mathbf{v}_\perp \rangle = 0$, and only non-adiabatic effects contribute to the evolution of $\langle f \rangle$. The equation obtained is the so-called *gyrokinetic equation* in the electrostatic limit

$$\boxed{\frac{\partial \langle f \rangle}{\partial t} + \left(v_\parallel \frac{\mathbf{B}}{B} + \mathbf{v}_E \right) \cdot \nabla_\mathbf{X} \langle f \rangle + \frac{q}{m} E_\parallel \frac{\partial}{\partial v_\parallel} \langle f \rangle = 0}$$

(6.25)

where $E_\parallel$ is the gyro-averaged parallel electric field which does not depend on particle velocity. One basic difference to the Vlasov equation is that here the coefficient of the second term is a function of the average space coordinate $\mathbf{X}$ because $\mathbf{B}$ and $\mathbf{v}_E$ are both functions of space. Therefore, in contrast to the Vlasov equation, this coefficient cannot be exchanged with the operator $\nabla_\mathbf{X}$. The non-commutativity is a result of the averaging procedure over the gyromotion. Writing the gyrokinetic equation in the form of a conservation equation

$$\frac{\partial \langle f \rangle}{\partial t} + \nabla_\mathbf{X} \cdot \left(v_\parallel \frac{\mathbf{B}}{B} + \mathbf{v}_E \right) \langle f \rangle + \frac{q}{m} \frac{\partial}{\partial v_\parallel} (E_\parallel \langle f \rangle) = \langle f \rangle \nabla_\mathbf{X} \cdot \left(v_\parallel \frac{\mathbf{B}}{B} + \mathbf{v}_E \right)$$

(6.26)

one recognizes that this average introduces a source term on the right-hand side of the equation which contributes to the evolution of the distribution function $\langle f \rangle$ but is itself a function of $\langle f \rangle$.

In a slightly more general but closely related way one finds another version of the averaged Vlasov equation, called the *drift kinetic equation*, for the average distribution function $f_d(v_\|, \mu, \mathbf{x}, t)$, where μ is the magnetic moment

$$\boxed{\frac{\partial f_d}{\partial t} + \nabla_\mathbf{x} \cdot (\mathbf{v}_d f_d) + \frac{\partial}{\partial v_\|} \left(\frac{F_\|}{m} f_d \right) = 0}$$

(6.27)

This equation is still only electrostatic, but accounts for the full particle drift

$$\mathbf{v}_d = \frac{v_\| \mathbf{B}}{B} + \mathbf{v}_E + \frac{\mathbf{F} \times \mathbf{B}}{q B^2}$$

(6.28)

since it includes all parallel and perpendicular forces (see Chap. 2)

$$F_\| = -\mu \nabla_\| B + q E_\|$$
$$\mathbf{F}_\perp = -\mu \nabla_\perp B - m v_\|^2 \frac{\mathbf{R}_c}{R_c^2} - m \frac{d\mathbf{v}_E}{dt}$$

(6.29)

Mirror effects and gradient and curvature drift in an inhomogeneous magnetic field are included in this form of the Vlasov equation which therefore is suitable to investigate the average behavior of trapped plasmas in the magnetospheric magnetic field.

6.3. Velocity Distributions

It is impossible to give graphic representations of the six-dimensional phase space distribution function, $f(\mathbf{x}, \mathbf{v}, t)$, as it varies in space, velocity, and time. But, usually, the most interesting property of the distribution function is its dependence on the velocity at a fixed position in configuration space. Many of the characteristic features of plasmas can be understood by knowing this velocity dependence. In the following we discuss a few examples of typical space plasma *velocity distribution functions*, $f(\mathbf{v})$, assuming the plasma to be spatially homogeneous and, in addition, stationary. Under these conditions the plasma does not change in time and does not exhibit spatial variations. In general, such a situation can be realized only when the plasma is in equilibrium; however, there are cases when the velocity distribution is of a form which is far from equilibrium. In such a case it must either be maintained by external means which inhibit relaxation of the distribution to its equilibrium form, or the time the distribution is observed is short with respect to the relaxation time.

Maxwellian Distributions

The general equilibrium velocity distribution function of a collisionless plasma is the *Maxwellian velocity distribution* or simply Maxwellian. A Maxwellian plasma is in thermal equilibrium which implies that it does not contain anymore free energy and, hence, there are no energy exchange processes between the particles in the plasma. It is then clear that the velocities the particles can assume must be distributed randomly around the average velocity. For a plasma at rest, the latter is zero, and the distribution of the velocities follow a simple Gaussian distribution of errors

$$g(\Delta x) = \left(\pi \langle \Delta x \rangle^2\right)^{-1/2} \exp\left(-\frac{(\Delta x)^2}{\langle \Delta x \rangle^2}\right) \tag{6.30}$$

Replacing Δx with one component of the velocity v_x, replacing the variance $\langle \Delta x \rangle$ with the average velocity spread $\langle v_x \rangle$, and multiplying with the average particle density, n, Eq. (6.30) yields the one-dimensional equilibrium velocity distribution function

$$f(v_x) = \frac{n}{\left(\pi \langle v_x \rangle^2\right)^{1/2}} \exp\left(-\frac{v_x^2}{\langle v_x \rangle^2}\right) \tag{6.31}$$

This distribution can be generalized to three dimensions by observing that $v^2 = v_x^2 + v_y^2 + v_z^2$. Hence, multiplying the three one-dimensional Maxwellians in the three directions one finds the full Maxwellian in an isotropic plasma

$$f(v) = \frac{n}{\left(\pi \langle v \rangle^2\right)^{3/2}} \exp\left(-\frac{v^2}{\langle v \rangle^2}\right) \tag{6.32}$$

Sometimes this function is written conveniently in the form

$$\boxed{f(v) = n \left(\frac{m}{2\pi k_B T}\right)^{3/2} \exp\left(-\frac{mv^2}{2k_B T}\right)} \tag{6.33}$$

with m denoting the particle mass and $k_B T$ the average thermal energy. The velocity spread, $\langle v \rangle = (2k_B T/m)^{1/2}$, can be identified as the thermal velocity, a relation which will be derived below in Sec. 6.5. Using the integral

$$\int_{-\infty}^{\infty} \exp(-x^2)\, dx = \sqrt{\pi} \tag{6.34}$$

it is easy to verify that the integral of the Maxwellian over the whole three-dimensional velocity space is n, the macroscopic number density. Hence, the Maxwellian velocity distribution tells us how the particle density, at a given point in space and time, in equilibrium is distributed over velocity space, depending on the average thermal energy of the particles. This function is sketched in the left-hand panel of Fig. 6.5. Its functional

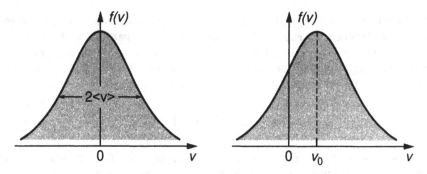

Fig. 6.5. Maxwellian and drifting Maxwellian velocity distributions.

form is symmetric with respect to the three velocity components and only depends on the magnitude v of the velocity, and its half-width gives the average velocity spread.

A simple further generalization of the equilibrium velocity distribution function is obtained by observing that in a plasma streaming at common velocity, $\mathbf{v}_0 = v_0 \hat{\mathbf{e}}_x$, in the x direction, the average velocity of the distribution function with respect to v_x is nonzero. Therefore, v_x must be replaced with $v_x - v_0$ or, more generally, $\mathbf{v}$ is replaced by $\mathbf{v} - \mathbf{v}_0$ to obtain

$$f(\mathbf{v}) = n \left(\frac{m}{2\pi k_B T} \right)^{3/2} \exp\left(-\frac{m(\mathbf{v} - \mathbf{v}_0)^2}{2 k_B T} \right) \tag{6.35}$$

for the drifting Maxwellian velocity distribution, which is sketched in the right-hand panel of Fig. 6.5.

Anisotropic Distributions

Not all velocity distributions are as simple as the isotropic (or drifting) Maxwellian. Already the drifting Maxwellian obeys some kind of asymmetry with respect to the zero point in velocity which is a kind of anisotropy. Especially the presence of magnetic fields introduces an anisotropy because it leads to different particle velocities parallel and perpendicular to the magnetic field. A particularly important case is that of gyrating particles. In this case the velocity distribution is independent of the angle of gyration, depending only on $v_\perp$ and $v_\parallel$. Because these two velocity components are independent, the equilibrium distribution can be modelled as the product of two Maxwellians (6.31) according to

$$f(v_\perp, v_\parallel) = \frac{n}{(\pi^3 \langle v_\perp \rangle^2 \langle v_\parallel \rangle)^{1/2}} \exp\left(-\frac{v_\perp^2}{\langle v_\perp \rangle^2} - \frac{v_\parallel^2}{\langle v_\parallel \rangle^2} \right) \tag{6.36}$$

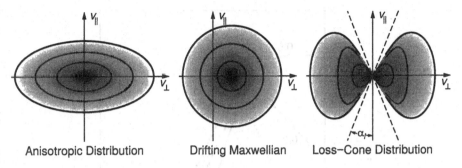

Fig. 6.6. Contours of constant f for typical anisotropic space plasma velocity distributions.

The resulting anisotropic distribution function is called *bi-Maxwellian distribution*. It accounts explicitly for the difference in the two average velocities $\langle v_\perp \rangle$, $\langle v_\parallel \rangle$ in the directions parallel and perpendicular to the magnetic field. Another representation can be based on Eq. (6.33)

$$f(v_\perp, v_\parallel) = \frac{n}{T_\perp T_\parallel^{1/2}} \left(\frac{m}{2\pi k_B}\right)^{3/2} \exp\left(-\frac{mv_\perp^2}{2k_B T_\perp} - \frac{mv_\parallel^2}{2k_B T_\parallel}\right) \tag{6.37}$$

Distributions of the form $f(v_\perp, v_\parallel)$, are the velocity distributions most often found in space plasmas. They are essentially two-dimensional and *gyrotropic velocity distributions*, which do not depend on the phase angle of the gyromotion and are often plotted as contour maps, i.e., curves of $f(v_\perp, v_\parallel) = \text{const}$, or grey-scale plots, where the grey-level encodes the phase space density. The left-hand side of Fig. 6.6 shows a sketch of the bi-Maxwellian (6.36) with $T_\perp > T_\parallel$, i.e., the average thermal energy of the particles perpendicular to the field greater than the average parallel energy. Instead of the circular contours of an isotropic Maxwellian, the contours are deformed into an elliptical shape.

The distribution in the middle of Fig. 6.6, with its circular contours displaced from the origin, is a *drifting Maxwellian* as described by Eq. (6.35). Here all particles drift with the same velocity perpendicular to the field, in addition to their thermal motion. Such a distribution would be representative for a plasma that is in thermal equilibrium, but convects perpendicular to the magnetic field under the action of an external electric field. In cases where the drift velocity is large compared to the thermal velocity such a distribution is called a *streaming distribution*. The corresponding perpendicular distribution function is of the same type as Eq. (6.35)

$$f(\mathbf{v}) = \frac{n}{T_\perp T_\parallel^{1/2}} \left(\frac{m}{2\pi k_B}\right)^{3/2} \exp\left(-\frac{m(\mathbf{v}_\perp - \mathbf{v}_{0\perp})^2}{2k_B T_\perp} - \frac{mv_\parallel^2}{2k_B T_\parallel}\right) \tag{6.38}$$

For example, a plasma drift in crossed electric and magnetic fields causes a drift velocity

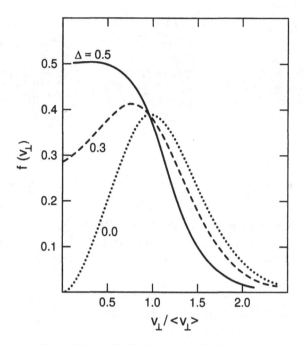

Fig. 6.7. Partially filled loss cone distributions.

$v_{0\perp} = v_E$. In an observer's frame the particle distribution has the form of Eq. (6.38).

In exactly the same way one can model a distribution which drifts along the magnetic field at velocity $v_{0\parallel}$

$$f(v_\parallel, v_\perp) = \frac{n}{T_\perp T_\parallel^{1/2}} \left(\frac{m}{2\pi k_B}\right)^{3/2} \exp\left(-\frac{mv_\perp^2}{2k_B T_\perp} - \frac{m(v_\parallel - v_{0\parallel})^2}{2k_B T_\parallel}\right) \qquad (6.39)$$

Such a distribution function is called a *parallel beam distribution*, a type of distribution function frequently encountered in the auroral magnetosphere, plasma sheet boundary layer, and in the foreshock region in front of the Earth's bow shock wave.

Loss Cone distributions

In the inner magnetosphere, where mirroring particles with high velocities parallel to the magnetic field may be lost to the ionosphere (see Sec. 3.2), an originally Maxwellian distribution will loose all particles inside the loss cone, α_ℓ. The right-hand side of Fig. 6.6 shows how such a *loss cone distribution* would look like. The form of the loss cone distribution depends heavily on the processes giving rise to the loss of particles. The simplest form is when the particles at the edge of the loss cone are cut out, but such distributions are quite unrealistic. A reasonable way is to model them via multiplication with some power of the perpendicular velocity $v_\perp^{2j} = v^{2j} \sin^{2j} \alpha$, describing an emptied loss cone near pitch angles $\alpha \approx 0$. Using a symmetric power $2j$ guarantees that the loss cone is symmetric for pitch angles $\alpha = 0$ and $\alpha = \pi/2$. This way one obtains, after normalization, the so-called *Dory-Guest-Harris loss cone distribution*

$$f(v_\|, v_\perp) = \frac{n}{A_j \langle v_\| \rangle \langle v_\perp \rangle^{4j}} \left(\frac{v_\perp}{\langle v_\perp \rangle} \right)^{2j} \exp\left(-\frac{v_\|^2}{\langle v_\| \rangle^2} - \frac{v_\perp^2}{\langle v_\perp \rangle^2} \right) \qquad (6.40)$$

and the constant is $A_j = \pi^{2j+1/2}(j!)^{2j}$. This distribution function looks complicated and has a very deep and absolutely empty loss cone. Sometimes the loss cone in the magnetosphere is not empty. In such a case it is more appropriate to use a loss cone distribution which accounts for a partial filling of the loss cone. The *partially-filled loss cone distribution* can be modelled by subtracting simple anisotropic Maxwellians

$$f(v_\|, v_\perp) = \frac{n}{(\pi^3 \langle v_\| \rangle^2 \langle v_\perp \rangle^4)^{1/2}} \exp\left(-\frac{v_\|^2}{\langle v_\| \rangle^2} \right) G(v_\perp, \Delta, \beta) \qquad (6.41)$$

The first part of this function is a parallel Maxwellian. The information about the loss cone is contained in the function $G(v_\perp, \Delta, \beta)$

$$G = \Delta \exp\left(-\frac{v_\perp^2}{\langle v_\perp \rangle^2} \right) + \frac{1-\Delta}{1-\beta} \left[\exp\left(-\frac{v_\perp^2}{\langle v_\perp \rangle^2} \right) - \exp\left(-\frac{v_\perp^2}{\beta \langle v_\perp \rangle^2} \right) \right] \qquad (6.42)$$

Here Δ and β are parameters chosen to fit the loss cone. In particular, $\Delta = 0$ describes an empty loss cone, the simplest form of the distribution equation (6.40), and $\Delta = 1$ reproduces a simple Maxwellian. Figure 6.7 gives examples of such partially filled loss cones. Since the dependence of the distribution function on $v_\|$ is Maxwellian, only the perpendicular distribution, i.e., the part depending on $v_\perp$ is shown. It is obvious from this figure that the loss cone gradually fills up with particles when Δ increases from 0 to 1. The constant β gives another freedom of changing the slope of the distribution inside the loss cone. It changes the average velocity of the subtracted Maxwellian component.

Energy Distributions

Considering the Maxwellian distribution equation (6.33) one realizes that the exponential depends on the ratio of two energies, the kinetic energy of the particles $mv^2/2$, and the average or thermal equilibrium energy $k_B T$. Hence, the equilibrium distribution function can be easily generalized to the inclusion of cases where the particles are localized in an external potential field. For an external electric field or potential, $\mathbf{E} = -\nabla \phi$, the potential energy is given by $U = -q\phi$, with q the charge of the particle. The total energy of the particle is then the sum of its kinetic and potential energies $W = mv^2/2 + U$, and the distribution function becomes

$$f(v) = n\left(\frac{m}{2\pi k_B T}\right)^{3/2} \exp\left(-\frac{W}{k_B T}\right)$$

(6.43)

This distribution can be written entirely in terms of the energy W as a variable by observing that the integral over the distribution must reproduce the density n. Hence,

$$f(W) = 2\left[\frac{2(W - U)}{m}\right]^{1/2} f(v)$$

(6.44)

with $f(v)$ given by Eq. (6.43). This distribution function is the *Boltzmann distribution*. It depends only on the particle energy.

Kappa and Power Law Distributions

Energy distributions are frequently measured in space. In most cases they do not resemble Boltzmann distributions but have more complicated shapes exhibiting long tails which strongly deviate from simple Maxwellians. Such tails can be modelled by *power law distributions*, where the distribution function varies like $f(W) \propto (W_0/W)^{-\kappa}$ with κ some constant power. This functional dependence is an approximation to a more general distribution, called the *kappa distribution*

$$f_\kappa(W) = n\left(\frac{m}{2\pi\kappa W_0}\right)^{3/2} \frac{\Gamma(\kappa + 1)}{\Gamma(\kappa - 1/2)}\left(1 + \frac{W^*}{\kappa W_0}\right)^{-(\kappa+1)}$$

(6.45)

Here W_0 is the particle energy at the peak of the distribution which can be related to the average thermal energy by $W_0 = k_B T(1 - 3/2\kappa)$. For $\kappa \gg 1$ the two are identical, and the distribution becomes a simple Maxwellian. For smaller $\kappa > 1$ the distribution possesses a high-velocity tail. Using $W^* = (\sqrt{W} - \sqrt{W_s})^2$, where W_s is a so-called shift energy, instead of the more simple $W = mv^2/2$ provides an additional parameter by which the distribution can be shifted in energy or velocity space, leaving sufficient freedom to fit measured energy distribution functions.

6.4.   Measured Distribution Functions

Distribution functions are probability densities in phase space. As such, the concept of a distribution function looks rather theoretical. However, there is another quantity, which is less theoretical and, more important, easy to measure, namely the particle flux.

Differential Particle Flux

There is a close relationship between the *differential particle flux*, $J(W, \alpha, \mathbf{x})$, per unit area at a given energy, pitch angle, and position and the particle phase space distribution $f(\mathbf{v}, \mathbf{x})$. A particle flux across a surface is given by the number density times the velocity component normal to the surface. Looked at differentially or, in other words, considering the particles found in a velocity interval dv coming from a solid angle $d\Omega$, the number density of particles with velocity v in a phase space volume element is $dn = f v^2 dv d\Omega$. Multiplying by v, one finds that the differential flux of particles with velocity v is given by

$$J(W, \alpha, \mathbf{x}) dW d\Omega = f(v_\parallel, v_\perp, \alpha, \mathbf{x}) v^3 dv d\Omega \qquad (6.46)$$

The left-hand side of this expression has been written in terms of the particle energy in the interval dW, simply because it is easier to measure the energy of particles in a certain interval than their individual velocities. Since $dW = mv dv$, the relation between the flux and the distribution function becomes simply

$$\boxed{J(W, \alpha, \mathbf{x}) = \frac{v^2}{m} f(v_\parallel, v_\perp, \alpha, \mathbf{x})} \qquad (6.47)$$

a very useful formula which directly relates the measured flux in a certain energy interval to the velocity distribution function of the measured particles.

Due to the factor v^2 even simple Maxwellians, which drop monotonically with increasing velocity if displayed as $f(v)$, exhibit a peak if plotted as $J(W)$. The particle energy at the peak of the particle flux, W_0, can be related to the average thermal energy by $W_0 = k_B T$. Figure 6.8 shows an example of the particle flux distribution for a Maxwellian and a kappa distribution.

Another quantity which is often used in space plasma physics is the *differential energy flux*. The latter is defined as the product of the differential particle flux times the particle energy. The energy flux drops off even less rapidly than the particle flux and this representation is thus often used to highlight features in the high-energy tail of a particle distribution. Especially in case of isotropic distributions, differential fluxes are often integrated over the solid angle $d\Omega$ and then called *omnidirectional differential flux*.

The differential fluxes form the basis of measurement of velocity distributions in space. Such distributions have been measured since the mid-sixtieth of this century in the solar wind and in the magnetosphere and have been used to obtain information about the plasma state in these regions. The instrumental technique is based on the measurement of a directed particle flux entering the narrow window of the instrument, generally a retarding potential analyzer, from a certain angular direction. One either has a large number of such windows and instruments distributed over a solid angle 4π, or one takes advantage of the rotation of the spacecraft to cover the full solid angle. Ions and electrons can be discriminated by applying positive or negative potentials. The former

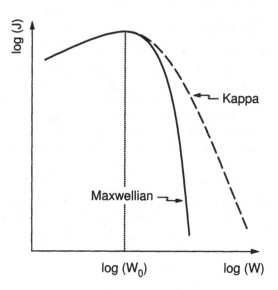

Fig. 6.8. Example of a kappa distribution.

prevent ions, the latter electrons from penetration. Selection with respect to particle energies takes place by other charged grids which deflect lower energy particles an prevent them from entering the instrument.

Particle Fluxes in Near-Earth Space

Measured velocity distribution functions in space are numerous. One can characterize the different regions in near-Earth space (see Sec. 1.2) by their characteristic long time averaged distribution functions. A number of such distributions is given in Fig. 6.9. This figure shows characteristic average omnidirectional differential ion and electron energy fluxes.

Figure 6.9 exhibits the great variety of particle distributions and energy fluxes in near-Earth space. The variation of the measured maximum fluxes covers about six orders of magnitude for the ions, and three orders of magnitude for electrons. The solar wind has the highest ion fluxes which are distributed over a narrow energy range close to an energy of 1 keV, the typical streaming energy of solar wind protons in the rest frame of the magnetosphere. These high fluxes identify to some extent the solar wind as the main particle and energy source in near-Earth space. Ion fluxes in the magnetosheath have been degraded by two orders of magnitude indicating that the solar wind has passed through the bow shock, slowed down and become heated. Slowing-down and heating is obvious from flux and energy decrease, as well as the increase of the energy spread in

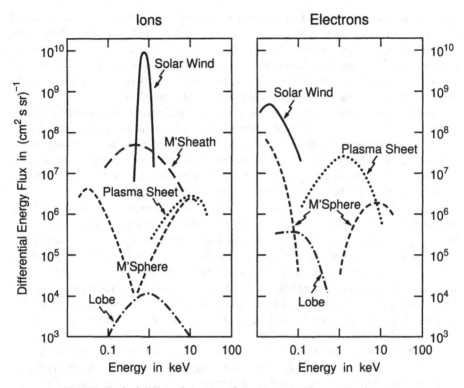

Fig. 6.9. Typical differential energy fluxes measured in near-Earth space.

the differential flux.

The tail lobe flux distributions found on the open polar cap field lines close to the magnetopause have an average energy of 1 keV, but very low density levels. The energy of the lobe fluxes is slightly higher than that of the magnetosheath fluxes, while the shape of the two distributions is the same. This implies that the lobe plasma is predominantly a tiny fraction of magnetosheath plasma which has gained a small amount of energy when adiabatically entering the open lobe field lines.

Inside the magnetosphere cold plasmaspheric and warm outer ring current plasma components are found. They have similar flux levels, but are clearly separated in their average energy. The plasma sheet fluxes are similar to the warm magnetospheric component. The low energy component has a different origin than the other two. It comes from the plasmasphere in the inner magnetosphere and therefore has no relation to the solar wind. On the other hand, the energetic plasma sheet distribution is clearly related to the magnetosheath plasma. It has a similar shape but lower fluxes and higher energy.

Hence, it has been energized by some process to about ten times the typical energy of the magnetosheath plasma during or after entry into the magnetosphere. The magnetospheric energetic component belongs to the energetic ring current and is part of the plasma sheet ions which perform at least a partial drift around the Earth. Clearly it is slightly more energetic than the former and has a lesser energy spread implying that the ring current ions are faster but colder than the plasma sheet plasma.

Similar considerations apply to the electron distributions. The streaming solar wind electrons are warm, considerably warmer than the solar wind ions, as observed from their large energy spread. On the other hand, they have lower streaming energy due to their low mass. When passing through the bow shock, their energy and temperature increases due to processes acting at the bow shock and turbulence in the magnetosheath, but to a lesser extend than found for the ions (not shown in figure).

Again the two magnetospheric components are well separated into cold and dense plasmaspheric electron fluxes and hot outer magnetospheric trapped electrons of solar wind origin which have been energized in the magnetosphere. These fluxes are considerably lower than the low energy plasmaspheric fluxes, indicating the dilute state of the external electron component. The plasma sheet electron fluxes have lower energies than the plasma sheet ions, reflecting both the lower magnetosheath electron temperature and lesser heating during their transfer from the magnetosheath into the plasma sheet. Finally the electron component in the lobe has both low fluxes and low energies, but the fluxes are higher than the lobe ion fluxes. Their origin is less obvious than that of the ions. Escaping polar ionosphere electrons may provide a significant contribution.

In addition to these average flux measurements many different types of particle distributions have been measured at different locations in the near-Earth space. The solar wind generally exhibits streaming Maxwellians, while in front of the bow shock electron and ion beam distributions dominate. In the bow shock one finds some kind of top-flattened electron distribution. On closed field lines in the magnetosphere various types of loss cone distributions of the energetic electron and ion components are found, while the plasma sheet plasma is well approximated by bi-Maxwellian distributions. Finally, in the lower magnetospheric auroral zone loss cone distributions dominate again but ion and electron beam distributions are sometimes observed.

6.5. Macroscopic Variables

One may ask oneself whether these distribution functions are practical in a more general sense than discussed in the previous section and particularly the last subsection. Measurement of a distribution function does not provide physical quantities as velocities and densities, but gives merely probabilities how many particles are found in a certain velocity or energy interval. Hence the question is how to find such macroscopic measurable quantities from a known distribution function. Fortunately, there is a unique answer to this question which follows from the very definition of a probability distribution. Given a probability distribution, a physical quantity related to the probability is defined as a certain *velocity moment* of this distribution.

Velocity Moments

The idea behind the procedure of calculating moments is simple. The distribution function depends on the velocity, on space, and on time. The physical macroscopic quantities like density, n, bulk flow velocity, v_b, average temperature, T, etc., do not depend on the particle velocities but only on space and time. Hence, to obtain a quantity which does not depend on velocity, one naturally would integrate over all velocities contributing to it. The distribution of particles with velocity is given by the distribution function $f(\mathbf{v}, \mathbf{x}, t)$. Therefore the corresponding integrals must be weighted by f before performing the integration. To find the i-th moment of the distribution function, the following integral must be calculated

$$\mathcal{M}_i(\mathbf{x}, t) = \int f(\mathbf{v}, \mathbf{x}, t) \, \mathbf{v}^i d^3 v \tag{6.48}$$

where $\mathbf{v}^i$ denotes the i-fold dyadic product, a tensor of rank i (see App. A.4). The number of moments which can be calculated from the distribution function is in principle infinite. However only the first few are of physical relevance. Their definition shows the usefulness of the distribution function which permits to calculate macroscopic and more familiar quantities by simple integrations.

Using the above definition, one can calculate the first few moments, $i = 0, 1, 2$, and identify important macroscopic quantities as variants of these moments. The *number density* is given by the zero-order moment

$$\boxed{n = \int f(\mathbf{v}) d^3 v} \tag{6.49}$$

The mean or *bulk flow velocity* v_b is naturally defined by the first-order moment

$$\boxed{\mathbf{v}_b = \frac{1}{n} \int \mathbf{v} f(\mathbf{v}) d^3 v} \tag{6.50}$$

The bulk velocity describes the macroscopic flow of the entire particle component in which each particle participates. It is an average flow velocity of the particle species or component under consideration. We have denoted it by the symbol $\mathbf{v}_b$ in order to distinguish it from the particle velocities. In later fluid applications where no confusion will be possible we will replace it by the conventional notation $\mathbf{v}$ of a velocity.

The *pressure tensor* is defined as the contribution of the fluctuation of the velocities of the ensemble from this mean velocity. Its calculation is based on the second-order moment

$$\mathbf{P} = m \int (\mathbf{v} - \mathbf{v}_b)(\mathbf{v} - \mathbf{v}_b) f(\mathbf{v}) d^3 v \qquad (6.51)$$

Since the two velocity product appearing in the pressure integral is a dyadic product, the pressure is a tensor.

Sometimes the next higher moment is also used to describe deviations from equilibrium. This moment is called the *heat tensor*

$$\mathbf{Q} = m \int (\mathbf{v} - \mathbf{v}_b)(\mathbf{v} - \mathbf{v}_b)(\mathbf{v} - \mathbf{v}_b) f(\mathbf{v}) d^3 v \qquad (6.52)$$

It is a third rank tensor or dyad. In itself it is not a very useful quantity, but its trace vector $\mathbf{q}$

$$\mathbf{q} = \frac{m}{2} \int (\mathbf{v} - \mathbf{v}_b) \cdot (\mathbf{v} - \mathbf{v}_b)(\mathbf{v} - \mathbf{v}_b) f(\mathbf{v}) d^3 v \qquad (6.53)$$

is the *heat flux vector* describes the transport of heat into a direction in the plasma which is not necessarily the direction of the mean flow.

Concept of Temperature

The pressure tensor consists of a trace and and the traceless off-diagonal part. The former gives, in an isotropic plasma, the isotropic pressure $p = n k_B T$, in an anisotropic plasma the anisotropic pressure. The traceless part contains the stresses in the plasma. The thermal pressure p can be used to define the temperature of the plasma component

$$T = \frac{m}{3 k_B n} \int (\mathbf{v} - \mathbf{v}_b) \cdot (\mathbf{v} - \mathbf{v}_b) f(\mathbf{v}) d^3 v \qquad (6.54)$$

a definition which identifies the temperature as a scalar quantity. This temperature is the *kinetic temperature*, a quantity which can formally be calculated for any type of distribution function and therefore is not necessarily a true temperature in the thermodynamic sense, which can only be calculated for plasmas in or close to thermal equilibrium. It rather is a measure of the spread of the particle distribution in velocity space. In addition, because each particle species may have its own distribution function, the kinetic temperatures of the plasma components may differ from each other. Also, in an anisotropic plasma the temperatures parallel and perpendicular to the magnetic field are in general different, because the particle distributions parallel and perpendicular have different shapes (see Sec. 6.3).

To demonstrate the physical meaning of Eq. (6.54) we calculate the kinetic temperature for the Maxwellian velocity distribution (6.32)

$$f(v) = \frac{n}{\left(\pi \langle v \rangle^2\right)^{3/2}} \exp\left(-\frac{v^2}{\langle v \rangle^2}\right) \tag{6.55}$$

in an isotropic plasma at rest, $v_b = 0$, and in thermal equilibrium. Performing the integration in Eq. (6.54) one finds that in the isotropic case the volume element becomes $d^3v = 4\pi v^2 dv$. It is shown in App. A.7 how the remaining integrals over velocity can be treated. One finds that the density and the factors π cancel, and the result is

$$T = \frac{m \langle v \rangle^2}{2k_B} \tag{6.56}$$

The thermal energy of the plasma is $k_B T$. For later convenience, we define the *thermal velocity* as $v_{th}^2 = \langle v \rangle^2 / 2$ or

$$v_{th} = \left(\frac{k_B T}{m}\right)^{1/2} \tag{6.57}$$

For the anisotropic bi-Maxwellian distribution we would have obtained a parallel, $T_\parallel = m v_{th\parallel}^2 / k_B$, and a perpendicular temperature, $T_\perp = m v_{th\perp}^2 / k_B$, instead of the isotropic temperature, T. In all these cases the temperature has a well-defined meaning, because the plasma is in thermal equilibrium even when the equilibrium may be anisotropic. For more complicated non-equilibrium conditions, however, the temperature loses its original meaning. As a kinetic temperature it merely contains information about the average kinetic energy in the plasma.

Concluding Remarks

The derivation of the Vlasov equation from the Klimontovich-Dupree equation is only one way of finding the kinetic equations of a plasma. Historically one has gone a different route starting from the Liouville equation of statistical mechanics and descending from it to the Vlasov equation. This is achieved by the construction of reduced distribution functions, where integration over the individual phase spaces of each particle gets rid of the coordinates of this particle and, hence, its individuality. The two approaches are entirely equivalent, but the Liouville equation approach is mathematically much more complicated. However, it is slightly more rigorous and may be preferred by theoretically oriented readers. We have sketched this approach in App. B.4.

Further Reading

More about the kinetic theory is found in the relevant chapters of the monographs listed below. The construction of the Vlasov equation using statistical mechanics is best described in [5]. Derivation of the Fokker-Planck equation can be found in many places. One particularly good reference for plasma physics is again [5], where the expressions for the averages over the fluctuations are explicitly given. The drift-kinetic equation in the form presented here can be found in [1]. Loss cone and anisotropic distribution functions are given in [2]. The principles of measurement of distribution functions together with a number of examples can be found [7].

[1] A. Hasegawa, *Plasma Instabilities and Nonlinear Effects* (Springer Verlag, Heidelberg, 1975).

[2] C. F. Kennel and M. Ashour-Abdalla, in *Magnetospheric Plasma Physics*, ed. A. Nishida (D. Reidel Publ. Co., Dordrecht, 1982), p. 245.

[3] Y. L. Klimontovich, *The Statistical Theory of Non-Equilibrium Processes in a Plasma* (MIT Press, Cambridge, 1967).

[4] E. M. Lifschitz and L. P. Pitaevskii, *Physical Kinetics* (Pergamon Press, Oxford, 1981).

[5] D. L. Montgomery and D. A. Tidman, *Plasma Kinetic Theory* (McGraw-Hill, New York, 1964).

[6] D. R. Nicholson, *Introduction to Plasma Theory* (Wiley & Sons Inc., New York, 1983).

[7] M. Schulz and L. J. Lanzerotti, *Particle Diffusion in the Radiation Belts* (Springer Verlag, Heidelberg, 1974).

7. Magnetohydrodynamics

Plasmas are most precisely described by particle distribution functions in phase space. In the previous chapter it has been shown that such distribution functions evolve according to kinetic equations, the Vlasov equation in the special case of a collisionless plasma. In many cases, however, it is not necessary to know the exact evolution of the distribution function but it is sufficient to determine the spatial and temporal development of the macroscopic moments of the distribution, such as densities, velocities and temperatures. In particular, for slow time variations the moments describe the state of the plasma good enough for most cases of interest. Clearly, it will be simpler to investigate their evolution than to determine that of the distribution function.

Since the macroscopic moments are quantities which one is already familiar from fluid and gas dynamics, the resulting theory falls into the domain of fluid theories. The aim of the present chapter is therefore to derive and discuss such a hydrodynamic theory for plasmas. This theory will be called *magnetohydrodynamics* because it is the fluid theory of electrically charged fluids subject to the presence of external and internal magnetic fields. However, magnetohydrodynamics is already a further approximation to a more general hydrodynamic theory, the *multi-fluid theory* of plasmas. In the following we are going to derive the multi-fluid equations first before proceeding to the one-fluid magnetohydrodynamic approximation to this theory. However, the discussion and application of the former will be delayed to later chapters, because of the greater transparency of the magnetohydrodynamic approach when applied to the plasma in the magnetosphere and solar wind.

7.1. Multi-Fluid Theory

Fluid theory is looking for evolution equations for the basic macroscopic moments, i.e., number density, $n_s(\mathbf{x}, t)$, bulk flow velocity, $\mathbf{v}_{b,s}(\mathbf{x}, t)$, pressure tensor, $\mathbf{P}_s(\mathbf{x}, t)$, and kinetic temperature, $T_s(\mathbf{x}, t)$, of the particle species s in a plasma. For a two-fluid plasma consisting of ions and electrons, we have $s = i, e$. Further, since no distinction is made anymore between the individual particle velocities, we will henceforth drop the subscript 'b' and understand that $\mathbf{v}_s$ is the mean flow velocity of species s.

The definition of the moments given in the previous chapter suggests that the evolution equations under question can be derived from the Vlasov equation by performing an appropriate integration with respect to velocity space. For the sake of a better understanding we will explicitly show how such fluid-like equations arise from the kinetic equation of a plasma.

Continuity Equation

In order to demonstrate how this procedure works, let us take the zero-order moment and integrate the Vlasov equation (6.20) over the entire velocity space

$$\int \left[\frac{\partial f_s}{\partial t} + \mathbf{v} \cdot \nabla_{\mathbf{x}} f_s + \frac{q_s}{m_s} (\mathbf{E} + \mathbf{v} \times \mathbf{B}) \cdot \nabla_{\mathbf{v}} f_s \right] d^3 v = 0 \qquad (7.1)$$

Since the velocity space volume element, $d^3 v$, does not depend on time, the time derivative can be exchanged with the integral to find that the first term in Eq. (7.1) becomes

$$\frac{\partial}{\partial t} \int f_s d^3 v = \frac{\partial n_s}{\partial t} \qquad (7.2)$$

where the definition of the zero-order moment, the particle density of species s, has been used. The second term is evaluated in exactly the same way observing that spatial and velocity coordinates are independent variables and, hence, the integration over $d^3 v$ and the differentiation $\nabla_{\mathbf{x}}$ can be exchanged. Since according to Eq. (6.50) the integral represents the particle flux density, $n_s \mathbf{v}_s$, the whole expression yields the divergence of the particle flux of species s

$$\nabla_{\mathbf{x}} \cdot \int \mathbf{v} f_s d^3 v = \nabla \cdot (n_s \mathbf{v}_s) \qquad (7.3)$$

where we dropped the index $\mathbf{x}$ on ∇ on the right-hand side because the particle flux density depends only on space. Finally, consider the last term in the integrated Vlasov equation. Applying the velocity gradient $\nabla_{\mathbf{v}}$ to the full integrand, $(\mathbf{E} + \mathbf{v} \times \mathbf{B}) f_s$, makes it a total differential. Integrating over the total differential gives its values at the boundaries in velocity space which are at infinity. Since no particle has infinite speed, the distribution function is zero here, and this integral vanishes. Now, we are left with

$$\int f_s \nabla_{\mathbf{v}} \cdot (\mathbf{E} + \mathbf{v} \times \mathbf{B}) \, d^3 v \qquad (7.4)$$

The mixed vector product in this expression vanishes because the magnetic field does not depend on particle velocity, and the product can be written as $\mathbf{B} \cdot (\nabla_{\mathbf{v}} \times \mathbf{v}) = 0$. Moreover, the electric field is a function of space only and the first term also vanishes.

Hence, the last integral in the integrated Vlasov equation does not contribute. Collecting the surviving terms, we find

$$\frac{\partial n_s}{\partial t} + \nabla \cdot (n_s \mathbf{v}_s) = 0 \qquad (7.5)$$

This is the continuity equation of the s-component fluid of particles in the plasma. Its physical meaning is that in the absence of any interaction processes which create or annihilate particles of this species, the particle number density, and also mass and charge density is conserved during the motion of the fluid.

Equation of Motion

The continuity equation is the first fluid equation of the multi-fluid plasma; it is the zero-order moment equation of the collisionless Vlasov equation. However, it is not an equation for the density alone but it couples the plasma density to the fluid velocity. Therefore, another equation is required for the velocity of the plasma. Since the latter is the first moment of the distribution function, this second fluid equation will naturally result from a first moment treatment of the Vlasov equation.

Let us dyadically multiply the Vlasov equation with the particle velocity $\mathbf{v}$ and integrate term by term with respect to the velocity

$$\int \mathbf{v} \left[\frac{\partial f_s}{\partial t} + \mathbf{v} \cdot \nabla_{\mathbf{x}} f_s + \frac{q_s}{m_s} (\mathbf{E} + \mathbf{v} \times \mathbf{B}) \cdot \nabla_{\mathbf{v}} f_s \right] d^3 v = 0 \qquad (7.6)$$

Applying the reasoning of the previous paragraph since also the phase space coordinate, $\mathbf{v}$, is independent of time, and again exchanging differentiation and integration, the first integral term results in

$$\frac{\partial}{\partial t} \int \mathbf{v} f_s d^3 v = \frac{\partial}{\partial t} (n_s \mathbf{v}_s) \qquad (7.7)$$

which is the temporal variation of the flux density of s-component fluid. Thus identification of the first term is trivial. However, the second term provides more serious difficulties, because it contains a dyadic form, $\mathbf{vv}$. As a first step let us again exchange the differentiation with respect to space and the integration over velocity which is permitted because both are independent coordinates, keeping in mind that because of the symmetry of the dyad $\mathbf{v}(\mathbf{v} \cdot \nabla_{\mathbf{x}}) = \nabla_{\mathbf{x}} \cdot (\mathbf{vv})$. Let us rewrite the dyadic form as

$$\mathbf{vv} = (\mathbf{v} - \mathbf{v}_s)(\mathbf{v} - \mathbf{v}_s) - \mathbf{v}_s \mathbf{v}_s + \mathbf{vv}_s + \mathbf{v}_s \mathbf{v} \qquad (7.8)$$

and introduce this into the second integral, with $\nabla_{\mathbf{x}}$ extracted out of the integral. The integral over the term resulting from the first product on the right-hand side of Eq. (7.8) is then after Eq. (6.51) identified as the fluid pressure tensor divided by the mass, $\mathbf{P}_s/m_s$. The integral over the term resulting from the second product on the right-hand side of Eq. (7.8) becomes simply $-n_s\mathbf{v}_s\mathbf{v}_s$ since the fluid bulk velocity $\mathbf{v}_s$ is independent of $\mathbf{v}$. Finally, the two remaining integrals reproduce twice the same value but with positive sign, $+2n_s\mathbf{v}_s\mathbf{v}_s$, so that the total sum is $n_s\mathbf{v}_s\mathbf{v}_s$. Combining all these terms, the second integral becomes

$$\nabla_\mathbf{x} \cdot \int \mathbf{v}\mathbf{v} f_s d^3v = \nabla \cdot (n_s\mathbf{v}_s\mathbf{v}_s) + \frac{1}{m_s}\nabla \cdot \mathbf{P}_s \tag{7.9}$$

The last integral of the Vlasov equation can be treated by the same method of applying the operator $\nabla_\mathbf{v}$ to the full integrand, i.e., transforming it into a total derivative with respect to $\mathbf{v}$, the integral over which vanishes because one integrates over all of velocity space, and subtracting the part which one has added to obtain the total derivative. In the remaining non-vanishing integral the operator $\nabla_\mathbf{v}$ is applied only to the velocity, $\nabla_\mathbf{v}\mathbf{v}$. The operator is the unit tensor, $\mathbf{I}$. As on p. 131, the mixed vector product vanishes and the electric field does not depend on velocity. Hence, this last integral becomes

$$\int f_s(\nabla_\mathbf{v}\mathbf{v}) \cdot (\mathbf{E} + \mathbf{v} \times \mathbf{B})\, d^3v = -n_s(\mathbf{E} + \mathbf{v}_s \times \mathbf{B}) \tag{7.10}$$

We can now add all non-vanishing integrals to obtain our final result

$$\boxed{\frac{\partial(n_s\mathbf{v}_s)}{\partial t} + \nabla \cdot (n_s\mathbf{v}_s\mathbf{v}_s) + \frac{1}{m_s}\nabla \cdot \mathbf{P}_s - \frac{q_s}{m_s}n_s(\mathbf{E} + \mathbf{v}_s \times \mathbf{B}) = 0} \tag{7.11}$$

This equation is the *momentum density conservation equation* of the s-component fluid of the plasma or the equation of motion of this fluid component. It is the equation for the fluid velocity which we looked for. It relates the fluid velocity to density and electromagnetic force acting on the fluid element, but not on the single particles anymore.

The momentum density equation has a close relationship to conventional hydrodynamics where it is known as the Navier-Stokes equation. Hence, the plasma momentum conservation equation is the Navier-Stokes equation including an electromagnetic Lorentz force acting on the charges in the plasma. The appearance of this force in the equation of motion couples the plasma fluid to the full set of electromagnetic equations and makes it very distinct from conventional hydrodynamics where the only forces acting on the fluid are pressure and viscous forces. The appearance of this force also couples all the charged plasma fluid components together. This is obvious from the fact that the electric and magnetic fields in the Lorentz force act on all charged components and, at the same time, all charged components do contribute to the electric and magnetic fields. It is therefore clear that in solving the equations for the motion of one plasma component one is unavoidably confronted with the problem of solving all the equations of motion for all plasma fluid components because neither of them are independent of each other and of the electric and magnetic fields.

Energy Equation

However, as we already expected, the equation of motion does not close the system of equations because the next higher order quantity, the pressure tensor, $\mathbf{P}_s$, appears in it, which again requires another equation determining its evolution. Such an equation is again expected to be found by calculation of the second-order moment equation from the Vlasov equation, i.e., multiplying the Vlasov equation by the second-order dyad $\mathbf{vv}$ and integrating over velocity space. This integration is much more involved than that which we have performed until now. Its result is another equation, the heat transfer or *energy density conservation equation*

$$\frac{3}{2} n_s k_B \left(\frac{\partial T_s}{\partial t} + \mathbf{v}_s \cdot \nabla T_s \right) + p_s \nabla \cdot \mathbf{v}_s = -\nabla \cdot \mathbf{q}_s - (\mathbf{P}'_s \cdot \nabla) \cdot \mathbf{v}_s \qquad (7.12)$$

where T_s is the temperature defined in Sec. 6.5 and p_s is the scalar pressure, both of which are related by the ideal gas equation, $p_s = n_s k_B T_s$. The quantity $\mathbf{q}_s$ is the heat flux vector and $\mathbf{P}'_s$ denotes the stress tensor part of the full pressure tensor, $\mathbf{P}_s$. The stress tensor part describes the *shear stress*, e.g., the transfer of y momentum by motion in the x direction. Naturally this equation again contains a new undetermined quantity, the heat flux which is a third-order moment and, hence, requires an additional expression. In most cases one can neglect the heat flux and truncate the system of basic equations in this way.

7.2. Equation of State

When calculating moment equations we have found that the fluid equations of a plasma form a hierarchy of ever increasing order where each order contains a next order quantity which must be determined from the next order equation. Such a procedure must be closed by truncation of the hierarchy at a certain level. The most common and simples way is assuming an *equation of state* for the pressure which makes the energy equation obsolete and avoids explicitly taking into account the transport of heat.

The actual form of the equation of state depends on the form of the pressure tensor, e.g., isotropic or anisotropic, and even more general on the behavior of the fluid condensed in the number and momentum density equations. The equations of state may differ and are indeed often different for different fluid components, especially electrons and ions.

Isotropic Pressure

If the pressure is taken to be isotropic, the pressure tensor becomes diagonal

$$\boxed{\mathbf{P}_s = p_s \mathbf{I}} \tag{7.13}$$

which reads in matrix notation

$$\mathbf{P}_s = \begin{pmatrix} p_s & 0 & 0 \\ 0 & p_s & 0 \\ 0 & 0 & p_s \end{pmatrix} \tag{7.14}$$

and only one such equation of state is needed. The actual form of this equation may differ drastically, but here we will give the equations for the two most important cases.

The most simple equation of state is that for an *isothermal* case, $T_s = $ const. Here one simply takes the ideal gas equation

$$p_s = n_s k_B T_s \tag{7.15}$$

Isothermal conditions can be applied when the temporal variations are so slow that the plasma has sufficient time to redistribute energy in order to maintain a constant heat bath temperature. Such conditions frequently apply to the global situation of the magnetosphere. When $T_s = T_{s0}$ is assumed constant, the pressure becomes proportional to the density of the species

$$\boxed{p_s = n_s k_B T_{s0}} \tag{7.16}$$

and the system of equations is truncated to a closed set.

In the other extreme, when the time variations are so fast that no susceptible heat exchange can take place the plasma evolves *adiabatically*. The change in temperature is then related in a simple way to the change in density. Intuitively this is clear because any gas will cool during a fast expansion of the volume and heat up when the volume is compressed. To find the adiabatic relation we set the right-hand side of Eq. (7.12) to zero. Using the continuity equation (7.5) in order to replace the divergence of the velocity by a time derivative of the density and applying the total time derivative $d/dt = \partial/\partial t + \mathbf{v} \cdot \nabla$, the heat transfer equation (7.12) can be cast into the form

$$\frac{3}{2} \frac{d(n_s k_B T_s)}{dt} - \frac{5}{2} k_B T_s \frac{dn_s}{dt} = 0 \tag{7.17}$$

an equation which is identical to

$$n_s \frac{dT_s}{dt} - \frac{2}{3} T_s \frac{dn_s}{dt} = 0 \tag{7.18}$$

which has the well known adiabatic solutions

$$T_s = T_{s0} \left(\frac{n_s}{n_{s0}} \right)^{\gamma - 1} \tag{7.19}$$

or, written in term of scalar pressure, using Eq. (7.15)

$$p_s = p_{s0} \left(\frac{n_s}{n_{s0}} \right)^{\gamma} \tag{7.20}$$

The *adiabatic index*, $\gamma = c_p/c_v = 5/3$, which is the ratio of the two specific heats at constant pressure and constant volume, is constant in a collisionless ideal isotropic plasma, and as long as there are no further interactions between the different species in the plasma the temperatures and densities of each species evolve according to the adiabatic law. The index γ can also be regarded as a *polytropic index*, comprising not only the adiabatic case, but also the isobaric or constant pressure, $\gamma = 0$, the isothermal or constant temperature, $\gamma = 1$, and the isometric or constant density, $\gamma = \infty$, cases.

Anisotropic Pressure

In anisotropic plasmas, the pressure tensor splits into parallel and perpendicular pressure

$$\mathbf{P}_s = p_{s\perp}\mathbf{I} + (p_{s\|} - p_{s\perp})\frac{\mathbf{BB}}{B^2} \tag{7.21}$$

which, in a coordinate system where the z axis is aligned with the magnetic field direction, reads in matrix notation

$$\mathbf{P}_s = \begin{pmatrix} p_{s\perp} & 0 & 0 \\ 0 & p_{s\perp} & 0 \\ 0 & 0 & p_{s\|} \end{pmatrix} \tag{7.22}$$

It is then not a priori clear that both parallel and perpendicular pressure evolve according to the same adiabatic laws while it is still a good approximation to use the ideal gas equation for both pressures

$$\begin{aligned} p_{s\|} &= n_s k_B T_{s\|} \\ p_{s\perp} &= n_s k_B T_{s\perp} \end{aligned} \tag{7.23}$$

If the adiabatic approximation is justified, one can use the general definition of the adiabatic index

$$\gamma = (d + 2)/d \tag{7.24}$$

where d is the degree of freedoms the particles of the plasma have. In monatomic $3d$ plasmas one recovers $\gamma = 5/3$. In $1d$ plasmas $\gamma = 3$, and in $2d$ plasmas $\gamma = 2$. The parallel pressure has $d = 1$, while the perpendicular has $d = 2$, and the adiabatic equations of state become

$$p_{s\parallel} = p_{s\parallel 0}\left(\frac{n_s}{n_{s0}}\right)^3$$

$$p_{s\perp} = p_{s\perp 0}\left(\frac{n_s}{n_{s0}}\right)^2 \tag{7.25}$$

However, the above reasoning clearly neglects the coupling between the two pressure components due to inhomogeneous magnetic fields. Most important, the above equations do not include any dependence on the magnetic field strength, whereas, for example, in a pure mirror geometry (see Sec. 2.5), the magnetic field strength determines the ratio between parallel and perpendicular pressures. Hence, these equations can only be applied in situations where the magnetic field is of minor importance.

Double-Adiabatic Invariants

A slightly better approximation to the equation of state of anisotropic plasmas can be found by assuming that the presence of a sufficiently strong magnetic field not only introduces the symmetry breaking between parallel and perpendicular pressures but also some kind of ordering. For instance, for the perpendicular motion of the particles we already know that under some rather weak conditions, when the variation of the plasma is slower than the gyration, the magnetic moment $\mu = mv_\perp^2/2B$ of the particles is conserved. This conservation clearly affects the evolution of the perpendicular energy of the particles and, hence, the perpendicular pressure.

Calculating the perpendicular temperature moment over an anisotropic Maxwellian and dividing by the magnetic field strength, one realizes that the result is the distribution function average or moment over the magnetic moment

$$\langle \mu \rangle = \frac{k_B T_\perp}{B} = \frac{p_\perp}{nB} \tag{7.26}$$

Because the ensemble averaged $\langle \mu \rangle$ must be conserved, the right hand side of this equation is a constant, and the perpendicular pressure evolves in proportionality to the magnetic field strength. The perpendicular adiabatic law can be read from this behavior as

$$\frac{d}{dt}\left(\frac{p_\perp}{nB}\right) = 0 \tag{7.27}$$

Since this expression results from magnetic moment conservation, no heat is transferred into the perpendicular direction. No adiabatic exponent enters this equation because the pressure is entirely determined by the magnetic field. Moreover, this relation couples magnetic field, perpendicular temperature and density together so that no simple adiabatic index exists. Instead it depends on either the field or the temperature.

Finding the parallel adiabatic equation of state is more involved because no such simple conservation equation exists as for the magnetic moment. One must rewrite the heat transfer equation for an anisotropic plasma, neglect all dissipative terms and inhibit parallel in addition to perpendicular heat transfer. This requirement is particularly strong for electrons of which we know that they easily escape along the magnetic field lines. Therefore the parallel adiabatic equation will from the very beginning impose a rather strong condition on the applicability to a plasma. One may apply it to ions, but the application to electrons will be limited.

One can show that the equation obtained for the parallel pressure $p_\parallel$ under the assumption of suppressed heat flow and no dissipation becomes

$$p_\perp \frac{dp_\parallel}{dt} + 2p_\parallel \frac{dp_\perp}{dt} + 5p_\perp p_\parallel \nabla \cdot \mathbf{v} = 0 \qquad (7.28)$$

Replacing the divergence of $\mathbf{v}$ with the help of the continuity equation (7.5) written in a different form

$$-n\nabla \cdot \mathbf{v} = \partial n/\partial t + \mathbf{v} \cdot \nabla n \qquad (7.29)$$

and using $d/dt = \partial/\partial t + \mathbf{v} \cdot \nabla$, Eq. (7.28) becomes, after rearranging some terms, a total time derivative

$$\frac{d}{dt}\left(\frac{p_\parallel p_\perp^2}{n^5}\right) = \frac{d}{dt}\left(\frac{p_\parallel B^2}{n^3}\right) = 0 \qquad (7.30)$$

This is the parallel equation of state. Again, the parallel pressure is a function of the magnetic field and the density.

Combining the two double-adiabatic or *Chew-Goldberger-Low equations* of state with the parallel and perpendicular ideal gas equations (7.23), it is easy to show that in this theory the parallel and perpendicular temperatures depend on the magnetic field strength according to

$$T_\perp \propto B$$
$$T_\parallel \propto (n/B)^2 \qquad (7.31)$$

Hence, for an increasing magnetic field strength the perpendicular temperature increases while the parallel temperature decreases. A streaming plasma where the magnetic field strength increases along the direction of the stream will therefore exhibit a growing temperature anisotropy with a higher temperature in the perpendicular than in the parallel direction. Such a situation is encountered in the magnetosphere and in the near-Earth magnetosheath.

To conclude this section we briefly discuss the behavior of the parallel and perpendicular adiabatic indices. Formally it is possible to define such indices in analogy to the adiabatic case by simply writing $p_\parallel \propto n^{\gamma_\parallel}$ and $p_\perp \propto n^{\gamma_\perp}$. Using the double-adiabatic equations of state, one finds that

$$\gamma_\perp = 1 + \frac{\ln(B/B_0)}{\ln(n/n_0)}$$
$$\gamma_\parallel = 3 - 2\frac{\ln(B/B_0)}{\ln(n/n_0)} \tag{7.32}$$

are functions of magnetic field and density. However, from Eq. (7.30) one can derive a condition on the parallel and perpendicular adiabatic indices

$$\gamma_\parallel + 2\gamma_\perp - 5 = 0 \tag{7.33}$$

which shows that the parallel and perpendicular adiabatic indices in a double-adiabatic plasma are not independent but closely related. It is sufficient to know one of the indices in order to determine the other one. Yet these quantities are not constants but spatially varying functions.

7.3. One-Fluid Theory

Plasmas consist of electrons of mass, m_e, and charge, $q_e = -e$, and ions of mass, m_i, and charge, $q_i = Ze$. Let us for simplicity assume that there is only one ion component present in the plasma. If these are protons, the atomic charge is $Z = 1$. For ease of use, let us repeat the fundamental equations of such a plasma, the continuity equation (7.5) and the equation of motion (7.11) for the s-component fluid

$$\frac{\partial n_s}{\partial t} + \nabla \cdot (n_s \mathbf{v}_s) = 0 \tag{7.34}$$

$$\frac{\partial (n_s \mathbf{v}_s)}{\partial t} + \nabla \cdot (n_s \mathbf{v}_s \mathbf{v}_s) = -\frac{1}{m_s}\nabla \cdot \mathbf{P}_s + \frac{n_s q_s}{m_s}(\mathbf{E} + \mathbf{v}_s \times \mathbf{B}) \tag{7.35}$$

which must be completed with the equations of state for the electron and ion components and with the set of Maxwell's equations, where we define the charges and currents by

$$\rho = e(n_i - n_e) \tag{7.36}$$
$$\mathbf{j} = e(n_i \mathbf{v}_i - n_e \mathbf{v}_e) \tag{7.37}$$

Quasineutrality in such a case is defined by a vanishing electric space charge $\rho = 0$, yielding $n = n_e = n_i$, which implies equal charge densities. For a current-free plasma,

$\mathbf{j} = 0$, the particle flux densities must be equal, $n_i \mathbf{v}_i = n_e \mathbf{v}_e$. This is not generally the case, since most plasmas are quasineutral but carry currents.

Sometimes it is convenient to neglect the difference between the particle species in a plasma and to consider the plasma as a conducting fluid carrying magnetic and electric fields and currents. In such a case the fluid field variables are some combinations of the densities and velocities of the single components. The resulting equations are called *magnetohydrodynamic equations* of a plasma. They can be derived from the above two-fluid equations by choosing as variables the following combinations

$$n = \frac{m_e n_e + m_i n_i}{m_e + m_i} \tag{7.38}$$

$$m = m_e + m_i = m_i \left(1 + \frac{m_e}{m_i} \right) \tag{7.39}$$

$$\mathbf{v} = \frac{m_i n_i \mathbf{v}_i + m_e n_e \mathbf{v}_e}{m_e n_e + m_i n_i} \tag{7.40}$$

for the fluid number density, n, the fluid mass, m, and the fluid velocity, $\mathbf{v}$.

Continuity Equation

It is now simple to derive the continuity equation for the total fluid. We multiply the two-fluid continuity equation (7.34) for ions by m_i and for electrons by m_e, add the two resulting equations, and make use of the above definitions in Eqs. (7.38) through (7.40) to obtain

$$\boxed{\frac{\partial n}{\partial t} + \nabla \cdot (n\mathbf{v}) = 0} \tag{7.41}$$

This equations represent the usual form of a *fluid continuity equation* and does not anymore discriminate between the different kinds of particles. The physical content of the continuity equation is that in a classical and nonrelativistic plasma, mass is conserved.

Equation of Motion

Constructing the momentum density conservation equation for the total fluid, or *fluid equation of motion*, is more difficult because of the appearance of the nonlinear terms, $n_s \mathbf{v}_s \mathbf{v}_s$, in Eq. (7.35). We therefore demonstrate explicitly how this equation is obtained. To be even more general, let us for convenience include a simple collisional term in Eq. (7.35), which may account for a momentum transfer between electrons and ions via some kind of friction, either collisional in the classical sense or due to anomalous collisions between the two kinds of particles. As demonstrated in Chap. 4, the presence of collisions gives rise to a non-vanishing electric field and resistive currents. As a consequence an Ohm's law exists in the plasma. The derivation of the momentum conservation equation for the fluid therefore necessarily results also in a generalized Ohm's law as a second material equation for the plasma, coupling the currents in the medium to the electromagnetic field.

The collisional term is defined by $\mathbf{R} = \mathbf{R}_{ie} = -\mathbf{R}_{ie}$. Because this term describes transfer of momentum from ions to electrons in the ion equation and from electrons to ions in the electron equation, conservation of the transferred momentum requires that the two terms are equal in magnitude but have different sign. Hence, the two momentum equations read

$$
\begin{aligned}
\frac{\partial(n_e \mathbf{v}_e)}{\partial t} + \nabla \cdot (n_e \mathbf{v}_e \mathbf{v}_e) &= -\frac{1}{m_e} \nabla \cdot \mathbf{P}_e - \frac{n_e e}{m_e}(\mathbf{E} + \mathbf{v}_e \times \mathbf{B}) + \frac{\mathbf{R}}{m_e} \\
\frac{\partial(n_i \mathbf{v}_i)}{\partial t} + \nabla \cdot (n_i \mathbf{v}_i \mathbf{v}_i) &= -\frac{1}{m_i} \nabla \cdot \mathbf{P}_i + \frac{n_i e}{m_i}(\mathbf{E} + \mathbf{v}_i \times \mathbf{B}) - \frac{\mathbf{R}}{m_i}
\end{aligned}
\tag{7.42}
$$

The equation of motion of the single-fluid plasma is obtained by adding these two equations and making use of the above definitions for m, n, $\mathbf{v}$, $\mathbf{j}$, and ρ. When multiplying the first equation by m_e, the second by m_i, and adding up, the two collisional terms cancel. The right-hand side becomes

$$
-\nabla \cdot (\mathbf{P}_e + \mathbf{P}_i) + e(n_i - n_e)\mathbf{E} + e(n_i \mathbf{v}_i - n_e \mathbf{v}_e) \times \mathbf{B} = -\nabla \cdot \mathbf{P} + \rho \mathbf{E} + \mathbf{j} \times \mathbf{B} \tag{7.43}
$$

where we have introduced the total pressure tensor, $\mathbf{P} = \mathbf{P}_e + \mathbf{P}_i$, and made use of the definitions of the space charge in Eq. (7.36) and of the current density in Eq. (7.37). The first term on the left-hand side of Eq. 7.42, after adding up and observing the definition of the fluid bulk velocity, $\mathbf{v}$, in Eq. (7.40) becomes

$$
\frac{\partial}{\partial t}(m_e n_e \mathbf{v}_e + m_i n_i \mathbf{v}_i) = \frac{\partial}{\partial t}(nm\mathbf{v}) \tag{7.44}
$$

In the second nonlinear term of Eq. 7.42 we take advantage of the smallness of the electron mass, $m_e \ll m_i$, and assume that the two densities are nearly equal. In this case $n_i \approx n_e$, and the dominant term in the sum $m_i n_i(\mathbf{v}_i \mathbf{v}_i + \mathbf{v}_e \mathbf{v}_e(m_e n_e/m_i n_i))$ is the first ion term. With this approximation, which is good for nearly quasineutral plasmas, the equation of motion becomes

$$
\boxed{\frac{\partial(nm\mathbf{v})}{\partial t} + \nabla \cdot (nm\mathbf{v}\mathbf{v}) = -\nabla \cdot \mathbf{P} + \rho \mathbf{E} + \mathbf{j} \times \mathbf{B}} \tag{7.45}
$$

This is the momentum conservation equation in magnetohydrodynamics.

Generalized Ohm's Law

The momentum conservation equation (7.45) contains the electric current density, $\mathbf{j}$, as a new variable. To close the system of equations, one therefore needs an additional expression for the evolution of $\mathbf{j}$. This equation is the *generalized Ohm's law* of a plasma. It is found by subtracting the two-fluid momentum equations (7.42). For practical reasons it is convenient to multiply the electron equation by m_i and the ion equation by m_e before subtraction. For small current densities, we can neglect quadratic terms in the velocities, i.e., the second terms on the left-hand sides of the momentum equations. It is then easy to show that the current density satisfies the following relation

$$\frac{m_e}{e}\frac{\partial \mathbf{j}}{\partial t} = \nabla \cdot \left(\mathbf{P}_e - \frac{m_e}{m_i}\mathbf{P}_i\right) - \left(1 + \frac{m_e}{m_i}\right)\mathbf{R}$$
$$+n_e e \left(1 + \frac{m_e n_i}{m_i n_e}\right)\left[\mathbf{E} + \left(\mathbf{v}_e + \frac{m_e n_i}{m_i n_e}\mathbf{v}_i\right) \times \mathbf{B}\right] \quad (7.46)$$

The right-hand side of this equation still contains the partial densities, velocities, and pressures. It is, however, possible to replace them by use of Eqs. (7.38) through (7.40). The algebra is simplified when neglecting terms with small mass ratios, $m_e/m_i \ll 1$, and assuming quasineutrality, $n \approx n_i \approx n_e$. Then only the electron pressure plays a role in the first term on the right-hand side of the above equation. Moreover, quasineutrality also simplifies the terms containing $\mathbf{E}$ and $\mathbf{B}$. In particular, it allows to omit the term containing $\mathbf{v}_i$. Hence, the above equation can be rewritten as

$$\frac{m_e}{e}\frac{\partial \mathbf{j}}{\partial t} = \nabla \cdot \mathbf{P}_e + ne(\mathbf{E} + \mathbf{v}_e \times \mathbf{B}) - \mathbf{R} \quad (7.47)$$

Before proceeding, we note briefly some interesting implications of this expression. The first one is that in the one-fluid theory thermal effects on the electric current density enter only through the electron pressure, while the macroscopic bulk fluid velocity is also affected by the ion pressure via the full pressure tensor entering Eq. (7.45). It is the changes in the electron partial pressure which modulate the current. Second, the Lorentz term $\mathbf{E} + \mathbf{v}_e \times \mathbf{B}$ on the right-hand side only contains the electron velocity $\mathbf{v}_e$. Hence, even in the one-fluid the electron fluid behaves different from the ion fluid and has a greater effect on the current. In particular, in the stationary ideal case the electron fluid is stronger frozen-in to the magnetic field than the ions.

Again omitting small mass ratio terms and assuming quasineutrality, one can find a fluid expression for the electron velocity, since under these assumptions Eq. (7.40) can be written as

$$\mathbf{v}_i = \mathbf{v} \quad (7.48)$$

Inserting this approximation into Eq. (7.37) one obtains

$$\mathbf{v}_e = \mathbf{v} - \frac{\mathbf{j}}{ne} \quad (7.49)$$

Inserting the above expression into Eq. (7.47) yields the following equation

$$\frac{m_e}{e}\frac{\partial \mathbf{j}}{\partial t} = \nabla \cdot \mathbf{P}_e + ne(\mathbf{E} + \mathbf{v} \times \mathbf{B}) - \mathbf{j} \times \mathbf{B} - \mathbf{R} \tag{7.50}$$

Finally, we remember that the friction term, $\mathbf{R}$, is proportional to the velocity difference of the two oppositely charged species, and is symmetric in the densities. The proportionality factor is the collision frequency, ν_c, times the electron mass

$$\mathbf{R} = m_e n^2 \nu_c(\mathbf{v}_i - \mathbf{v}_e) \tag{7.51}$$

The factor in front of the brackets on the right-hand side includes similar quantities as the plasma resistivity, $\eta = m_e \nu_c / ne^2$, introduced in Sec. 4.2, thus permitting to write

$$\mathbf{R} = \eta ne \mathbf{j} \tag{7.52}$$

Hence, the generalized Ohm's law of a single-fluid magnetohydrodynamic plasma becomes after rearrangement of the different terms

$$\boxed{\mathbf{E} + \mathbf{v} \times \mathbf{B} = \eta \mathbf{j} + \frac{1}{ne}\mathbf{j} \times \mathbf{B} - \frac{1}{ne}\nabla \cdot \mathbf{P}_e + \frac{m_e}{ne^2}\frac{\partial \mathbf{j}}{\partial t}} \tag{7.53}$$

This equation is an important expression in several respects. First one recognizes that in a plasma the simple Ohm's law derived in Chap. 4 becomes considerably more complicated. In addition to the resistive term, $\eta \mathbf{j}$, it contains the anisotropic electron pressure term, a Lorentz force term $\mathbf{j} \times \mathbf{B}$ which is often called *Hall term* and even in a collisionless plasma gives rise to contributions transverse to both the current and the magnetic field. Finally, the generalized Ohm's law turns out to contain the time variation of the current which can be interpreted as the contribution of electron inertia to the current flow.

From Eq. (7.53) it also is obvious that in an ideally conducting magnetohydrodynamic fluid with $\eta = 0$ the convective approximation or frozen-in condition

$$\mathbf{E} = -\mathbf{v} \times \mathbf{B} \tag{7.54}$$

requires additional assumptions. Vanishing electron pressure gradients and slow time variations of current density are necessary to neglect the two corresponding terms in the generalized Ohm's law. But even under these conditions neglecting the Lorentz force term, $\mathbf{j} \times \mathbf{B}$, is more difficult to justify. Actually, using Eq. (7.49) in Eq. (7.53), one can readily show that

$$\mathbf{E} = -\mathbf{v}_e \times \mathbf{B} \tag{7.55}$$

Hence, only the electron fluid is frozen to the magnetic field, while the motion of the ion fluid may deviate from the field. Nevertheless, when the transverse currents are small, the ideal magnetohydrodynamic condition is frequently applied to space plasmas. In the solar wind and the magnetosphere for slow variations, negligible pressure gradients and weak currents it is often satisfied very well.

Energy Conservation Equation

For completeness, we add the *energy conservation equation* of magnetohydrodynamics. It follows by multiplying the momentum conservation equation (7.45) by the fluid velocity, $\mathbf{v}$, and manipulating the resulting expression with the help of the continuity equation into the form

$$\frac{\partial}{\partial t}\left[nm\left(\tfrac{1}{2}v^2 + w\right) + \frac{B^2}{2\mu_0}\right] = -\nabla \cdot \mathbf{q} \qquad (7.56)$$

The right-hand side of this expression is the divergence of the magnetohydrodynamic energy or heat flux density vector

$$\mathbf{q} = \left(\frac{v^2}{2} + w + \frac{p + \mu_0^{-1}B^2}{nm}\right)nm\mathbf{v} - \frac{\mathbf{B}}{\mu_0}\left(\mathbf{v} + \frac{\mathbf{j}}{ne}\right)\cdot\mathbf{B} - \frac{\eta\mathbf{j} \times \mathbf{B}}{\mu_0} + \frac{jB^2}{\mu_0 ne} + \frac{m_e\mathbf{B}}{\mu_0 ne^2} \times \frac{\partial\mathbf{j}}{\partial t}$$

$$(7.57)$$

Of the dissipative terms in Ohm's law we have taken into account only the resistivity and electron inertia in deriving this form of $\mathbf{q}$. The quantity w is the free internal energy density or *enthalpy* of the fluid, and p is the isotropic pressure. The left-hand side of Eq. (7.56) is the local time derivative of the total energy density which is the sum of the kinetic and internal energy densities to which the pressure of the magnetic field has been added. On the other hand, the energy flux contains the convective losses of kinetic energy, internal energy, and temperature, Joule heating due to resistivity and electron inertia, and magnetic field energy flux related to the divergence of the Lorentz force term. If all these losses vanish, the divergence of $\mathbf{q}$ is zero, and the total energy density including kinetic, inner, and magnetic energy does locally not change in time.

Entropy Equation

In Eq. (7.56) the conservation of energy has been written in the form of a heat conduction equation. This is not the only possible way to express energy conservation. Several other forms have been proposed in the literature. For example, one may introduce the entropy, S, and write Eq. (7.56) as a continuity equation of the entropy

$$\boxed{\frac{dS}{dt} = \frac{\partial S}{\partial t} + \mathbf{v} \cdot \nabla S = 0} \qquad (7.58)$$

To calculate the entropy change in an ideal isotropic gas from the measurements of temperature and density, one can use

$$\boxed{\Delta S = R_0 \ln\left[\frac{T^{1/(\gamma-1)}}{n}\right]} \qquad (7.59)$$

where R_0 is the ideal gas constant (cf. App. A.1) and γ is the adiabatic index (see App. A.6, where also a formula for anisotropic plasmas is given).

Equation (7.58) replaces the energy conservation and heat transfer equations insofar as the entropy is the fundamental state quantity of any thermodynamic system. It is defined differentially as the ratio $dS = dQ/T$ of the differential amount of heat dQ generated in the gas divided by the temperature T. Since the entropy can only grow or stay constant, the differential is always positive or vanishes when no heat is produced. Heat production requires dissipation. Hence, in a dissipationless medium like ideal magnetohydrodynamics the entropy is conserved. This is the content of Eq. (7.58). But if the plasma is dissipative, heat is generated, and the right-hand side of Eq. (7.58) is different from zero. Then one must add a Joule heating term, $\mathbf{j} \cdot \mathbf{E}$, to the right-hand side, where the electric field is being replaced with the help of the general Ohm's law (see p. 141).

Magnetic Tension and Plasma Beta

The Lorentz force or Hall term, $\mathbf{j} \times \mathbf{B}$, appearing in the magnetohydrodynamic equation of motion and the generalized Ohm's law introduces a new effect which is specific for magnetohydrodynamics and which we will discuss separately. It is the effect of the *magnetic tension* on a conducting magnetohydrodynamic fluid.

For slow variations, when the displacement current in the plasma can be neglected, the first Maxwell equation (2.3) can be used to rewrite

$$\mathbf{j} \times \mathbf{B} = -\frac{1}{\mu_0}\mathbf{B} \times (\nabla \times \mathbf{B}) \tag{7.60}$$

Applying some vector algebra to the right-hand side, this expression can be written as

$$\boxed{\mathbf{j} \times \mathbf{B} = -\nabla \left(\frac{B^2}{2\mu_0}\right) + \frac{1}{\mu_0}\nabla \cdot (\mathbf{BB})} \tag{7.61}$$

The first term on the right-hand side of this equation corresponds to a pressure term with the *magnetic pressure* defined as

$$p_B = \frac{B^2}{2\mu_0} \tag{7.62}$$

This pressure simply adds to the thermal pressure of the plasma. The second term is a consequence of the vector product of current and magnetic field and thus of the vector character of the magnetic field. It is the divergence of a *magnetic stress tensor*, $\mathbf{BB}/\mu_0$. The magnetic field introduces a magnetic stress in the plasma, which contributes to tension and torsion in the conducting fluid.

The concept of magnetic pressure can also be used to define another useful quantity. Starting from the fluid equation of motion (7.45) and assuming quasineutrality, which cancels the $\rho\mathbf{E}$ term, equilibrium conditions, $d\mathbf{v}/dt = 0$, and using Eq. (7.60), we are left with

$$\nabla \cdot \mathbf{P} = -\frac{1}{\mu_0}\mathbf{B} \times (\nabla \times \mathbf{B}) \qquad (7.63)$$

This equation forms together with those Maxwell's equation describing the divergence and the curl of $\mathbf{B}$ a closed set and is often called *magnetohydrostatic equation*. It tells us that in equilibrium the particle pressure gradient is balanced by magnetic tension.

If we neglect the off-diagonal or stress terms in the two pressure tensors, e.g., for cases where the particle pressure is nearly isotropic and the magnetic field is approximately homogenous, and use Eq. (7.61), we can approximate Eq. (7.63) by

$$\nabla \left(p + \frac{B^2}{2\mu_0} \right) = 0 \qquad (7.64)$$

Hence, in an equilibrium, isotropic, and quasineutral plasma the total pressure is a constant. Under these conditions one can define a *plasma beta* parameter as the ratio of thermal and magnetic pressure

$$\boxed{\beta = \frac{2\mu_0 p}{B^2}} \qquad (7.65)$$

In anisotropic plasmas where the pressure splits into parallel and perpendicular components one frequently uses parallel and perpendicular plasma beta

$$\begin{aligned} \beta_\parallel &= \frac{2\mu_0 p_\parallel}{B^2} \\ \beta_\perp &= \frac{2\mu_0 p_\perp}{B^2} \end{aligned} \qquad (7.66)$$

The β parameters measure the relative importance of particle and magnetic field pressures. A plasma is called a *low-beta plasma* when $\beta \ll 1$ and a *high-beta plasma* for $\beta \approx 1$ and greater. Both types of plasmas are encountered in near-Earth space.

Equation of State

The above equations are not complete until one adds appropriate equations for the pressure tensor components. Neglecting all dissipative effects, except for the electrical resistivity, the pressure tensors may have up to two independent components, $p_\perp$ and $p_\parallel$, which require additional equations. Under the assumption that the medium behaves like an ideal gas, this pressure equation becomes the ideal gas equation $p = nk_B T$, or two equations of state in the anisotropic case, providing the connection between pressure, temperature, and density.

The equation determining the behavior of the temperature is the energy or heat conduction equation. For an isotropic magnetohydrodynamic fluid it is sufficient to derive an overall energy conservation equation. But this equation will contain the unknown heat flux and must be truncated. If the heat flux is neglected, the energy equation becomes an equation for the temperature, as has been discussed in Sec. 7.1. In many cases, however, one can avoid to solve this equation assuming either isothermal conditions with $T = $ const or adiabaticity or, more general, polytropicity with $p \propto n^\gamma$ (or the various anisotropic equivalents of this expression), thereby closing the full set of magnetohydrodynamic relations.

7.4. Stationarity and Equilibria

Stationarity implies absence of any time variations which mathematically means that partial and sometimes even total time derivatives are set to zero. Stationarity also implies that the state of the plasma persists for long time. It is thus an equilibrium state and one sometimes speaks of plasma equilibria. Finding the equilibrium state of a fluid plasma requires solution of the time-independent fluid equations for the special case under consideration. Under given boundary conditions this may become a formidable task. But it is possible to draw some more general conclusions about the plasma behavior.

Boltzmann's Law

Let us first treat a rather simple case, which involves only the electron fluid. Consider the stationary electron momentum conservation equation (7.11) for scalar pressure and absence of an external magnetic field. Setting the convective derivative to zero, it can be written as

$$\nabla p_e = -n_e e \mathbf{E} \tag{7.67}$$

demonstrating that the electrons are in equilibrium with the electric field. The electric field can be represented as the gradient of an electric potential, $\mathbf{E} = -\nabla \phi$. Now assuming that the electron temperature is constant, which is reasonable since under stationary conditions one expects that the plasma had sufficient time to achieve an isothermal state, the above equation becomes with $p_e = n_e k_B T_e$

$$\nabla \left(\ln n_e - \frac{e\phi}{k_B T_e} \right) = 0 \tag{7.68}$$

The solution of this equation is the *Boltzmann law* which relates the stationary electron density to the electrostatic potential as

$$\boxed{n_e = n_0 \exp\left(\frac{e\phi}{k_B T_e} \right)} \tag{7.69}$$

where n_0 is the average electron density. The interpretation of this law is that in a stationary electric field, present in a plasma and maintained by external means, the electron density is necessarily spatially inhomogeneous and changes exponentially with the local electrostatic potential. The electron fluid reacts very sensitively to an electric field. According to the attraction exerted on the negative electrons by positive electric potentials the electron density assumes its maximum where the electric potential attains its highest positive value. Application of this law is possible in all cases when electric fields are present in the plasma in the direction parallel to the magnetic field and when the time variations are so slow that electron motions can be neglected.

Diamagnetic Drift

The next conclusion can be obtained in full generality even for the multi-fluid case. Let us return to the fluid momentum conservation equation (7.11) for the s-component fluid. Let us also assume stationary conditions so that the convective derivative terms can be dropped. For simplicity we also assume that the pressure tensor, $\mathbf{P}_s$, is anisotropic in the form of Eq. (7.21). Then the fluid equation of motion reduces to

$$q_s n_s (\mathbf{E} + \mathbf{v}_s \times \mathbf{B}) = \nabla p_{s\perp} + \nabla \cdot \left[(p_{s\parallel} - p_{s\perp}) \frac{\mathbf{BB}}{B^2} \right] \qquad (7.70)$$

Taking the cross-product of this equation with $\mathbf{B}/B^2$ and rearranging the different terms, we obtain for the stationary drift velocity of the s-component fluid

$$\mathbf{v}_s = \frac{\mathbf{E} \times \mathbf{B}}{B^2} + \frac{1}{q_s n_s B^2} \mathbf{B} \times \nabla p_{s\perp} + \frac{1}{q_s n_s B^2} \mathbf{B} \times \nabla \cdot \left[(p_{s\parallel} - p_{s\perp}) \frac{\mathbf{BB}}{B^2} \right] \qquad (7.71)$$

In a manner analogous to single particle theory, this expression defines the drift of the s-component fluid of the plasma across the magnetic field. As one immediately recognizes, the first term on the right hand side is nothing else but the $\mathbf{E} \times \mathbf{B}$ drift of the fluid which is the effect of the Lorentz transformation. The second and third terms are, however, entirely new and did not appear in single particle theory.

The second term describes a drift perpendicular to the magnetic field and to the transverse gradient (because only $\nabla_\perp$ survives the cross-product with the magnetic field) of the perpendicular pressure $p_{s\perp}$. Its dependence on the pressure, which is a moment of the distribution function and thus an average variable of the plasma, indicates that this drift arises due to a collective behavior of the plasma.

Consider a plasma of gyrating particles of one species. All particles gyrate in the same direction around the field. Consequently, at each point in a homogeneous plasma there would be exactly the same number of particles having exactly same but oppositely directed transverse velocities, resulting from particles which are displaced by just one gyroradius across the magnetic field, so that the average velocity would be zero. In a

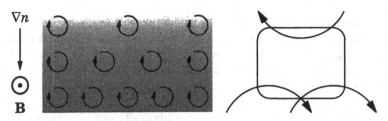

Fig. 7.1. Origin of the diamagnetic drift, assuming positive ions.

non-uniform plasma, the change in transverse pressure can be either due to a gradient in density or a gradient in transverse temperature.

The presence of a transverse density gradient introduces an asymmetry since in the direction of decreasing particle density there are less particles gyrating and, hence, not sufficient oppositely directed velocities to average the transverse velocity out. Consequently there is an excess of transverse particle gyration velocity perpendicular to the density gradient which remains and simulates a gross fluid drift motion (see Fig. 7.1). If the plasma temperature changes across the magnetic field, decreasing temperature implies smaller transverse gyroradii and velocities which are unable to make the average velocity zero. Hence, both density and temperature gradients contribute to a transverse *diamagnetic fluid drift*

$$\mathbf{v}_{dia,s} = \frac{\mathbf{B} \times \nabla_\perp p_{s\perp}}{q_s n_s B^2} \tag{7.72}$$

across the magnetic field. Because this velocity depends on the charge of the fluid particles, differently charged fluid components will drift into opposite directions, thus giving rise to an effective drift current flow in the plasma. In a quasineutral electron-ion plasma this *diamagnetic current* becomes

$$\mathbf{j}_{dia} = \frac{\mathbf{B} \times \nabla_\perp p_\perp}{B^2} \tag{7.73}$$

where $p_\perp = p_{e\perp} + p_{i\perp}$ is the total perpendicular pressure. Any plasma containing transverse density or pressure gradients carries such diamagnetic currents. They are called diamagnetic since they diminish the external magnetic field.

In an isotropic stationary plasma this diamagnetic drift is the only relevant drift term. In the anisotropic case the third term on the right-hand side of Eq. (7.71) comes into play. It is a little more involved to treat this term, since it contains derivatives of the type $\nabla \cdot (\mathbf{BB}/B^2)$. These can be replaced by using the fact that the derivative of a unit vector is equal to the outer (negative sign) normal, $\mathbf{n}$, divided by the radius of curvature,

R_c. Hence, the third term in Eq. (7.71) becomes

$$\mathbf{v}_{c,s} = -\frac{(p_{s\parallel} - p_{s\perp})}{q_s n_s B^2 R_c} \mathbf{B} \times \mathbf{n} \qquad (7.74)$$

This is a fluid drift velocity which depends on the pressure difference and is non-zero only in a curved magnetic field. It thus resembles the curvature drift but this time it is caused by the collective effects in the pressure anisotropy. Its direction is the same as the diamagnetic drift. It is interesting to note that this fluid drift may be negative as well as positive depending on the sign of the pressure anisotropy. The anisotropic curvature current resulting from this drift is given by

$$\mathbf{j}_{dia,c} = -\frac{p_\parallel - p_\perp}{B^2 R_c} \mathbf{B} \times \mathbf{n} \qquad (7.75)$$

with $p_\parallel$ the total parallel pressure. This current, for large parallel pressure, is negative but can change its sign when the perpendicular pressure becomes large. It may therefore amplify or weaken the effect of the isotropic diamagnetic current.

Neutral Sheet Current

A typical example of a diamagnetic current is the neutral sheet current in the geomagnetic tail which divides the tail into northern and southern lobes with their stretched magnetic field lines. In the southern lobe the field lines extend from the southern polar cap and point anti-sunward, while in the northern lobe they come from the distant tail pointing sunward and ending in the northern polar cap.

This stretching of the otherwise approximately dipolar terrestrial magnetic field can be accounted for by a diamagnetic current flowing across the magnetospheric tail from dawn to dusk (see Sec. 1.3 and Fig. 1.6). Such a current transports positive charges from dawn to dusk and negative charges from dusk to dawn across the tail and, because of its stationarity and its macroscopic magnetic effect, cannot be anything else but a diamagnetic current. Its cause must therefore be a gradient in the plasma pressure perpendicular to the current layer pointing from north to south in the upper (northern) half and from south to north in the lower (southern) half of the current layer. Hence, the current layer is a concentration of dense and hot plasma which is called the *neutral sheet* because of the weak magnetic field it contains.

Spacecraft measurements have revealed that the neutral sheet in the geomagnetic tail contains a quasineutral ion-electron plasma of roughly 1–10 keV temperature and a density of about 1 cm^{-3}. The transverse magnetic field in the neutral sheet is not zero but rather weak, of the order of 1–5 nT. Due to the weak magnetic field, the plasma beta parameter has typically rather high values, $\beta \approx 100$. The main current sheet has a typical thickness of 1–2 R_E and the maximum current density is of the order of some nA/m^2.

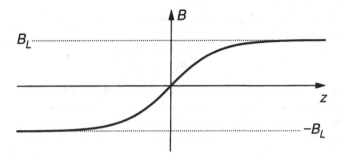

Fig. 7.2. Harris model of the neutral sheet magnetic field.

But especially during disturbed times and before substorm onset, the current sheet can be much thinner and the current density much higher.

The width of the current layer must be larger than an ion gyroradius. To find an estimate of the pressure gradient length in the neutral sheet, L_p, we perform a dimensional analysis. Since the neutral sheet current is a diamagnetic current, it is approximated as

$$j = \frac{2p_\perp}{L_p B_L} \tag{7.76}$$

where $p_\perp$ is the perpendicular plasma pressure in the neutral sheet and B_L is the lobe magnetic field strength. From Ampère's law this current is equal to the curl of the lobe magnetic field

$$j = \frac{B_L}{\mu_0 L_B} \tag{7.77}$$

where L_B is the variation length of the lobe magnetic field. Hence, one finds that

$$\frac{L_p}{L_B} = \frac{2\mu_0 p_\perp}{B_L^2} \tag{7.78}$$

This ratio is something like the plasma beta, but with respect to the lobe magnetic field and the neutral sheet plasma pressure. In equilibrium, when the magnetosphere is undistorted, this ratio is about one. But in the interior of the neutral sheet we have $\beta \gg 1$ and the pressure has nearly flat profile, while outside the neutral sheet, where β is small, the pressure drops steeply towards the lobe.

A very simple theoretical model of the neutral layer, which empirically accounts for these observational properties, is the so-called *Harris sheet*. It assumes that the magnetospheric tail has a simple geometry with the magnetic field pointing in the sunward x direction in the northern lobe and in the $-x$ direction in the southern lobe and varies in the z direction, perpendicular to the plasma sheet equatorial plane, according to a hyperbolic tangent function

$$B_x = B_L \tanh(z/L_B) \tag{7.79}$$

where B_L is again the lobe magnetic field strength and L_B is assumed to be constant.

This model, which is shown in Fig. 7.2, neglects any northward B_z component or inclination of the magnetic field in the neutral sheet. Since the measured magnetic field inside the neutral sheet has also a northward B_z component, typically of the order of 1–5 nT, the Harris model can be taken only as a crude approximation to real tail field.

Field-Aligned Currents

The various perpendicular drift currents in a plasma add up to a total perpendicular current density $\mathbf{j}_\perp$ which in an inhomogeneous and possibly time-varying plasma is not necessarily divergence-free. Under slowly variable conditions the requirement of closed current circuits necessarily leads to the generation of field-aligned currents in a way similar to that discussed in connection with the generation of field-aligned currents in the ionosphere (see Sec. 5.4). Since from Ampére's law under quasi-stationary conditions the divergence of total current must vanish

$$\nabla \cdot \mathbf{j} = \nabla \cdot (\mathbf{j}_\perp + \mathbf{j}_\parallel) = 0 \tag{7.80}$$

one obtains

$$\nabla \cdot \mathbf{j}_\parallel = -\nabla \cdot \mathbf{j}_\perp \tag{7.81}$$

Introducing a coordinate, s, along the magnetic field, this expression can be rewritten in terms of the scalar parallel current density

$$\frac{\partial}{\partial s}\left(\frac{j_\parallel}{B}\right) = -\frac{\nabla \cdot \mathbf{j}_\perp}{B} \tag{7.82}$$

The divergence of the perpendicular current contributing to the right-hand side, most notably the diamagnetic neutral sheet current, serves as source of the field-aligned currents in the magnetosphere. They close in the ionosphere as discussed in Secs. 5.5 and 5.7. For $B = B(s)$ Eq. (7.82) can be integrated to obtain

$$j_\parallel(s) = -\int_0^s \frac{\nabla \cdot \mathbf{j}_\perp(s')}{B(s')}\, ds' \tag{7.83}$$

for the field-aligned current density at a given position along a magnetic field line.

Polarization Drifts

We have so far assumed stationarity and thus neglected any time dependence of the fluid velocity. If we assume that such time variations exists, but with a time scale much longer than the gyration period, one can add the term $m_s n_s d\mathbf{v}_s/dt$ to the right-hand side of Eq. (7.70), where the fluid velocity of a particular species, $\mathbf{v}_s$, is now understood as the slow drift velocity of the fluid. The dominant term in this drift velocity is the electric convection fluid drift and one recovers the guiding center *polarization drift* velocity (see Sec. 2.3)

$$\mathbf{v}_{P,s} = \frac{1}{\omega_{gs} B} \frac{d\mathbf{E}_\perp}{dt} \qquad (7.84)$$

In contrast to the gradient and curvature particle drifts, the polarization drift turns out to survive the transition from single particles to fluids. It is one of the important fluid drifts. Since ω_{gs} carries the sign of the charge and also depends on the mass, the polarization drift term also leads to *polarization current* or *inertial current* in the plasma.

In addition to the electric polarization drift, time variations in the pressure gradient term, i.e., the first term on the right-hand side of Eq. (7.70), yield a further contribution to the polarization drift. Assuming that the magnetic field is stationary and that the fluid temperature is constant, we obtain the *density polarization drift* term as

$$\mathbf{v}_{Pn,s} = -\frac{k_B T_s}{q_s \omega_{gs} B} \nabla_\perp \frac{d \ln n_s}{dt} \qquad (7.85)$$

This drift is a pure fluid drift and caused by slow variations in the perpendicular gradient of the plasma density. It is a collective effect of the fluid which does not exist for single particles. Since the signs of ω_{gs} and q_s cancel, ions and electrons drift in the same direction. But they drift with different velocities because of their different mass and, possibly, different temperature, and thus add to the polarization current.

Another polarization drift can be obtained for the one-fluid case from the generalized Ohm's law (7.53). This law contains the electron inertial term in the form of a time derivative of the current density $\mathbf{j}$. Vector multiplication of Ohm's law with $\mathbf{B}/B^2$ and solving for the fluid velocity yields the *current polarization drift*

$$\mathbf{v}_{Pc} = \frac{m_e}{ne^2 B^2} \mathbf{B} \times \frac{\partial \mathbf{j}}{\partial t} \qquad (7.86)$$

This drift does neither depend on the charge nor the mass of the species and does not contribute to a current. It is a pure plasma flow, where a time variation of the perpendicular current density sets the whole plasma into motion in a direction which is both perpendicular to the magnetic field and to the current.

Force-Free Fields

The ideal magnetohydrodynamic equilibrium equations are obtained by making the transition to the stationary case while at the same time dropping all dissipations. Would we not drop the dissipative terms, it can be trivially realized that the final long-time stationary state would be the null state with maximum temperature, no motion and no fields because dissipation would have destroyed the field. Hence, in a medium with dissipation only time dependent states of physical significance are of interest. In an ideal magnetohydrodynamic fluid, however, we can consider the conditions for equilibrium by returning to the basic equations. The equation of motion for a quasineutral fluid

$$mn\frac{d\mathbf{v}}{dt} = -\nabla \cdot \mathbf{P} + \mathbf{j} \times \mathbf{B} \qquad (7.87)$$

suggests that equilibrium can be established whenever the right-hand side of this equation vanishes and the plasma becomes force-free because all forces cancel.

One particularly interesting case, frequently discussed in magnetohydrodynamic applications to stars as the sun, is the case when the pressure force is neglected. Then the force-free condition reduces to a vanishingly small Lorentz force

$$\mathbf{j} \times \mathbf{B} = 0 \qquad (7.88)$$

The vanishing of the Lorentz force implies that the currents in the plasma flow entirely parallel to the magnetic field. The currents are field-aligned and this alignment with the field applies not only to the external field but to the total field. Hence, the current flows parallel to the sum of the external field and the field generated by the current itself. A fully general representation of the same condition on the current would then to write the current as

$$\mu_0\mathbf{j} = \alpha_L\mathbf{B} \qquad (7.89)$$

where $\alpha_L(\mathbf{x})$ is a scalar to make sure that the current is really parallel to the field. This proportionality factor is called *lapse field*. It may depend on position, but is constant in time. Thus one can use the stationary Ampère's law, $\nabla \times \mathbf{B} = \mu_0\mathbf{j}$, to find that the condition of force-free magnetohydrodynamics can be written as

$$\mathbf{B} \times (\nabla \times \mathbf{B}) = 0 \qquad (7.90)$$

The condition shows that force-free fields must have a complicated helical structure, which is intuitively clear because the parallel current generates a helical field around the external guiding field. Using Eq. (7.89), Ampère's law can be written as

$$\nabla \times \mathbf{B} = \alpha_L\mathbf{B} \qquad (7.91)$$

or taking the divergence of both sides of this equation and using $\nabla \cdot \mathbf{B} = 0$

$$\mathbf{B} \cdot \nabla\alpha_L = 0 \qquad (7.92)$$

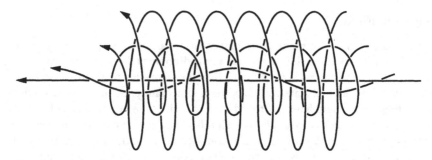

Fig. 7.3. Typical force-free field and current configuration.

This is a condition on the lapse field, which shows that α_L is constant along the magnetic field lines and varies only in the perpendicular direction. Hence, the force-free field-aligned current is also proportional to the total self-consistent magnetic field.

Figure 7.3 shows a schematic force-free magnetic field configuration which can be imagined schematically. The originally straight magnetic field, when exposed to a sufficiently strong field-aligned current which carries its own solenoidal field, is deformed. Due to the superposition of the straight and solenoidal fields, the resulting magnetic field configuration winds up to become helical in order to compensate for the Lorentz force.

Equation 7.92 has a very interesting analogy. It is well-known that in a magnetized plasma which convects with velocity $\mathbf{v}_E = \mathbf{E} \times \mathbf{B}/B^2$ the electric convection potential is constant along the magnetic field and, hence, satisfies the same equation as the lapse function α_L. It is therefore tempting, to identify α_L with an electric convection field potential. This is a quite general interpretation, because the requirement of a force-free field is identical to the requirement that the Lorentz force $\mathbf{E} + \mathbf{v} \times \mathbf{B}$ vanishes. This in turn is just the condition that the plasma convects at velocity $\mathbf{v}_E$.

We may thus interpret α_L as an equivalent electric potential which must be constant along the magnetic field. If α_L itself is constant throughout the whole space this condition is identically satisfied, corresponding to the special case that the electric field in the plasma vanishes. In all other cases of varying α_L the plasma must necessarily undergo convective motion at the velocity

$$\mathbf{v}_L = \frac{\mathbf{B} \times \nabla \alpha_L}{B^2} \tag{7.93}$$

showing that all general force-free field configurations can exist only in convecting plasmas, and a quiescent motionless force-free field can only exist when α_L is uniform. In this case one finds by taking the curl of Eq. (7.91) that the force-free magnetic field satisfies the Helmholtz equation

$$\nabla^2 \mathbf{B} + \alpha_L^2 \mathbf{B} = 0 \tag{7.94}$$

In the near-Earth plasma environment force-free fields are of little importance, since the strong guiding field of the Earth does not permit for force-free conditions. The field-aligned currents flowing in the inner magnetosphere are far too weak to affect the external magnetospheric field effectively. However, force-free fields have been successfully used to model magnetic fields in the distant magnetospheric tail and in the atmospheres of stars as the Sun.

Concluding Remarks

The magnetohydrodynamic equations derived in this chapter are simplifications of the more general multi-fluid equations which on their own are moments of the kinetic equation. Each of the steps leading to these sets of equations is based on simplifying assumptions which limit their applicability. In concluding this chapter, it is useful to review some of the conditions which have to be observed when applying the set of magnetohydrodynamic equations summarized below.

Validity of Magnetohydrodynamics

The most basic condition of validity of magnetohydrodynamics is found when remembering that in magnetohydrodynamics no distinction is made between the different components of the plasma. This approximation requires that time-scales of variation of the fluid and fields must be longer than the characteristic time-scale of the heaviest particle component. Hence, the characteristic frequency, ω, of any change must be smaller than the ion cyclotron frequency

$$\omega < \omega_{gi} \tag{7.95}$$

Because of the same reason, the characteristic length scale, L, where magnetohydrodynamics becomes applicable must be longer than the ion gyroradius

$$L > r_{gi} \tag{7.96}$$

Magnetohydrodynamics is therefore restricted to both, very low frequencies and large spatial scales. At such low frequencies one can in most cases ignore the displacement current term in the first Maxwell equation.

The above basic conditions are very important and generally used when deriving low-frequency waves from kinetic theory. One should, however, remark that these conditions are merely derived from the equation of motion of the fluid. Further conditions follow when considering Ohm's law and the induction equation. The equation of state is unaffected by the conditions. It is thus more precise than magnetohydrodynamics requires.

The *general induction law* of magnetohydrodynamics follows, in the same way as described in Sec. 5.1, from a combination of Faraday's law and the generalized Ohm's law (7.53), which serves to replace the electric field

$$\boxed{\frac{\partial \mathbf{B}}{\partial t} = \nabla \times \left(\mathbf{v} \times \mathbf{B} - \frac{1}{ne} \mathbf{j} \times \mathbf{B} + \frac{1}{ne} \nabla \cdot \mathbf{P}_e - \frac{m_e}{ne^2} \frac{\partial \mathbf{j}}{\partial t} - \eta \mathbf{j} \right)} \qquad (7.97)$$

As discussed earlier, the first term on the right-hand of this generalized induction equation (7.97) describes frozen-in fields and plasmas. The other terms tend to break the frozen-in state. If they can all be neglected, one ends up with ideal magnetohydrodynamics. For the pressure gradient term to be neglected one compares it with the magnitude of the first term, $|\mathbf{v} \times \mathbf{B}|$, on the right and finds dimensionally

$$\frac{p_e}{neVBL_p} = \frac{n_e k_B T_e}{neVBL_p} = \frac{m_e v_{the}^2}{2eVBL_p} = \frac{r_{ge}}{L_p} \frac{v_{the}}{V} \qquad (7.98)$$

where L_p is the gradient scale length of the electron pressure and V and B are average velocity and magnetic field strength, respectively. Hence, when the pressure gradient scale divided by the electron gyroradius is much larger than the ratio of electron to flow velocity, the pressure term can be ignored. This is valid for sufficiently cold plasmas with small $\beta \ll 1$.

The dimensional ratio of the second-last to the last term on the right-hand side of the induction law is obtained by replacing the $\partial/\partial t$ term by the characteristic frequency of change, ω, in the second-last term

$$\frac{m_e \omega}{ne^2 \eta} = \frac{\omega}{v_c} \qquad (7.99)$$

where use has been made of the definition of the resistivity $\eta = m_e v_c / ne^2$. Hence, the inertial term in the induction equation can be neglected whenever the characteristic frequency of change is smaller than the electron-ion collision frequency. This condition is well met in resistive magnetohydrodynamics. However, in time-varying ideal magnetohydrodynamics it poses severe problems and holds strictly only for stationary cases.

A similar problem is encountered when considering the Hall-term. Forming the dimensional ratio of the second to the last term in the induction law, one finds

$$\frac{B}{ne\eta} = \frac{eB}{m_e v_c} = \frac{\omega_{ge}}{v_c} \qquad (7.100).$$

the ratio of electron gyrofrequency to collision frequency. In most of the cases the former is much higher than the latter. It is thus clear that neglecting the Hall term in the ideal magnetohydrodynamic case is rather delicate and is, in general, not permitted, unless it can be justified by additional assumptions. Only for strongly collisional plasmas in weak magnetic fields the Hall term may be dropped. In the light of this discussion the relevant induction equation becomes, when dropping the pressure and resistive terms

$$\frac{\partial \mathbf{B}}{\partial t} = \nabla \times \left(\mathbf{v} \times \mathbf{B} - \frac{1}{ne}\mathbf{j} \times \mathbf{B} - \frac{m_e}{ne^2}\frac{\partial \mathbf{j}}{\partial t} \right) \qquad (7.101)$$

Dividing the two last terms yields dimensionally the ratio of the characteristic frequency to the electron gyrofrequency

$$\frac{m_e \omega}{eB} = \frac{\omega}{\omega_{ge}} \qquad (7.102)$$

This ratio is always much smaller than one and the inertial term can usually be neglected in comparison to the Hall term. Finally, the dimensional ratio of the Hall and convection terms can be written in the form

$$\frac{j}{neV} = \frac{B}{\mu_0 L_B neV} = \frac{c}{V}\frac{c}{\omega_{pe}L_B}\frac{\omega_{ge}}{\omega_{pe}} \qquad (7.103)$$

where L_B is the characteristic length over which the magnetic field varies. The first term, the ratio of the light to the flow velocity is large, typically of the order of 1000. The second ratio is the ratio of the *electron inertial length*, c/ω_{pe}, to the magnetic field scale length, which is a small number. The last ratio is the electron cyclotron to plasma frequency ratio. In strong magnetic fields it is large and the Hall term becomes important. Only in dense plasmas, where this ratio becomes small, the Hall term is negligible against the convection term.

The induction equation (7.101), neglecting the presumably small electron inertia term, allows another interesting conclusion. Because of the relation (7.49) between the current and the flow velocities, the simplified induction equation becomes

$$\frac{\partial \mathbf{B}}{\partial t} = \nabla \times (\mathbf{v}_e \times \mathbf{B}) \qquad (7.104)$$

In this form the induction equation shows that in collisionless magnetohydrodynamics with Hall currents not neglected, the electrons are the only plasma component which is frozen to the magnetic field. Hence, the magnetic field is moving with the electron fluid while the plasma flow can, in general, deviate from the motion of the field lines. Only when Hall currents become small or can be ignored, the frozen-in concept of plasma physics applies to the magnetohydrodynamic flow itself.

Summary of Magnetohydrodynamic Equations

In this summary of magnetohydrodynamic equations the time derivative of the electric field is understood to be so slow that the validity of the magnetohydrodynamic approximation is not violated. Then the last Maxwell equation, $\nabla \cdot \mathbf{E} = \rho/\epsilon_0$, is not needed for closing the system of equations and can be used to determine the space charges maintained in the plasma. The system of equations given below has to be complemented by an energy equation or an equation of state. Usually one assumes adiabatic conditions and uses $p \propto n^\gamma$ with $\gamma = 5/3$. However, as discussed in Sec. 7.2, other forms of this equation may equally be valid in certain situations.

$$\frac{\partial n}{\partial t} + \nabla \cdot (n\mathbf{v}) = 0$$

$$\frac{\partial (nm\mathbf{v})}{\partial t} + \nabla \cdot (nm\mathbf{v}\mathbf{v}) = -\nabla \cdot \mathbf{P} + \rho \mathbf{E} + \mathbf{j} \times \mathbf{B}$$

$$\mathbf{E} + \mathbf{v} \times \mathbf{B} = \eta \mathbf{j} + \frac{1}{ne}\mathbf{j} \times \mathbf{B} - \frac{1}{ne}\nabla \cdot \mathbf{P}_e + \frac{m_e}{ne^2}\frac{\partial \mathbf{j}}{\partial t}$$

$$\nabla \times \mathbf{B} = \mu_0 \mathbf{j} + \mu_0 \epsilon_0 \frac{\partial \mathbf{E}}{\partial t}$$

$$\nabla \times \mathbf{E} = -\frac{\partial \mathbf{B}}{\partial t}$$

$$\nabla \cdot \mathbf{B} = 0$$

Further Reading

A thorough description of the equations of state is given in [5]. Magnetohydrodynamics is discussed in many textbooks, but in most cases from the point of view of a conducting fluid, which applies more to the interior of the Earth than to its plasma environment. A general reference is [3], but this book misses the generalized Ohm's law, which is given in [1] and [2]. For other applications of magnetohydrodynamics see [4].

[1] F. F. Chen, *Introduction to Plasma Physics and Controlled Fusion, Vol. 1* (Plenum Press, New York, 1984).

[2] N. G. van Kampen and B. U. Felderhof, *Theoretical Methods in Plasma Physics* (North-Holland Publ. Co., Amsterdam, 1967)

[3] L. D. Landau and E. M. Lifshitz, *Electrodynamics of Continuous Media* (Pergamon Press, Oxford, 1975).

[4] E. N. Parker, *Cosmical Magnetic Fields* (Clarendon Press, Oxford, 1979).

[5] G. L. Siscoe, in *Solar-Terrestrial Physics*, eds. R. L. Carovillano and J. M. Forbes (D. Reidel Publ. Co., Dordrecht, 1983), p. 11.

8. Flows and Discontinuities

Plasma equilibria frequently arise in plasma flows. The dominant plasma flow in near-Earth space is the solar wind. It interacts with the bodies of the solar system, planets, moons, and comets. This interaction depends on the properties of these bodies. It is particularly strong when they are magnetized as is the case for Earth, Mercury, Jupiter, Saturn, Uranus and Neptune. Such an interaction results in the formation of a magnetosphere with a thin boundary, a magnetopause, representing a sudden transition from the solar wind plasma to the planetary magnetic field region. Transition layers like the magnetopause are called discontinuities.

8.1. Solar Wind

The *solar wind* is the high-speed particle stream continuously blowing out from the solar corona into interplanetary space, extending far beyond the orbit of the Earth and terminating somewhere in interstellar space after having hit the weakly ionized interstellar gaseous medium around 160 AU (1 Astronomical Unit = $1.50 \cdot 10^{11}$ m; see App. A.1). Near the Earth's orbit at 1 AU, the solar wind velocity typically ranges between 300–1400 km/s. The value of 500 km/s given in Sec. 1.2 is the most probable value of the solar wind velocity. It corresponds to an about 4-day particle flight from the Sun to the Earth. Streams with velocities of less than 400 km/s are known as low-speed, those with velocities exceeding 600 km/s are called high-speed solar wind streams.

Since the solar atmosphere, where the solar wind originates, is a quiescent low-speed region of low temperature near 6000 K, the solar wind must be accelerated to its high interplanetary streaming velocities in the *solar corona*. The corona is a hot region with a temperature of about $1.6 \cdot 10^6$ K and a density near $5 \cdot 10^{17}$ cm^{-3}. The strength of the magnetic field in the corona is not well known, but at its bottom it might be of the order of some 10^{-2} T, decaying away with increasing distance. Most of the dynamics of the solar wind is determined by this field. When the field is closed with both foot points on the Sun, the solar atmosphere is trapped. But in regions, where the field lines are stretched out into interplanetary space, the so-called *coronal holes*, the solar atmospheric plasma can flow out. These are the source regions for the solar wind.

159

Solar Coronal Outflow

Neglecting the minor influence of the magnetic field, the expansion of the solar wind from a coronal hole can be treated as a one-dimensional radial hydrodynamic flow problem. Radial flux conservation requires

$$\frac{d}{dr}(r^2 n v) = 0 \tag{8.1}$$

where v is the purely radial flow velocity and n is the radially variable coronal plasma density. Stationary momentum conservation yields

$$v\frac{dv}{dr} = -\frac{GM_\odot}{r^2} - \frac{1}{nm_i}\frac{dp}{dr} \tag{8.2}$$

where G is the gravitational constant, $M_\odot$ the solar mass, and p the thermal pressure of the coronal plasma. Instead of an energy law one assumes a simple polytropic relation

$$\frac{d}{dr}\left(\frac{p}{n^\gamma}\right) = 0 \tag{8.3}$$

where adiabatic conditions correspond to $\gamma = 5/3$ and isothermal and turbulent flows have $\gamma = 1$. Intermediate values, $1 < \gamma < 5/3$, indicate the presence of coronal heating. The latter two equations can be integrated to yield

$$\frac{v^2}{2} + \frac{\gamma}{\gamma - 1}\frac{p}{nm_i} - \frac{GM_\odot}{r} = K \tag{8.4}$$

where K is a constant. Identifying the term p/nm_i as the *sound velocity*, c_s

$$\boxed{c_s^2 = \frac{\gamma k_B T}{m_i}} \tag{8.5}$$

permits to eliminate the explicit dependence on pressure and radial velocity from Eq. (8.4) by defining a *sonic Mach number*, M_s, as the ratio of flow to sound velocity

$$\boxed{M_s^2 = \frac{m_i v^2}{\gamma k_B T}} \tag{8.6}$$

Introducing the gravitational potential of the Sun at its surface at radius $R_\odot$

$$\Phi_\odot = GM_\odot/R_\odot \tag{8.7}$$

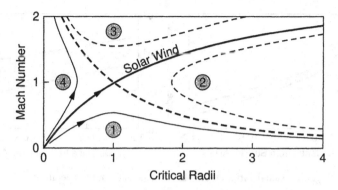

Fig. 8.1. The four types of outflow solutions in the (M,R) plane.

and integrating Eq. (8.1) with the help of Eqs. (8.4) and (8.6), yields for $R = r/R_\odot$.

$$\left(K + \frac{\Phi_\odot}{R}\right)\frac{M_s^2 - 1}{m^2}R\frac{dM_s^2}{dR} = \left(1 + (\gamma - 1)\frac{M_s^2}{2}\right)\left(4K + \frac{3\gamma - 5}{\gamma - 1}\frac{\Phi_\odot}{R}\right) \quad (8.8)$$

It is impossible to solve this equation analytically, but one can draw some qualitative conclusions. For $|M_s| = 1$ the left-hand side vanishes. Since the right-hand side must also vanish, the polytropic index must lie in the range $1 < \gamma < 5/3$, which requires coronal heating. For such γ the *sonic point*, $M_s = 1$, lies at the critical radial distance

$$R_c = \frac{\Phi_\odot}{4K}\frac{5 - 3\gamma}{\gamma - 1} \quad (8.9)$$

For the sonic point to be found outside the solar surface $R_c > 1$ and thus

$$K < \frac{\Phi_\odot}{4}\frac{5 - 3\gamma}{\gamma - 1} \quad (8.10)$$

Solutions satisfying this result describe coronal plasma outflows which at $R = R_c$ pass from flow speeds smaller than the sound velocity to speeds larger than the sound velocity. These are very special solutions, which divide the space of possible solutions into four categories.

The two heavier curves in Fig. 8.1 correspond to the two solutions passing through the point $M_s(R_c) = 1$. Only two of the four regions in the (M_s, R_c) plane correspond to physical solutions. These are regions (1) and (4). In region (1) the solar wind accelerates from low flow velocity yet subsequently decelerates, never reaching supersonic velocities while escaping from the corona into interplanetary space to large distances. In region (4) it starts at low velocities and increases to supersonic speeds but cannot leave the solar corona but falls back into the solar atmosphere. Since the solar wind is a high-speed stream, it must correspond to the critical solution passing through the sonic point.

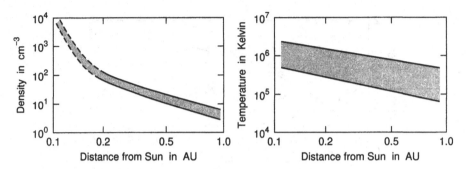

Fig. 8.2. Radial variation of solar wind density and temperature.

Properties of the Solar Wind

Once the solar wind has been formed, its plasma expands into the interplanetary space. This expansion is a radial propagation of the solar wind plasma fluid away from the Sun during which it must become diluted and at the same time cool down. Measurements have shown that the solar wind density outside the solar corona up to a distance of a few AU decreases approximately as $n(r) \propto r^{-2}$ (see Fig. 8.2). Close to the Earth, the solar wind is a fully ionized plasma consisting of electrons, protons, and α-particles. Since the abundance of He^{2+} is only about 3% of the total density of about 5 cm^{-3}, their presence can typically be neglected.

The decrease in the temperature is less severe (see Fig. 8.2). It decreases by roughly a factor of 20 from coronal temperatures of a few million Kelvin to electron and proton temperatures at 1 AU of about $T_e \approx 1.4 \cdot 10^5$ K and $T_i \approx 1.2 \cdot 10^5$ K, respectively. The large uncertainty of the temperature indicated in Fig. 8.2 is caused by fluctuations in the corona and by a yet unknown solar wind heating mechanism. Although the temperatures of electrons and protons are not very different, they have drastically different thermal velocities at 1 AU, with $v_{the} \approx 1500$ km/s and $v_{thi} \approx 35$ km/s. The electron thermal velocity is typically larger than the flow speed, while the ion thermal velocity is always very small compared to the flow speed.

Additional insight into the nature of the solar wind is obtained by defining, in analogy to Eq. (8.5), the 'magnetic sound' or *Alfvén velocity*, v_A, where twice the magnetic pressure, B^2/μ_0, appears instead of the thermal pressure

$$v_A^2 = \frac{B^2}{\mu_0 n m_i} \qquad (8.11)$$

The Alfvén velocity is an important plasma parameter and, as will be shown in Chap. 9, it is the fundamental speed at which magnetic signals in a plasma can be transported

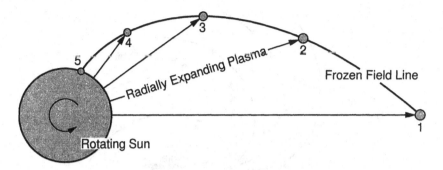

Fig. 8.3. Spiral form of interplanetary field line due to solar rotation.

by waves. Using the typical densities given above and an interplanetary magnetic field amplitude of about 5 nT, we find that with sound and Alfvén velocities of only about 30–50 km/s, the solar wind is a supersonic and super-Alfvénic flow.

Interplanetary Magnetic Field

One important property of the solar wind is its state of magnetization. This state can be described by considering plasma beta, the ratio of thermal to magnetic energy densities given in Eq. (7.65). Near 1 AU the solar wind magnetic field is of the order of 5 nT, so that $\beta \approx 1 - 30$ is large. The solar wind flow entirely determines the behaviour of the field, which fact has an important consequence for its structure. The radial outflow of the wind from the solar corona transports the field into interplanetary space, while its footpoint remain anchored in the solar atmosphere. Because of the 27-day solar rotation period, the *interplanetary magnetic field* will not maintain the form it had in the solar corona. Figure 8.3 shows schematically what happens to a field line which is frozen into the solar wind plasma flow and is transported radially outward with a constant radial velocity while the Sun rotates. As a result of the combined motion of outflow and rotation the magnetic field line becomes bent into an *Archimedian spiral* form.

At 1 AU this spiral makes an angle of approximately 45° to the Earth-Sun line, so that it hits the Earth from the late morning direction. The magnetic field strength at this position is about 5–10 nT. As depicted in Fig. 8.4, the direction of the interplanetary magnetic field in the ecliptic plane is divided into sectors according to solar and anti-solar directions. The boundaries between sectors of opposing field direction are regions of zero magnetic field and thus current sheets. To reproduce the direction of the inter-planetary magnetic field, the solar wind current sheet must be tilted with respect to the ecliptic plane at the sector boundaries, while high above and below the ecliptic plane it turns into horizontal direction. It looks like a ballerina skirt sketched in Fig. 8.5.

Taking into account of the azimuthal speed of the Earth of about 30 km/s, the so-

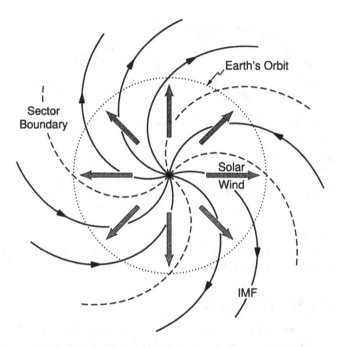

Fig. 8.4. Sector structure of interplanetary magnetic field.

lar wind hits the Earth magnetosphere with an *aberration* of typically about 5° from the radial direction. This angle decreases the higher the solar wind speed becomes. In occasional high-speed solar wind streams having velocities near and larger 1000 km/s, the solar wind direction is practically radial.

For completeness, we note that the solar wind is by no means a quiet laminar flow. It is subject to many variations in density, flow velocity, temperature, pressure, magnetic field strength and magnetic field direction. These variations do in part originate in the solar atmosphere and are swept along with the solar wind into interplanetary space. However, a large part of them is produced in the solar wind itself. Such variations are either waves which propagate in the solar wind, or they are localized disturbances like narrow boundaries.

Interaction with Obstacles

Because of its supersonic and super-Alfvénic velocity, the solar wind flow is subject to shocking if there is an obstacle in its path. Such obstacles are planets, comets, and asteroids, but the most interesting obstacles are the magnetized planets with their extended

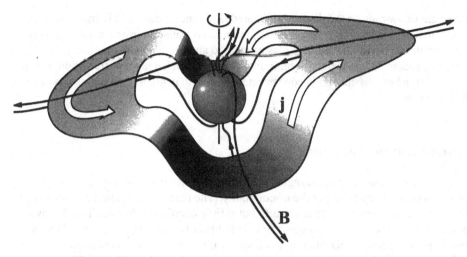

Fig. 8.5. Three-dimensional topology of the solar wind current sheet.

magnetospheres. These magnetospheres enhance the cross-section of a planet by a large factor. For the Earth this factor is about 150. One therefore expects that an extended standing *bow shock* exist in front of such planets.

In addition to the creation of a bow shock, the interaction with magnetized planets leads to the formation of another thin boundary, a *magnetopause*. In order to understand the nature of these phenomena we must investigate the various possibilities for the evolution of thin boundaries in a plasma. We will develop the theory of such boundaries in the next section.

8.2. Fluid Boundaries

Equilibria between plasmas of different properties often give rise to the formation of boundaries separating the two adjacent plasma regions. In general these boundaries are not imposed from the outside, but evolve inside the plasma due to the interaction of the plasmas on both sides, forming narrow layers called *discontinuities*. Plasmas share the capability for the evolution of discontinuities with ordinary hydrodynamics, yet the much greater richness of plasma processes offers more possibilities for discontinuities. Since the properties of an ideal magnetohydrodynamic plasma must be maintained on both sides of the discontinuity, not all kinds of discontinuities can be supported in a given plasma state. Instead the discontinuities will group into different classes. In the following we determine these classes and discuss their properties.

It is convenient to start from ideal one-fluid magnetohydrodynamics, assuming that the plasma flow outside the discontinuity is ideal, in the sense that it is free of dissipation. There is, however, no need for dissipation to be excluded from the interior of the discontinuity. Transition from one plasma state to another typically requires some dissipative process. Because the discontinuity represents such a layer of transition, it will be a region where dissipation is concentrated while it is negligible on both sides of the discontinuity.

General Jump Conditions

Discontinuities separate plasma regions of different properties, with the properties possibly changing abruptly across the discontinuity. The fields and plasma parameters are not independent across the discontinuity but satisfy certain conditions. This behavior is similar to the behavior of electromagnetic fields at boundaries, where the dielectric properties of optical media change. In all such cases the boundaries are assumed to be infinitesimally thin. In optics the boundary is thin with respect to the wave length of light and thick with respect to the spacing of the molecules in the crystal structure, while magnetohydrodynamic discontinuities are thin with respect to the scale lengths of the fluid parameters, but thick with respect to Debye length and ion gyroradius.

The conditions at the outermost edges of the discontinuities can be derived from ideal one-fluid magnetohydrodynamics. The conventional method to find the boundary conditions is based on an integration of the conservation laws across the discontinuity boundary in order to determine the differences between the field and fluid quantities on both sides of the discontinuity. It is therefore convenient to write the ideal one-fluid equations for a quasi-neutral plasma derived in Sec. 7.3 in the form of conservation laws

$$\frac{\partial n}{\partial t} + \nabla \cdot (n\mathbf{v}) = 0 \tag{8.12}$$

$$\frac{\partial (nm\mathbf{v})}{\partial t} + \nabla \cdot (nm\mathbf{v}\mathbf{v}) = -\nabla \cdot \left(\mathbf{P} + \frac{B^2}{2\mu_0} \mathbf{I} \right) + \frac{1}{\mu_0} \nabla \cdot (\mathbf{B}\mathbf{B}) \tag{8.13}$$

$$\frac{\partial \mathbf{B}}{\partial t} = \nabla \times (\mathbf{v} \times \mathbf{B}) \tag{8.14}$$

$$\nabla \cdot \mathbf{B} = 0 \tag{8.15}$$

where use has been made of $\mu_0 \mathbf{j} = \nabla \times \mathbf{B}$ for slowly variable fields, the ideal Ohm's law, $\mathbf{E} = -\mathbf{v} \times \mathbf{B}$, has been assumed, and space charges have been neglected, $\rho \mathbf{E} = 0$. These equations have to be complemented by an energy equation and by appropriate equations of state for the independent components of the pressure tensor. A particularly useful form of the energy conservation equation is following from Eq. (7.56) and the first two terms on the right-hand side of Eq. (7.57)

$$\nabla \cdot \left\{ nm\mathbf{v} \left[\tfrac{1}{2}v^2 + w + \frac{1}{nm} \left(p + \frac{B^2}{\mu_0} \right) \right] \quad \frac{1}{\mu_0} (\mathbf{v} \cdot \mathbf{B}) \mathbf{B} \right\} = 0 \qquad (8.16)$$

Moreover, we assume in most of the following text that the pressure tensor is isotropic, since an anisotropy introduces only a slight modification of the final results.

For a thin discontinuity the plasma, the only important changes occur perpendicular to the discontinuity boundary, while parallel to the discontinuity the plasma remains uniform. The discontinuity surface can be described by a two-dimensional function, $S(\mathbf{x}) = 0$. The normal vector of the discontinuity, $\mathbf{n}$, is defined as

$$\mathbf{n} = -\frac{\nabla S}{|\nabla S|} \qquad (8.17)$$

It is a unit vector perpendicular to the discontinuity and directed outward. The vector derivative has one single component in the direction of $\mathbf{n}$, $\nabla = \mathbf{n} \, (\partial/\partial n)$.

Choosing a certain point on the discontinuity surface, S, and taking the line integral of any field quantity along a rectangular box centered at this point from medium 1 to medium 2 and back to medium 1, the two sides of the box tangential to the surface do not contribute to the integral because the plasma properties do not change along to the discontinuity. On the other hand, the two integrations along the normal direction contribute twice due to their opposite sense. Hence, any closed line integral of a quantity X crossing S reduces to

$$\oint_S \frac{dX}{dn} dn = 2 \int_1^2 \frac{dX}{dn} dn = 2(X_2 - X_1) = 2[X] \qquad (8.18)$$

The quantity $[X]$ is the jump of X when crossing the boundary. Integrating the conservation laws in the above-prescribed way and observing that an integral over a conservation law vanishes, one can divide by 2 and replace the vector operations by the prescriptions

$$\nabla X \;\rightarrow\; \mathbf{n}[X]$$
$$\nabla \cdot \mathbf{X} \;\rightarrow\; \mathbf{n} \cdot [\mathbf{X}] \qquad (8.19)$$
$$\nabla \times \mathbf{X} \;\rightarrow\; \mathbf{n} \times [\mathbf{X}]$$

Boundaries in a plasma may flow with the medium or even relative to the medium. In both cases it is possible to transform to a system moving with the discontinuity at its local speed $\mathbf{U}$. Because of the Galileian invariance, it is sufficient to replace the time derivative by

$$\partial/\partial t = -\mathbf{U} \cdot \nabla = -\mathbf{U} \cdot \mathbf{n} \, (\partial/\partial n) \qquad (8.20)$$

while transforming the fluid velocity according to $\mathbf{v}' = \mathbf{v} - \mathbf{U}$. In the co-moving frame the discontinuity is stationary so that the time derivative can be dropped and the conservation laws in Eqs. (8.12) through (8.15) are written in terms of $\mathbf{v}'$. In the following we will drop the prime keeping in mind that viewed from the non-moving frame the velocity is to be replaced by the difference between the plasma velocity and the speed of displacement of the discontinuity.

Rankine-Hugoniot Conditions

With the help of the above definitions for use of the vector derivative, dropping the prime on the velocity, and setting $\mathbf{P} = p\mathbf{I}$, the conservation laws across the discontinuity can be replaced by *jump conditions* across the discontinuity

$$\mathbf{n} \cdot [n\mathbf{v}] = 0$$

$$\mathbf{n} \cdot [nm\mathbf{vv}] + \mathbf{n}\left[p + \frac{B^2}{2\mu_0}\right] - \frac{1}{\mu_0}\mathbf{n} \cdot [\mathbf{BB}] = 0$$

$$[\mathbf{n} \times \mathbf{v} \times \mathbf{B}] = 0 \tag{8.21}$$

$$\mathbf{n} \cdot [\mathbf{B}] = 0$$

An additional condition follows from the conservation of energy

$$\left[nm\mathbf{n} \cdot \mathbf{v}\left\{\frac{v^2}{2} + w + \frac{1}{nm}\left(p + \frac{B^2}{\mu_0}\right)\right\} - \frac{1}{\mu_0}(\mathbf{v} \cdot \mathbf{B})\mathbf{n} \cdot \mathbf{B}\right] = 0 \tag{8.22}$$

where the internal enthalpy, w, is related to the scalar pressure by

$$w = \frac{c_v p}{n k_B} \tag{8.23}$$

and the specific heat, c_v, is the typically the same on both sides of the discontinuity.

Paying for the moment no attention to the energy conservation, the last line in the above system of equations (8.21) shows the familiar result that the normal component of the magnetic field is continuous across any surface so that its jump vanishes

$$[B_n] = 0 \tag{8.24}$$

Similarly, from the first condition in Eq. (8.21) one finds that the mass flow normal to the discontinuity is always constant and thus

$$[nv_n] = 0 \tag{8.25}$$

Splitting between the normal and tangential components of the fields and making use of the last two relations one finds the remaining jump conditions

$$nmv_n[v_n] = -\left[p + \frac{B^2}{2\mu_0}\right]$$

$$nmv_n[\mathbf{v}_t] = \frac{B_n}{\mu_0}[\mathbf{B}_t] \qquad (8.26)$$

$$B_n[\mathbf{v}_t] = [v_n\mathbf{B}_t]$$

where the subscript t denotes the tangential component of the corresponding vector. For a given equation of state these equations are the boundary conditions which must be satisfied across any discontinuity. These equations are known under the name *Rankine-Hugoniot equations*. They contain the all basic properties of discontinuities in ideal magnetohydrodynamics.

To determine which classes of discontinuities are possible, the system of nonlinear jump conditions can be rewritten in a quasi-linear form by defining averages

$$\langle X \rangle = \tfrac{1}{2}(X_1 + X_2) \qquad (8.27)$$

With this definition it is easy to prove that the jump of a product is

$$[AB] = \tfrac{1}{2}\left([A]\langle B\rangle + \langle A\rangle[B]\right) \qquad (8.28)$$

It is then convenient to define the specific volume, $V = (nm)^{-1}$, and the constant normal mass flux, $F = nmv_n$. These definitions allow to rewrite the system of jump conditions

$$F[V] - [v_n] = 0$$

$$F[\mathbf{v}] + \mathbf{n}[p] + \mu_0^{-1}\mathbf{n}\langle\mathbf{B}\rangle \cdot [\mathbf{B}] - \mu_0^{-1}B_n[\mathbf{B}] = 0$$

$$F\langle V\rangle[\mathbf{B}] + \langle\mathbf{B}\rangle[v_n] - B_n[\mathbf{v}] = 0 \qquad (8.29)$$

$$[B_n] = 0$$

$$F\left\{[w] + \langle p\rangle[V] + (4\mu_0)^{-1}[V][\mathbf{B}_t]^2\right\} = 0$$

Understanding the average quantities as given and the jumps as the unknowns, these are eight equations for nine jumps. Adding an expression for the enthalpy or an equation of state to relate the pressure to the fields, the system of equations becomes linear and homogeneous in the jumps. Its determinant must vanish to allow for a solution. It is possible, though tedious, to write the determinant in factorized form

$$F\left(F^2 - \frac{B_n^2}{\mu_0\langle V\rangle}\right)\left\{F^4 + F^2\left(\frac{[p]}{[V]} - \frac{\langle\mathbf{B}\rangle^2}{\mu_0\langle V\rangle}\right) - \frac{B_n^2}{\mu_0\langle V\rangle}\frac{[p]}{[V]}\right\} = 0 \qquad (8.30)$$

This is a seventh-order equation for the normal mass flux, F, across the discontinuity. It consists of three factors which can be put to zero independently, demonstrating that in magnetohydrodynamics three different classes of discontinuities may evolve. The first class of discontinuities is determined by the condition

$$\boxed{F_{\mathrm{I}} = 0}$$

(8.31)

corresponding to zero mass flow across the discontinuity. We will find that two independent types of discontinuities, *tangential discontinuities* and *contact discontinuities*, satisfy this condition. The second class attributes a finite value to the mass flux across the discontinuity

$$\boxed{F_{\mathrm{II}} = \pm \frac{B_n}{\sqrt{\mu_0 \langle \mathcal{V} \rangle}}}$$

(8.32)

This value depends only on the normal component of the magnetic field, which is continuous, and on the average density of the fluid. The class of discontinuities corresponding to this particular condition is called *rotational discontinuities*. For the third class the expression in the second parenthesis of Eq. (8.30) must vanish

$$\boxed{F_{\mathrm{III}}^4 + F_{\mathrm{III}}^2 \left(\frac{[p]}{[\mathcal{V}]} - \frac{\langle \mathbf{B} \rangle^2}{\mu_0 \langle \mathcal{V} \rangle} \right) - \frac{B_n^2}{\mu_0 \langle \mathcal{V} \rangle} \frac{[p]}{[\mathcal{V}]} = 0}$$

(8.33)

This factor contains the jumps of pressure and the specific volume and allows for a variety of different conditions. It is a biquadratic equation which has two independent solutions

$$F_{\mathrm{III}}^2 = -\frac{1}{2} \left(\frac{[p]}{[\mathcal{V}]} - \frac{\langle \mathbf{B} \rangle^2}{\mu_0 \langle \mathcal{V} \rangle} \right) \pm \sqrt{\frac{\Delta}{4}}$$

(8.34)

These solutions are called *shocks* or *shock waves*. It can be shown that the discriminant

$$\Delta = \left(\frac{[p]}{[\mathcal{V}]} - \frac{\langle \mathbf{B} \rangle^2}{\mu_0 \langle \mathcal{V} \rangle} \right)^2 + \frac{4 B_n^2}{\mu_0 \langle \mathcal{V} \rangle} \frac{[p]}{[\mathcal{V}]} > 0$$

(8.35)

is always positive. For shocks to evolve, $F_{\mathrm{III}}^2 > 0$, and thus the negative sign in front of the square root must be ignored. Shocks do always exist, when the first term in the above solution is also positive. This is the case for

$$\frac{\langle \mathbf{B} \rangle^2}{\mu_0 \langle \mathcal{V} \rangle} > \frac{[p]}{[\mathcal{V}]}$$

(8.36)

a condition which is trivially satisfied if the sign of the ratio $[p]/[\mathcal{V}]$ is negative. Hence, in all cases when the pressure and specific volume change oppositely across the shock, shocks in a magnetohydrodynamic fluid become possible. When pressure and specific volume vary the same way, however, the existence of shocks is more restricted. Since the pressure always increases across a shock, the latter case can be realized only for a *rarefaction shock*, where the density decreases during the shock transition.

8.3. Discontinuities

The general jump conditions lead to three different families of discontinuities. For each of these an explicit set of jump conditions can be specified.

Contact and Tangential Discontinuities

The first family of discontinuities is characterized by zero normal mass flow and thus $v_n = 0$. Such a situation can be realized in two different cases. In the first the magnetic field has a non-vanishing but continuous component normal to the discontinuity surface, $B_n \neq 0$, $[B_n] = 0$. In this case the second condition in Eq. (8.26) requires that the tangential magnetic field vector is continuous across the discontinuity. Armed with this knowledge, it is easy to show that the third jump condition (8.26) demands continuity of the tangential velocity vector, while the first jump condition (8.26) reduces to the continuity of the pressure. The only quantity which can experience a change across the discontinuity is the plasma density, while all other quantities are continuous. This type of discontinuity is called *contact discontinuity*, because two plasmas are attached to each other at the discontinuity and tied by the normal component of the magnetic field such that they flow together at the same tangential speed. Contact discontinuities satisfy the relations

$$
\begin{aligned}
[p] &= 0 \\
[\mathbf{v}_t] &= 0 \\
[B_n] &= 0 \\
[\mathbf{B}_t] &= 0
\end{aligned}
\tag{8.37}
$$

Since the pressure remains constant across the discontinuity, any change in density must be balanced by a change in temperature. However, a temperature difference between both sides of the discontinuity should rapidly be dispersed by electron heat flux along the magnetic field, and contact discontinuities should usually not persist for long.

The second and more interesting case has a magnetic field purely tangential to the discontinuity, with zero normal component, $B_n = 0$. Since $v_n = 0$, the second and third of the Rankine-Hugoniot conditions (8.26) are trivially satisfied for any jumps in the tangential velocity and magnetic field. The only nontrivial condition which survives is the first, requiring the continuity of the total pressure across the discontinuity

$$
\left[p + \frac{B^2}{2\mu_0} \right] = 0
\tag{8.38}
$$

Such a discontinuity is a surface of total pressure balance between the two contacting plasmas with no mass or magnetic flux crossing the discontinuity from either side, while

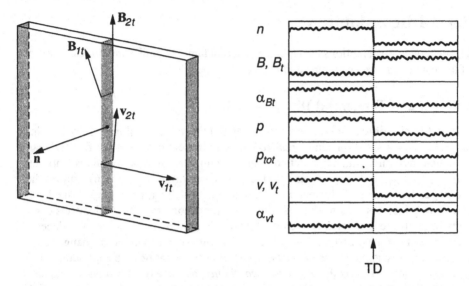

Fig. 8.6. Changes of magnetic field and plasma moments across a tangential discontinuity.

all other quantities can experience arbitrary changes. It is called *tangential discontinuity*, because both the plasma flow and the magnetic field are tangential to but discontinuous at the discontinuity. Typical changes of magnetic field, density, pressure, and bulk velocity across a tangential discontinuity are sketched in Fig. 8.6. The left part of the figure shows that the tangential magnetic and velocity vectors may arbitrarily change their magnitudes, B_t, v_t, and directions, α_{Bt}, α_{vt} across the discontinuity. The right-hand part of the figure shows how plasma and field quantities would change for a spacecraft crossing a tangential discontinuity.

Rotational Discontinuities

Assuming a finite mass flow across the discontinuity, $nv_n \neq 0$, but a continuous normal flow velocity, $[v_n] = 0$, yields discontinuities of family II. Because of the continuity of nv_n there can be no jump in the plasma density across the discontinuity. But the condition $F_{II} \neq 0$ requires that a non-vanishing normal flow is possible only when the magnetic field also has a non-vanishing normal component, $B_n \neq 0$. The first condition in Eq. (8.26) with vanishing left-hand side requires the continuity of the total pressure, like for tangential discontinuities. However, the second and third condition (8.26) show that the tangential components of the velocity and the magnetic field can only change together. The second condition together with Eq. (8.32) yields an appropriate jump condition for the tangential components. Applying the rule (8.28) to resolve the bracket in the third Rankine-Hugoniot condition (8.26)

$$B_n[\mathbf{v}_t] = v_n[\mathbf{B}_t] \qquad (8.39)$$

reveals another interesting property. Since v_n and B_n are constant, the tangential velocity and magnetic field vectors must rotate together across the discontinuity, but without changing their magnitudes. As a simple consequence, that magnetic field strength and thermal pressure are each continuous across the discontinuity. Summarizing, discontinuities of the second family obey the following jump conditions

$$
\boxed{
\begin{aligned}
[n] &= 0 \\
[p] &= 0 \\
[v_n] &= 0 \\
[B_n] &= 0 \\
[B^2] &= 0 \\
\left[\mathbf{v}_t - \frac{\mathbf{B}_t}{\sqrt{nm\mu_0}}\right] &= 0
\end{aligned}
}
\qquad (8.40)
$$

Such a discontinuity is called *rotational discontinuity* and, as sketched in Fig. 8.7, is a region where the tangential flow and magnetic field rotate by some arbitrary angle. The last jump condition actually relates the jump in the tangential component of the flow velocity to the jump of the tangential component of the Alfvén velocity defined in Eq. (8.11)

$$\boxed{[\mathbf{v}_t] = \left[\frac{\mathbf{B}_t}{\sqrt{nm\mu_0}}\right]} \qquad (8.41)$$

At a rotational discontinuity the jump in the tangential flow velocity is exactly equal to the jump in the tangential Alfvén velocity, a fact which shows that rotational discontinuities are closely related to the transport of magnetic signals across a boundary from one medium to another. Since the density in rotational discontinuities must be constant, the jump in the tangential Alfvén velocity arises only from the jump in the tangential magnetic field.

The constant normal component of the flow velocity is naturally related to the normal component of the Alfvén velocity. It can be found from the constancy of F_{II}, observing that the continuity of the density implies that the specific volume is also continuous

$$\boxed{v_n = \frac{B_n}{\sqrt{nm\mu_0}}} \qquad (8.42)$$

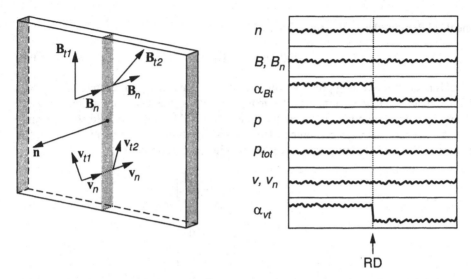

Fig. 8.7. Changes of magnetic field and plasma moments across a rotational discontinuity.

This latter equation is often called *Walén relation* and can be used to determine the normal component of the flow velocity.

   Since plasma pressure and density must each be constant across a rotational discontinuity, the temperature of the plasmas on both sides of a rotational discontinuity must be the same. This implies in turn that rotational discontinuities do not lead to an increase in entropy. The three discontinuities encountered until now all are reversible. In the case of the former two there was no mixing of the plasma and reversibility was trivial. But in the case of the rotational discontinuity mixing of two different plasmas takes place and reversibility comes in as a surprise.

Effect of Pressure Anisotropy

So far we have considered cases when the plasma pressure is isotropic. It is possible to generalize the results obtained to anisotropic plasmas with the pressure given by Eq. (7.21). The inclusion of pressure anisotropy affects, in the first place, the equation of motion and the second jump condition in Eq. (8.21), which now reads

$$\mathbf{n} \cdot [nm\mathbf{vv}] + \mathbf{n} \left[p_\perp + \frac{B^2}{2\mu_0} \right] - \frac{1}{\mu_0} \mathbf{n} \cdot \left[\mathbf{BB} \left(1 - \frac{\mu_0(p_\parallel - p_\perp)}{B^2} \right) \right] = 0 \qquad (8.43)$$

Since tangential discontinuities have $B_n = 0$, anisotropy has only a minor effect on the conditions for tangential discontinuities. Only in the jump condition for the total pressure the isotropic pressure is replaced by the perpendicular pressure

$$\left[p_\perp + \frac{B^2}{2\mu_0} \right] = 0 \tag{8.44}$$

It is considerably more involved to determine the anisotropic jump conditions for rotational discontinuities and only a few important points are mentioned here. The Walén relation is modified such that the normal component of the flow is equal to the modified normal Alfvén velocity in Eq. (8.42)

$$v_n = \left\{ \left(\frac{B_n^2}{nm\mu_0} \right) \left(1 - \frac{\mu_0(p_\parallel - p_\perp)}{B^2} \right) \right\}^{1/2} \tag{8.45}$$

on both sides of the discontinuity. Since the tangential magnetic fields can have arbitrary directions in an anisotropic plasma, the two tangential magnetic field vectors and the vector normal to the rotational discontinuity are in general not coplanar. Moreover, the density may also have a non-vanishing jump, $[n] \neq 0$, and pressure equilibrium holds only for the total pressure, $[p + B^2/2\mu_0] = 0$.

Entropy Changes

It is instructive to investigate the behavior of the entropy at discontinuities. Consider the stationary ideal heat conduction or entropy conservation equation (7.58)

$$\mathbf{v} \cdot \nabla S = 0 \tag{8.46}$$

In incompressible media with $\nabla \cdot \mathbf{v} = 0$, this equation can be rewritten as

$$\nabla \cdot (\mathbf{v} S) = 0 \tag{8.47}$$

or in the form of a jump condition as

$$[v_n S] = 0 \tag{8.48}$$

In the more general case of compressible media, the right-hand side does not vanish, showing that the compressibility of the fluid acts as a source of entropy

$$\nabla \cdot (\mathbf{v} S) = S \nabla \cdot \mathbf{v} \tag{8.49}$$

At a one-dimensional discontinuity the last equation can be rewritten as

$$[v_n S] = \int dn\, S \frac{dv_n}{dn} = \int S dv_n \tag{8.50}$$

Hence, discontinuities with constant normal velocity, $[v_n] = 0$, conserve entropy. All discontinuities discussed so far belong to this class of *isentropic discontinuities* with no increase of entropy when crossing from upstream to downstream regions. Discontinuities in compressible media with a non-vanishing jump in the normal flow speed will necessarily lead to an increase in entropy across the discontinuity and therefore represent irreversible changes in the state of the plasma across the discontinuity. These types of discontinuities are discussed next.

8.4. Shocks

The third discontinuity family is characterized by non-vanishing normal fluxes, $n v_n \neq 0$. For this family the plasma moves across the discontinuity as in the case of rotational discontinuities, but the density is discontinuous.

Intermediate Shocks

In fact, Eq. (8.33) suggests that $[\mathcal{V}] \neq 0$, because otherwise this condition reduces to

$$[p]\left(F_{\mathrm{III}}^2 - \frac{B_n^2}{\mu_0 \langle \mathcal{V} \rangle}\right) = 0 \qquad (8.51)$$

The latter equation has two solutions. Either

$$[p] = 0 \qquad (8.52)$$

which implies that the pressure is continuous, or

$$\left(F_{\mathrm{III}}^2 - \frac{B_n^2}{\mu_0 \langle \mathcal{V} \rangle}\right) = 0 \qquad (8.53)$$

showing that under the special condition of continuous density, a family III discontinuity resembles a rotational discontinuity. Indeed, the latter condition is identical with condition (8.32), which the normal flux satisfies at rotational discontinuities. In this very special case and with $[p] \neq 0$ one speaks of an *intermediate shock*. On the other hand, when both factors in the above condition are simultaneously zero, one recovers the ordinary rotational discontinuities. Rotational discontinuities form a subclass of intermediate shocks with continuous pressure and no increase in entropy.

True Shocks

In all other cases Eq. (8.33) possesses two pairs of conjugate solutions for non-vanishing jumps in pressure, specific volume, and density, implying that the plasma flow across the discontinuity switches from one thermodynamic state to another. Of these four solutions only the three which yield $F_{III}^2 > 0$ are physically relevant. These are one solution for $[p]/[\mathcal{V}] > 0$, and two solutions for $[p]/[\mathcal{V}] < 0$. For $[p] > 0$ there is only one solution with $[\mathcal{V}] > 0$ or $[n] < 0$, which corresponds to a dilutive transition in a *rarefaction shock*, while there are two solutions with $[n] > 0$ corresponding to compressive transitions. Since the pressure always increases, rarefaction shocks are always accompanied by an increase in temperature across the shock, indicating that inside the shock transition region the plasma is heated. This is an irreversible process which increases the entropy.

However, entropy increases are not restricted to rarefaction shocks, but appears in all family III discontinuities. All shocks are accompanied by a change in entropy across the discontinuity and thus irreversible. Because the plasma moves across the discontinuity at continuous normal flux, the finite jump in the density implies that the normal velocity will be changed in the opposite way across the discontinuity. This is most easily observed from the continuity of the normal flow, $[F_{III}] = [nv_n] = 0$, a condition which can be rewritten using Eq. (8.28)

$$[v_n] = -\frac{\langle v_n \rangle}{\langle n \rangle}[n] \tag{8.54}$$

Coplanarity

Knowing that $v_n \neq 0$, the two last conditions of Eq. (8.26) suggest that for shocks with $B_n \neq 0$ the jump in the tangential magnetic field vector is parallel to the jump in the tangential velocity vector. Eliminating $[\mathbf{v}_t]$ from (8.26) one obtains

$$[v_n \mathbf{B}_t] = \frac{B_n^2}{\mu_0 F_{III}}[\mathbf{B}_t] \tag{8.55}$$

Hence, the cross-product of the right- and left-hand sides must vanish

$$[\mathbf{B}_t] \times [v_n \mathbf{B}_t] = 0 \tag{8.56}$$

When resolving the brackets in the above expression, one obtains

$$(v_{n1} - v_{n2})(\mathbf{B}_{t1} \times \mathbf{B}_{t2}) = 0 \tag{8.57}$$

Since $[v_n] \neq 0$, the upstream and downstream tangential components of the magnetic field on both sides of the shock must be parallel to each other. Hence, the upstream and downstream tangential magnetic field vectors are coplanar with the shock normal vector, they all lie in the same plane normal to the shock. This *coplanarity theorem* implies that the magnetic field across the shock has a two-dimensional geometry. The same also holds for the bulk velocity. It is coplanar with the shock normal and has a two-dimensional geometry, yet a different one.

Jump Conditions

To proceed further we need to include the energy transport across the shock, since the shock itself is a region where heat and entropy are produced. Assuming that the plasma behaves like an ideal gas and that all variations proceed so fast that adiabatic conditions can be assumed, the internal enthalpy is

$$w = \frac{p}{nm(\gamma - 1)} \tag{8.58}$$

with γ being the polytropic index. With these assumptions in mind, one can bring the conservation of energy (8.22) into the form

$$\left[v_n \left(\frac{nmv^2}{2} + \frac{\gamma p}{\gamma - 1} + \frac{B_t^2}{\mu_0} \right) - \frac{B_n \mathbf{v} \cdot \mathbf{B}}{\mu_0} \right] = 0 \tag{8.59}$$

Using the remaining Rankine-Hugoniot conditions (8.26) and realizing that

$$\left[\left(\mathbf{v}_t - \frac{B_n \mathbf{B}_t}{\mu_0 F_{\mathrm{III}}} \right)^2 \right] = 0 \tag{8.60}$$

the latter expression can be brought into the following form

$$\frac{[v_n^2]}{2} + \frac{\gamma[pv_n]}{(\gamma - 1)F_{\mathrm{III}}} + \frac{[v_n B_t^2]}{\mu_0 F_{\mathrm{III}}} - \frac{B_n^2}{\mu_0^2 F_{\mathrm{III}}^2} \frac{[B_t^2]}{2\mu_0} = 0 \tag{8.61}$$

or, when splitting the jumps of the products into products of jumps and averages,

$$[p]\langle v_n \rangle + [v_n]\gamma \langle p \rangle + \frac{\gamma - 1}{4\mu_0}[v_n][B_t^2] = 0 \tag{8.62}$$

Since bulk velocity and magnetic field each obey the coplanarity theorem, one can write Eq. (8.55) in scalar form

$$[v_n]\langle B_t \rangle + [B_t]\langle v_n \rangle = \frac{B_n^2}{\mu_0 F_{\mathrm{III}}}[B_t] \tag{8.63}$$

and find from momentum conservation, i.e., the second Rankine-Hugoniot condition (8.26),

$$B_n[v_t] = \frac{B_n^2}{nm\mu_0}[B_t]$$

$$F_{\mathrm{III}}[v_n] = -\left[p + \frac{B_t^2}{2\mu_0} \right] \tag{8.64}$$

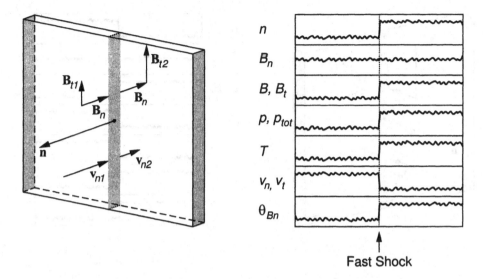

Fig. 8.8. Changes of magnetic field and plasma moments across a fast shock.

where $B_n^2/nm\mu_0$ is the square of the normal component of the Alfvén velocity given by the Walén relation (8.42). Equations (8.62) through (8.64) form a closed set of equations for the shock jump conditions and are the general Rankine-Hugoniot conditions for shocks, which are valid in an isotropic and adiabatic one-fluid plasma.

Fast and Slow Shocks

We can obtain some general results when eliminating the jump in the normal velocity $[v_n]$ from Eqs. (8.62) and (8.64). This way we obtain a relation between the jumps in plasma and magnetic pressures

$$\left(\frac{\langle v_n \rangle}{\gamma - 1} - H\right)[p] = \frac{H}{\mu_0 F_{\text{III}}}[B_t^2] \tag{8.65}$$

where the quantity H is defined as

$$F_{\text{III}} H = \frac{[B_t^2]}{4\mu_0} + \frac{\gamma \langle p \rangle}{\gamma - 1} \tag{8.66}$$

Since the pressure always increases across a shock transition from the undisturbed to the disturbed and heated plasma behind the shock, one has

$$[p] > 0 \tag{8.67}$$

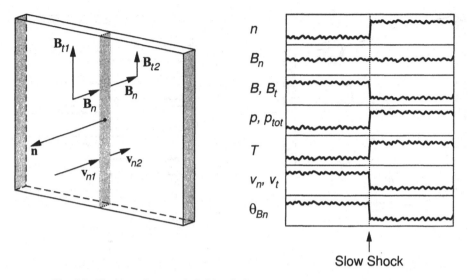

Fig. 8.9. Changes of magnetic field and plasma moments across a slow shock.

One can therefore distinguish between two different cases of shock waves. The first type of shocks is characterized by an increase in the Alfvén speed or, correspondingly, an increase in the magnetic pressure

$$[B_t^2] > 0 \qquad (8.68)$$

Such shocks satisfy the condition

$$\langle v_n \rangle > (\gamma - 1)H \qquad (8.69)$$

and are called *fast shocks* (see Fig. 8.8). The second type experiences a decrease in magnetic pressure when passing through the shock from the undisturbed to the disturbed medium

$$[B_t^2] < 0 \qquad (8.70)$$

Shocks of this type satisfy the opposite condition for the normal average velocity

$$\langle v_n \rangle < (\gamma - 1)H \qquad (8.71)$$

and are called *slow shocks* (see Fig. 8.9).

As can be recognized by comparing Figs. 8.8 and 8.9 or the two panels on the right-hand side of Fig. 8.11 further below, across fast shocks the magnetic field increases and is tilted toward the shock surface, while across slow shocks it decreases and bends toward the shock normal. The average normal velocity is always positive, because plasma is flowing across the shock from the undisturbed into the disturbed and heated medium. Hence, slow shocks can exist only when the average pressure is large enough to satisfy

$$\gamma \langle p \rangle - \frac{\gamma - 1}{4\mu_0} \left(B_{t1} B_{t2} - [B_t^2] \right) > 0 \tag{8.72}$$

The two types of shocks can be distinguished by the behaviour of the tangential magnetic field pressures across the shock (see Figs. 8.8 and 8.9). In addition, one finds that fast flows will cause fast shocks to evolve, while slow flows may give rise to slow shocks.

Mach Numbers

A shock may develop when the fluid velocity exceeds the *magnetosonic speed*

$$\boxed{c_{ms}^2 = c_s^2 + v_A^2} \tag{8.73}$$

of the fluid, which replaces the sound velocity (8.5) in an ordinary fluid. In this case the plasma flow is called super-magnetosonic and, in analogy to ordinary fluids, a non-moving obstacle will give rise to the evolution of a plasma shock front. One defines a *magnetosonic Mach number*, M_{ms}, by relating the fluid velocity to the magnetosonic speed

$$\boxed{M_{ms} = \frac{v}{c_{ms}}} \tag{8.74}$$

The condition for the evolution of a shock wave in a plasma then becomes that

$$M_{ms} > 1 \tag{8.75}$$

Whenever this condition is satisfied and the plasma flow is distorted due to the presence of a non-moving object, a shock front will develop across which the fluid quantities will jump discontinuously and the super-magnetosonic flow will become retarded to a sub-magnetosonic flow.

8.5. Bow Shock

The most famous example of a plasma shock is the Earth's bow shock. It develops as a result of the interaction of the Earth's magnetosphere with the supersonic solar wind. The magnetosphere is a blunt obstacle at rest, which brakes the solar wind flow. Figure 8.10 shows the parabolically shaped surface of the bow shock, across which the solar wind velocity decreases from super-magnetosonic to sub-magnetosonic. It divides the

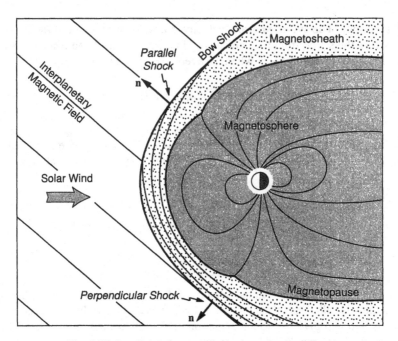

Fig. 8.10. Parallel and perpendicular bow shock regions.

solar wind flow into two regions, the undisturbed solar wind in the region upstream of the bow shock and the disturbed *magnetosheath* flow on the downstream side.

The bow shock is an ideal object to study the properties of shocks. Since the solar wind is a high-Mach number stream with $M_{ms} \approx 8$, the bow shock is a fast magnetosonic shock. The density and the magnetic field increase when crossing from the solar wind into the magnetosheath. As determined experimentally, both quantities jump by about a factor of 4.

However, the shock exists only over a limited region of space in front of the Earth because the Mach number is defined by the solar wind velocity component normal to the shock, $v_n = v_{sw} \cos \theta$. The condition $M_{ms} > 1$ is satisfied only as long as the angle $\theta < \arccos M_{ms}^{-1}$. For $M_{ms} \approx 8$ the maximum angle between the solar wind velocity and the shock normal up to which the bow shock exists is $\theta_{max} \approx 80°$. Hence, the bow shock forms a spatially restricted shield in front of the magnetosphere and undergoes a transition from a high-Mach number shock at its nose to a low Mach number shock at its flanks.

High-Mach number shocks having $M_{ms} > M_c$ are called *supercritical*. They behave differently from low Mach number shocks with $M_{ms} < M_c$, which are *subcritical*. The *critical Mach number*, M_c, is conventionally defined as the Mach number for which

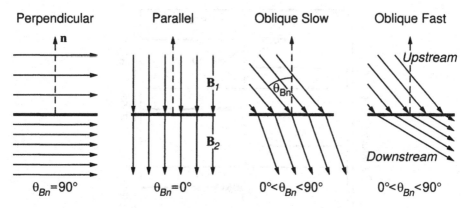

Fig. 8.11. Four possible geometries of shock normal and magnetic field.

the flow velocity downstream of the shock equals the downstream sound velocity so that the downstream magnetosonic Mach number is equal to unity. Solving the shock jump conditions under this restriction and under the assumption that the magnetic field is tangential to the shock yields a value of $M_c = 2.7$. This value decreases, however, for oblique magnetic field directions. At the bow shock it has been found that an average critical Mach number $1 < M_c < 2$ is more appropriate than the above theoretical value. Hence, the majority of observed bow shock transitions are supercritical.

Parallel and Perpendicular Shocks

Another distinctive difference between different parts of the bow shock can be realized from Fig. 8.10, namely the direction of the magnetic field with respect to the shock normal. For a normal Archimedian spiral form of the interplanetary field, the shock normal on the morning side of the bow shock is parallel to the direction of the interplanetary magnetic field while on the evening side the interplanetary magnetic field and the shock normal are orthogonal. Depending on the value of the *shock normal angle*, θ_{Bn}, shocks can be classified as *parallel shocks* ($\theta_{Bn} = 0°$), as *perpendicular shocks* ($\theta_{Bn} = 90°$) or as *oblique shocks* ($0° < \theta_{Bn} < 90°$). One also speaks of quasi-perpendicular or quasi-parallel shocks if the shock normal angle does not deviate too far from the perpendicular or parallel direction, respectively. The three possible cases are illustrated in Fig. 8.11 (together with the oblique slow mode shock geometry).

The distinction between the two shock directions is physically relevant. Strictly parallel shocks have their magnetic field directed along the shock normal and since B_n must be continuous, the magnetic field is not affected by the presence of the shock. But this case is never realized in real systems. Realistic parallel shocks are always quasi-parallel and react also magnetically. Any small deviation of the magnetic field direction

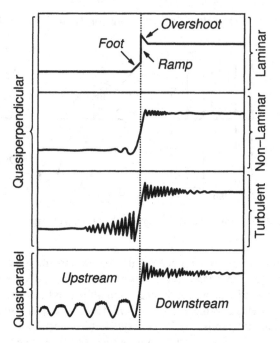

Fig. 8.12. Typical magnetic shock profiles.

from being perpendicular to the shock front results in a strong effect on the magnetic field, since the magnetic field is rotated out of coplanarity by sound waves radiated inside the shock into all directions tangential to the shock front. Such a distortion causes local disturbances which result in short wavelength oscillations of the magnetic field. The shock becomes turbulent. In addition, the generation of the new out-of coplanarity magnetic component turns a parallel shock into a quasi-perpendicular one close to the shock ramp. A magnetized shock always manages to turn the magnetic field locally quasi-perpendicular, even if far upstream of the shock the magnetic field was parallel.

Figure 8.12 shows characteristic magnetic shock profiles. The typical perpendicular shock profile consists of upstream and downstream regions connected by a steep *shock ramp*. Perpendicular shocks usually possess a *shock foot* region in front of the ramp, where the magnetic field gradually rises. In addition, the shock ramp generally shows a magnetic *shock overshoot* before settling at the average magnetic field strength behind the shock. For oblique magnetic fields the shock starts exhibiting oscillatory behaviour, which gradually becomes turbulent. Parallel shocks are highly oscillatory, up to large distances in front of the shock. This region is called *foreshock*, since here the upstream medium becomes notified of the shock's presence.

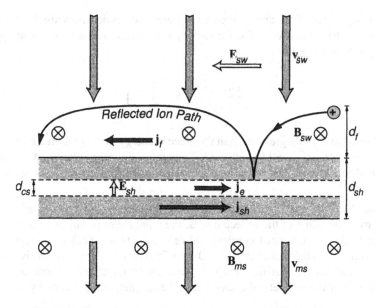

Fig. 8.13. Ion reflection and acceleration at a perpendicular shock.

Shock Currents

The jump $[B_t] \neq 0$ in the tangential magnetic field across the shock indicates that the bow shock itself is a current layer with an internal surface current density, j_{sh}, which accounts for the change in the magnetic field. It can be estimated from

$$j_{sh} = \frac{[B_t]}{\mu_0 d_{sh}} \qquad (8.76)$$

where d_{sh} is the shock width. This current increases the magnetic field strength behind the shock. It should, in principle, partially cancel the magnetic field in front of the shock, but this has not been observed. Instead one finds a slight increase in magnetic field strength in the shock foot region, as indicated in Fig. 8.12, due to the appearance of reflected ions in front of the shock.

Solar wind ions and electrons encountering the compressed magnetic field perpendicular shock will have different gyroradii. The ions can penetrate deeper into the field than the electrons. This difference in penetration depth will generate a charge separation electric field in the shock normal direction, pointing toward the sun. Such a field will reflect a number of ions back into the solar wind, while it attracts and captures electrons. It is given by

$$\epsilon_0 E_{sh} = e(n_{i,sh} + n_{e,sh})d_{cs} \qquad (8.77)$$

where d_{cs} is the width of the charge separation layer and the densities are the densities of the particles inside the shock. The ion density is given as $n_i = n_e - n_{ir}$, with n_{ir} the density of the reflected ions

$$E_{sh} = \frac{en_{e,sh}d_{cs}}{\epsilon_0}\left(1 - \frac{n_{ir}}{n_{e,sh}}\right) \qquad (8.78)$$

Hence, all ions having energies less than the electric energy in the potential drop across the shock, $e\phi = eE_{sh}d_{cs}$, will be reflected. They return into the solar wind in front of the shock and perform another gyration in the solar wind magnetic field as shown in Fig. 8.13.

Because the solar wind in the shock frame carries a convection electric field, which lies in the gyration plane of the reflected ions, the reflected ions will be accelerated in this electric field to about twice the solar wind velocity. These reflected and accelerated ions carry a current in the foot region, j_f. Only a fraction of solar wind ions is actually reflected, but this fraction carries a current which closes the shock current in front of the shock and over-compensates the decrease in the magnetic field caused by the shock current.

The magnetic overshoot in the shock profile is related to another current layer inside the shock transition region. This current is a pure electron drift current (see Fig. 8.13). It results from the presence of the charge separation field, E_{sh}, inside the shock. As argued above this field is restricted to a narrow layer, narrower than the ion gyroradius in the compressed shock magnetic field, but wider than the electron gyroradius. Therefore the electrons may perform an electric $E \times B$ drift motion in the crossed electric and magnetic fields within this layer while the ions are not affected. This drift gives rise to an electron current, j_e, flowing in the same direction as the shock current, j_{sh}, and amplifying it locally, thereby causing the magnetic overshoot.

8.6. Magnetopause

As described in Sec. 5.1, the fully ionized and magnetized solar wind plasma cannot mix with the terrestrial magnetic flux tubes. Instead, it will deviate from its original direction and will, by its dynamical pressure, compress the terrestrial field and confine it into a small region of space, the magnetosphere. During this interaction a narrow boundary layer evolves, the *magnetopause*. This layer is a discontinuity which must be different from the bow shock because the plasma flow behind the bow shock is subsonic. Actually, to first order, the magnetopause can be regarded as a tangential discontinuity.

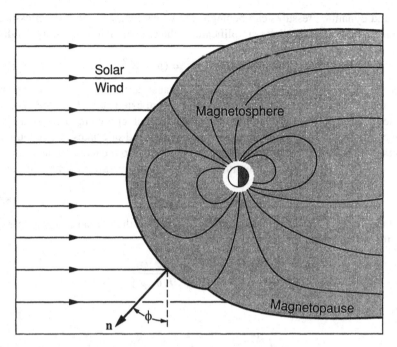

Fig. 8.14. Geometry of the Earth's magnetopause.

Magnetopause Shape

As a tangential discontinuity the magnetopause is a surface of total pressure equilibrium between the solar wind-magnetosheath plasma and the geomagnetic field confined in the magnetosphere. The weakness of the solar wind magnetic field allows, in a first approximation, to neglect the contribution of the interplanetary field to the pressure balance. In addition, since the main energy of the solar wind flow is stored in the bulk flow of the ions and not in the thermal pressure, it is sufficient to take into account only the solar wind dynamic ram pressure

$$p_{dyn} = n_{sw} m_i v_{sw}^2 \qquad (8.79)$$

This equation is valid for ideally specular reflection of the oncoming solar wind particles at the magnetopause boundary as shown in Fig. 8.14. The dynamic pressure exerted on the terrestrial field is proportional to the number density and the total change in energy of the particles during their turn-around. The latter is twice the dynamic solar wind ion energy, since the tiny contribution of the electrons can neglected. If the particles are not really specularly reflected, one must include an efficiency coefficient κ on the right-hand side of the above expression. On the other hand, inside the magnetosphere the plasma thermal and dynamic pressures can be neglected when compared with the pressure of the geomagnetic field. With these simplifications the tangential discontinuity condition (8.38) becomes

$$2\mu_0 \kappa n_{sw} m_i (\mathbf{n} \cdot \mathbf{v}_{sw})^2 = (\mathbf{n} \times \mathbf{B})^2 \tag{8.80}$$

Here $\mathbf{n}(r, \theta, \phi)$ is the outer normal to the magnetopause surface. It depends on all three spatial directions, because the magnetopause is a complicated curved surface. The left-hand side of Eq. (8.80) selects the normal solar wind velocity component as the only relevant component for the interaction, while the right-hand side takes into account that the magnetospheric field has no component perpendicular to the magnetopause.

Denoting the function describing the magnetopause surface in spherical coordinates as

$$S_{mp}(r, \theta, \varphi) = 0 \tag{8.81}$$

the outer normal to the magnetopause can be expressed as the normalized negative gradient of the surface function

$$\mathbf{n}(r, \theta, \varphi) = -\frac{\nabla S_{mp}(r, \theta, \varphi)}{|\nabla S_{mp}(r, \theta, \varphi)|} \tag{8.82}$$

Inserting into Eq. (8.80) yields

$$2\mu_0 \kappa n_{sw} m_i \left(\frac{\nabla S_{mp}}{|\nabla S_{mp}|} \cdot \mathbf{v}_{sw} \right)^2 = \left(\frac{\nabla S_{mp}}{|\nabla S_{mp}|} \times \mathbf{B} \right)^2 \tag{8.83}$$

The above equation contains the complicated structure of the magnetospheric magnetic field near the magnetopause. It also contains the three-dimensional derivatives of the unknown surface function, S_{mp}, and their second powers. Hence, it is a second-order three-dimensional nonlinear partial differential equation for S_{mp} and the solution can be found only by numerical methods.

One can, however, find a simple solution at the nose of the magnetopause, where the solar wind speed reduces to zero, the so-called *stagnation point*. Here the magnetopause is symmetrical in the angular coordinates so that all the angular derivatives vanish. The magnetospheric field is perpendicular to the ecliptic plane, and the solar wind velocity is in the ecliptic. Denoting the *stand-off distance* of the magnetopause from the Earth's center by R_{mp} and assuming that the magnetospheric field is dipolar (see Sec. 3.1), pressure equilibrium can be written as

$$n_{sw} m_i v_{sw}^2 = \frac{K B_E^2}{2\mu_0 R_{mp}^6} \tag{8.84}$$

Here B_E is the magnetic field at the surface of the Earth, and the constant K accounts for both κ and the deviation of the magnetic field from its dipolar value at R_{mp}. Rewriting

Mercury	Earth	Jupiter	Saturn	Uranus	Neptune
1.4	10	75	20	20	25

Table 8.1. Planetary stagnation point distances in planetary radii

this equation yields the stand-off distance in Earth radii

$$R_{mp} = \left(\frac{K B_E^2}{2\mu_0 n_{sw} m_i v_{sw}^2} \right)^{1/6}$$

(8.85)

as the sixth root of the ratio of the magnetic dipole energy at the Earth's surface to the dynamic solar wind energy density. Taking $n_{sw} = 5\,\mathrm{cm^{-3}}$, $v_{sw} = 400\,\mathrm{km/s}$, $B_E = 3.1 \cdot 10^4\,\mathrm{nT}$, and assuming $K = 2$, one finds $R_{mp} = 9.9\,R_E$. Since the stand-off distance changes as the sixth root of the values involved, it is not very sensitive to variations in the solar wind dynamic pressure. Under quiet conditions the magnetospheric nose or solar wind stagnation point is found at about $10\,R_E$.

Equation (8.85) is independent of the specific terrestrial situation. It is valid for any dipolar magnetic field which interacts with a weakly magnetized plasma stream. It can therefore be applied to any other magnetosphere, ranging from the magnetospheres of the planets to magnetospheres of stars and pulsars interacting with stellar winds or interplanetary gases. Table 8.1 collects the theoretical stagnation point distances for the magnetized planets of our solar system.

A similar conclusion as for the nose distance of the magnetopause can be drawn for the distance of the magnetopause at its flanks. At the flanks the solar wind flow is tangential to the magnetopause, $v_n = 0$, and the ram pressure of the solar wind vanishes. In the pressure equilibrium between the non-magnetized solar wind and the dipolar magnetospheric field the so far neglected thermal pressure, $p_{sw} = \gamma n_{sw} k_B T_{sw}$, comes into play at this point, yielding

$$R_{mpf} = \left(\frac{K B_E^2}{2\mu_0 \gamma n_{sw} k_B T_{sw}} \right)^{1/6}$$

(8.86)

for the geocentric distance of the magnetopause flanks in units of Earth radii. For a solar wind temperature of about $1.3 \cdot 10^5\,\mathrm{K}$, $\gamma = 5/3$, and the values used above, this distance becomes about $R_{mpf} \approx 1.8\,R_{mp}$, roughly two times the distance of the subsolar point. Observations of the shape of the magnetopause have shown that the magnetopause at the dawn-dusk meridian is found at about a distance of $14\,R_E$, slightly less than its theoretical distance. In addition these observations show that the dawn and dusk magnetopause

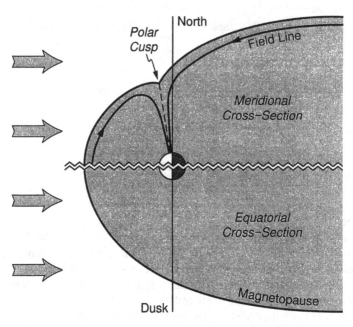

Fig. 8.15. Magnetopause cross-sections and cusp.

still experiences a non-vanishing normal flow velocity component, $v_n \neq 0$. In other words, the magnetosphere at dawn and dusk is still inflating and the radius of the magnetosphere still increases, when going from the dayside through the dawn-dusk meridian to the nightside magnetosphere. Only much farther downstream tail the magnetospheric boundary becomes approximately parallel to the flow. The reason for such a behavior can be found in the global magnetospheric current system.

Equation (8.83) has been solved numerically to obtain the shape of the magnetopause. Surprisingly, in the meridional plane there is no continuous solution connecting the dayside magnetopause to the nightside magnetopause. Figure 8.15 shows the calculated magnetopause cross-section in the equatorial and meridional planes. While in the equatorial plane the magnetopause is a smooth curve extending from the dayside into an open tail, at high latitudes the tangent to the magnetopause is discontinuous at one location. This point has been identified as the *polar cusp* and arises from the special geometry of the dipolar geomagnetic field. From this point onward the magnetic field lines are turned around to the tail as a consequence of their interaction with the solar wind. However, the field lines do not experience any discontinuity at the polar cusp, but simply change their topology from dayside-like to tail-like as indicated by the field line included in the figure.

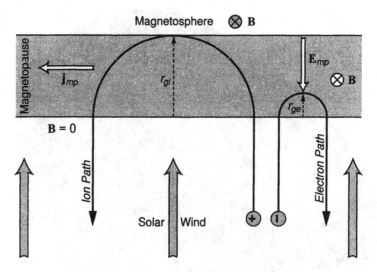

Fig. 8.16. Specular reflection off the magnetopause.

Magnetopause Current

Separating the solar wind from the magnetospheric magnetic field and being a surface across which the magnetic field strength jumps from its low interplanetary value to the high magnetospheric field strength, the magnetopause represents a surface current layer. The origin of this current can be understood from Fig. 8.16.

Specularly reflected ions and electrons hitting the magnetospheric field inside the magnetopause boundary will perform half a gyro-orbit inside the magnetic field before escaping with reversed normal velocity from the magnetopause back into the magnetosheath. The thickness of the solar wind-magnetosphere transition layer under such idealized conditions becomes of the order of the ion gyroradius, $r_{gi} = v_{sw}/\omega_{gi}$. Electrons also perform half gyro-orbits, but with much smaller gyroradii. The sense of gyration inside the boundary is opposite for both kinds of particles leading to the generation of a narrow surface current layer. This current provides the additional magnetic field, which compresses the magnetospheric field into the magnetosphere and at the same time annihilates its external part. It is a diamagnetic current caused by the perpendicular density gradient at the magnetopause.

The current density inside the magnetopause can be estimated to about 10^{-6} Am^{-2}. The total the current flowing in the magnetopause is of the order of 10^7 A. In the equatorial plane the magnetopause current flows from dawn to dusk, as shown schematically in Fig. 8.17 (see also Fig. 1.6). It closes on the tail magnetopause, where it splits into northern and southern parts flowing across the lobe magnetopause from dusk to dawn.

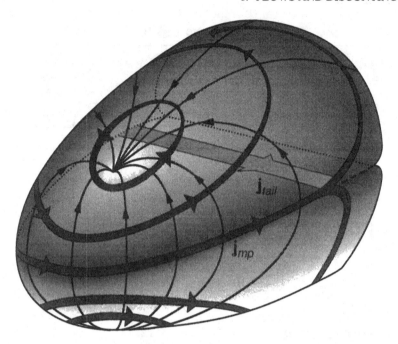

Fig. 8.17. Three-dimensional geometry of magnetopause currents.

The tail magnetopause current is additionally fed by the cross-tail neutral sheet current which flows from dawn to dusk.

Magnetosheath Flow

Knowing the three-dimensional shape of the magnetopause and bow shock, one is in the position to calculate the properties of the flow in the magnetosheath surrounding the magnetosphere. This can be done in several degrees of sophistication. The simplest one is to neglect the contribution of the magnetic field, solving the jump conditions across the bow shock and calculating the flow in the magnetosheath by assuming ideal hydrodynamic conditions and the condition of tangential flow. This is not fully realistic, but for a simple gasdynamic shock the Rankine-Hugoniot conditions simplify considerably. In particular, the shock distance from the blunt magnetospheric body for such a shock satisfies the condition

$$R_{bs} = \left(1 + 1.1 \frac{n_{sw}}{n_{bs}} \right) R_{mp} \tag{8.87}$$

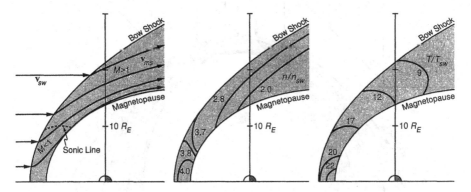

Fig. 8.18. Magnetosheath stream lines and density and temperature isocontours for $M_s = 8$.

where n_{bs} is the magnetosheath density adjacent to the shock ramp. From gasdynamic shock theory it follows that this density is at its maximum about $n_{bs} \approx 4\,n_{sw}$, yielding a distance of about $R_{bs} \approx 1.3\,R_{mp}$, both in good agreement with observations.

The results of such gasdynamic calculations depend on the polytropic index of the solar wind plasma and on its Mach number. As displayed in Fig. 8.18, for weak magnetic fields the magnetosphere bends the flow lines into an azimuthal direction, with the flow lines closer together at the magnetopause. At a certain distance from the subsolar point the nozzle effect of the magnetosheath causes the flow to again make the transition from subsonic to supersonic flow. The isodensity contours indicate compression of the magnetosheath plasma in a region close to the stagnation point at the nose of the magnetosphere. Outside this region the plasma is still compressed, but gradually becomes more dilute toward the flanks of the magnetopause. The temperature behind the shock is enhanced showing the generation of entropy in the course of the solar wind shocking process. Close to the stagnation point this enhancement is more than a factor of 20 and is still significant near the flanks of the magnetopause where the flow has cooled adiabatically.

Neglecting the magnetic field can be justified only if it is so small that it merely reacts passively to the interaction of the flow with the magnetosphere. If this is the case, the magnetic field is simply convected along the magnetosheath flow in a manner that it stays tangential to the magnetopause and satisfies the ideal magnetohydrodynamic conditions

$$\nabla \cdot \mathbf{B}_{ms} = 0$$
$$\nabla \times (\mathbf{v}_{ms} \times \mathbf{B}_{ms}) = 0 \tag{8.88}$$

The flow drapes the field around the magnetopause (see Fig. 8.19) and at the same time transports it downstream to the nightside. There is some compression of the field in the

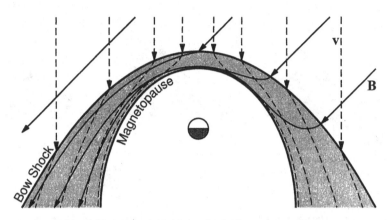

Fig. 8.19. Magnetic field draping in the magnetosheath.

magnetosheath similar to the closer positioning of the streamlines of the flow (see Fig. 8.18), but there is no reaction of the field on the flow in this model. Interestingly, for a more parallel direction of the interplanetary magnetic field the draping occurs only on that side of the magnetosphere side where the bow shock is quasi-perpendicular. Since the bow shock is a fast shock the magnetic field lines in the magnetosheath are refracted away from the shock normal. Behind the quasi-parallel part of the bow shock this refraction pulls the field lines away from the stagnation point thereby generating a region of lower magnetic field strength in the magnetosheath between the bow shock and the magnetopause. Usually this region is found on the early morning side of the magnetosheath.

The magnetic field draping has two earlier neglected effects. Firstly, due to the compression of the field related to the draping the magnetic field pressure increases. Ultimately this effect will lead to a breakdown of the gasdynamic model. The enhanced magnetic field pressure inside a compressed magnetosheath flux tube near the stagnation point will squeeze the magnetosheath plasma out of this tube into the flank-side magnetosheath. This effect has been observed and called *plasma depletion*. It effectively dilutes the magnetosheath plasma near the nose below its theoretical density. The effect is mainly observed when the magnetic fields in the magnetosheath and magnetosphere are nearly parallel to each other.

Reconnection

A much more important effect occurs when the magnetosheath magnetic fields has a southward component. As introduced in Secs. 5.1 and 5.2, in such a case *reconnection* or *merging* sets in at the magnetopause between the contacting antiparallel mag-

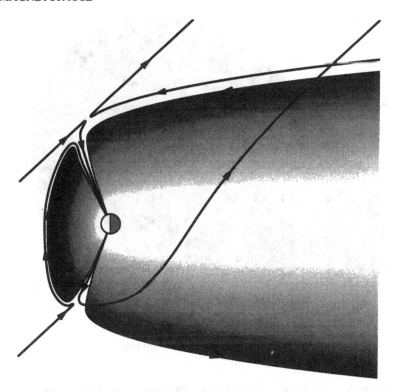

Fig. 8.20. Cusp merging for northward outer field.

netosheath and magnetospheric magnetic field lines. Magnetic reconnection is one of the most important though poorly understood processes in space plasma physics. The principle of reconnection is the merging of antiparallel magnetic field lines at a magnetic X-line, where the two fields annihilate each other (see Fig. 5.3). The mechanism of this process is related to an instability, one of the subjects of our companion volume, *Advanced Space Plasma Physics*.

As sketched in Fig. 5.4, when reconnection occurs, some magnetosheath field lines become connected with some of the magnetospheric field lines and are convectively transported tailward. The important point is that in the reconnection region the nature and topology of the magnetopause change fundamentally. While the magnetopause still maintains to be a surface of total pressure equilibrium, it looses the property of a tangential discontinuity and becomes locally a rotational discontinuity with a non-vanishing normal magnetic component, $B_n \neq 0$, generated in the reconnection process. For such a discontinuity the normal flux is also non-zero, $F_{\mathrm{II}} \neq 0$. Matter from the magnetosheath can get free access to the magnetosphere along the normal magnetic field component to

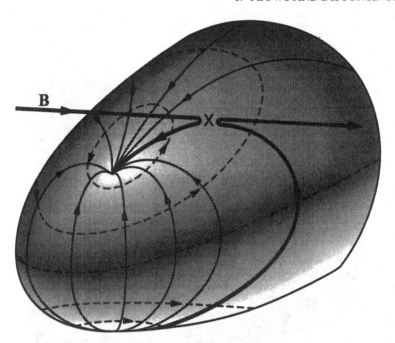

Fig. 8.21. High-latitude merging for ecliptic outer field.

inject magnetosheath plasma into the outer magnetospheric region thereby creating a broad boundary layer adjacent to the magnetospheric side of the magnetopause.

There is no need for the magnetic field in the magnetosheath to be directed southward for the onset of reconnection. Near the dayside stagnation point it is sufficient to have a southward component. Reconnection may occur between this component and a fraction of the magnetospheric magnetic field with the remaining field playing the role of a so-called guiding field. Moreover, non-southward magnetospheric fields can merge with the magnetospheric field anywhere along the magnetopause where the fields become opposite to each other. A northward magnetosheath field component may merge in this way with magnetospheric field lines on the magnetopause at latitudes higher than the latitude of the polar cusp and may lead to displacements and motions of the polar cusp as shown in Fig. 8.20. Also, a normal Archimedian spiral interplanetary magnetic field, with no component perpendicular to the ecliptic plane, may merge with parts of the high-latitude magnetic field diverging from the cusp (see Fig. 8.21) thereby causing asymmetric reconnection at the magnetopause.

All these types of reconnection may happen simultaneously at different places of the magnetopause, where the magnetopause looses its global character of a tangential discontinuity and becomes a surface, which in many places is perforated and magnetically connected to interplanetary space. This allows plasma to flow into the magneto-sphere and feeds energy into the magnetosphere and ultimately causes its various violent variations like the magnetospheric substorms described in Sec. 5.6.

Concluding Remarks

It should be noted that the general Rankine-Hugoniot conditions obtained in this chapter are of more far reaching importance than claimed so far. They are valid in a plasma as long as the typical scales of interest are much larger than the width of the discontinuity. In such a case one divides the plasma into the large-scale region outside the discontinuity, where the jump conditions hold, and into its interior, where the more complicated dissipative processes take place. There may, however, exist cases when the dissipative processes inside the discontinuity affect the behavior of the plasma outside the transition region so strongly that even the outside region cannot be considered as ideal. Then the Rankine-Hugoniot conditions hold only approximately, and the discontinuity treatment has to be based on a more precise many-fluid or even kinetic theory.

Kinetic theory must also be applied to investigate the real interior structure of discontinuities and shocks and the conditions under which they may develop. These questions have been ignored in the present chapter, but will be returned to in a later section.

Further Reading

The classical treatment of the ideal magnetohydrodynamic jump conditions can be found in [2]. A thorough description of the equations of state is given in the tutorial article [3]. A comprehensive tutorial of the gasdynamic theory of the flow around the magnetosphere has been developed in [5]. Figure 8.18 is based on calculations presented in that publication. A lot of useful information about magnetic reconnection can be found in [1], and the physics of the magnetopause is exhaustively treated in [4]. Finally, a good introduction on bow shock physics is given in [6].

[1] E. W. Hones, Jr (ed.), *Magnetic Reconnection in Space and Laboratory Plasmas* (American Geophysical Union, Washington, 1984).

[2] L. D. Landau and E. M. Lifshitz, *Electrodynamics of Continuous Media* (Pergamon Press, Oxford, 1975).

[3] G. L. Siscoe, G. L., in *Solar-Terrestrial Physics*, eds. R. L. Carovillano and J. M. Forbes (D. Reidel Publ. Co., Dordrecht, 1983), p. 11.

[4] B. U. Ö Sonnerup, M. Thomson, and P. Song (eds.), *Physics of the Magnetopause* (American Geophysical Union, Washington, 1995).

[5] J. R. Spreiter, A. Y. Alskne, and A. L. Summers, in *Physics of the Magnetosphere*, eds. R. L. Carovillano, J. F. McClay, and H. R. Radoski (D. Reidel Publ. Co., Dordrecht, 1968), p. 301.

[6] R. G. Stone and B. T. Tsurutani (eds.), *Collisionless Shocks in the Heliosphere: A Tutorial Review* (American Geophysical Union, Washington, 1985).

9. Waves in Plasma Fluids

In a plasma there are many reasons for the evolution of time-dependent effects. The high temperatures required to produce a plasma imply that the plasma particles are in fast motion. Such motions generate microscopic charge separations and currents and therefore temporally changing electric and magnetic fields. Hence, it is quite natural to expect that electric and magnetic fluctuations are typical for a plasma, even in its stationary state. Absolutely quiescent plasmas do not exist. Just due to the thermal motion of the particles in a plasma every plasma in equilibrium contains a certain level of fluctuations, which depends entirely on the temperature of the plasma and is therefore called *thermal fluctuation level*. The thermal spectrum of a plasma can be calculated as the balance between the generation of the thermal fluctuations and the reabsorption and dissipation of these fluctuations, but calculations of this kind require quantum theoretical methods which are outside the scope of this book.

In addition to these unavoidable fluctuations, any plasma will react to a violent distortion of its state imposed by outer means. All such disturbances may be thought of as a superposition of linear waves onto the quiescent plasma state which propagate across the plasma in order to transport the energy of the distortion and to communicate it to the entire plasma volume. Such plasma waves have been measured in many different frequency ranges. Figure 9.1 indicates that their frequencies may be as low as several Millihertz and as high as several tens of Kilohertz. Conventionally, this range is subdivided into ultra-low (ULF), extremely-low (ELF) and very-low frequency (VLF) waves.

But plasma waves are not generated at random. In order to exist, any disturbance must satisfy at least two conditions. First, it must be a solution of the appropriate equations of the plasma. Therefore the number of modes propagating in the plasma will not be continuous but discrete. Secondly, we can speak of a wave only if its amplitude exceeds the level of the thermal fluctuations always present in a plasma. The second condition sets a limit on the initial disturbance causing the waves. If it has an amplitude lower than the thermal noise level and if no mechanism acts to amplify the disturbance in the plasma, this disturbance does not affect the plasma and there is no wave.

In the present and the following chapter we investigate the consequences of the first condition, i.e., we consider the discrete modes which can propagate in a plasma. Since there are several different plasma models available, the number and properties of the

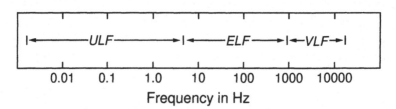

Fig. 9.1. Ranges of ultra-low, extremely-low, and very-low frequency waves.

wave modes depends on the chosen approximation to the kinetic plasma model. The present chapter investigates wave propagation in plasma fluids. In neglecting the second condition we automatically assume that the thermal noise level is much smaller than the wave amplitude. Hence, the plasma is assumed to be sufficiently cold. On the other hand, we will not deal with nonlinear effects in this chapter. Therefore the wave amplitudes are assumed to be small enough to allow any disturbance to be represented as superposition of plane waves.

With these remarks in mind, we can represent any wave disturbance, $A(x, t)$, in the plasma by plane waves, i.e., by its Fourier components. If the disturbance itself is a plane wave, it consists only of one Fourier component

$$A(x, t) = A(k, \omega) \exp(i k \cdot x - i\omega t) \qquad (9.1)$$

where the amplitude, $A(k, \omega)$, is a function of the wave vector, k, and the frequency, ω. This representation allows to define the phase and the group velocity of the wave

$$v_{ph} = \omega k / k^2 \qquad (9.2)$$

$$v_{gr} = \partial\omega/\partial k \qquad (9.3)$$

The phase velocity is always parallel to the wave vector, k, and shows the direction of wave propagation. The group velocity may deviate from this direction and describes the speed and direction of the energy flow in the wave.

9.1. Waves in Unmagnetized Fluids

As a first example and to introduce the concept of plasma waves let us consider an unmagnetized plasma consisting of equal numbers of electrons and ions. Two kinds of waves can propagate in such a plasma. The first kind are electromagnetic waves similar to waves in vacuum. Due to the presence of charges which respond to the electric and magnetic field of the waves, the properties of these waves will be modified. The second kind of waves are internal plasma oscillations, which do not exist in the vacuum, but are a specific property of the plasma. We will treat the second type of waves first.

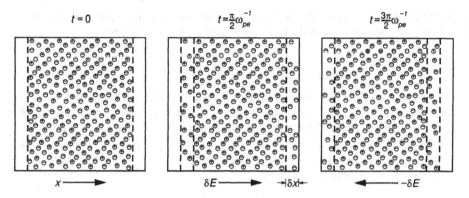

Fig. 9.2. Oscillation of a column of electrons at the plasma frequency.

Langmuir Oscillations

Consider a plasma where the ions are fixed while the electrons may undergo small trans-
lations relative to the ions. Such an assumption is reasonable if the timescale of the electron translation is so short that the ions cannot follow the electron motion because of their
large inertia. In other words we consider high-frequency electron oscillations in which
the ions do not participate.

Now take a column of electrons and displace this column with respect to the ions by
a short distance, δx, in the x direction (see Fig. 9.2). Such a displacement causes an electric field, δE, also pointing in x direction and exerting a force, $-e\delta E$, on each electron
which tries to pull the electron back to its mother ion in order to preserve quasineutrality. For the whole column of density n_e this means that the time variation of the density
distortion, δn, will be given by the electron fluid continuity equation

$$\frac{\partial \delta n}{\partial t} = -n_e \frac{\partial \delta v_{e,x}}{\partial x} \tag{9.4}$$

as the spatial derivative of the electron velocity disturbance, $\delta v_{e,x}$. The distortion of the
velocity is found from the electron momentum conservation as

$$\frac{\partial \delta v_{e,x}}{\partial t} = -\frac{e}{m_e}\delta E \tag{9.5}$$

and the electric field caused by all the displaced electrons satisfies Poisson's law

$$\frac{\partial \delta E}{\partial x} = -\frac{e}{\epsilon_0}\delta n \tag{9.6}$$

It is now simple to derive an equation for the disturbance of the density. Take the time
derivative of the first of the above equations, replace the time derivative of the velocity

disturbance in the resulting expression by the second of the above equation, and eliminate the spatial derivative with the help of the third equation. The result is

$$\frac{\partial^2 \delta n}{\partial t^2} + \frac{n_e e^2}{m_e \epsilon_0} \delta n = 0 \tag{9.7}$$

This is a linear equation for the variation of the density, which has the form of a linear oscillator equation. Clearly, the coefficient of the second term must have the dimension of an inverse time squared. This time is proportional to the characteristic period of the oscillation of the electron column around the equilibrium position of the ion column. The solution of the above equation is found by taking $\delta n \propto \exp(-i\omega t)$, where $\omega = \omega_{pe}$ is the angular frequency of the oscillation

$$\omega_{pe}^2 = \frac{n_e e^2}{m_e \epsilon_0} \tag{9.8}$$

Hence, the electrons will perform an oscillation around the position of the ions with the electron *plasma frequency*, ω_{pe}, already given in Sec. 1.1.

Langmuir Waves

The plasma oscillation is somewhat artificial, since the electrons are not at rest but have different velocities and will react differently to the attempt to displace them from there instantaneous positions. To account for this effect one must introduce the adiabatic variation of the electron thermal pressure, $\delta p_e = \gamma_e k_B T_e \delta n_e$, into the electron momentum conservation equation. Let us, for simplicity, assume that the electron temperature is constant. Then the linearized equation of motion of the displaced electron fluid

$$\frac{\partial \delta v_{e,x}}{\partial t} = -\frac{e}{m_e} \delta E - \frac{\gamma_e k_B T_e}{m_e n_e} \frac{\partial \delta n}{\partial x} \tag{9.9}$$

replaces Eq. (9.5). Eliminating once more δE and $\delta v_{e,x}$ now yields another more precise equation for the variation of density

$$\frac{\partial^2 \delta n}{\partial t^2} - \frac{\gamma_e k_B T_e}{m_e} \frac{\partial^2 \delta n}{\partial x^2} + \omega_{pe}^2 \delta n = 0 \tag{9.10}$$

This equation differs from the former one in the appearance of the second partial derivative with respect to x. Therefore it is of the form of a wave equation and can be solved by introducing the plane wave ansatz for the variation of the electron density, $\delta n \propto \exp(-i\omega t + ikx)$, into Eq. (9.10), which yields a relation between the angular frequency, ω, and the wavenumber, k

$$\omega_l^2 = \omega_{pe}^2 + k^2 \gamma_e v_{the}^2 \tag{9.11}$$

where we used the electron thermal velocity, $v_{the} = (k_B T_e/m_e)^{1/2}$ defined in Eq. (6.57). This is the *Langmuir dispersion relation*. It determines the dependence of the frequency of the Langmuir waves on wavenumber. The interesting point is that the thermal motion of the electrons leads to a dispersion of the electron plasma oscillations by introducing a wavenumber dependence into the wave frequency. This dependence drops out only if the electrons have zero temperature or for zero wavenumber, $k = 0$. In both cases one recovers the plasma oscillations. However, for finite temperatures or $k \neq 0$ the oscillations start propagating across the plasma and turn into travelling electrostatic waves which are oscillations of the electric field propagating through the plasma. The limit of vanishing wavenumber is of particular interest. Because k is inversely related to the wavelength, $\lambda = 2\pi/k$, the wavenumber becomes zero for infinitely long waves. Langmuir oscillations are thus Langmuir waves of very long wavelength.

Ion-Acoustic Waves

So far we have neglected the contribution of the ions and have considered very-high frequency electron oscillations. At lower frequencies the ion motion comes into play and it becomes necessary to take into account the ion equation of motion, in addition to the electron equations. On the other hand, in a first approach electron inertia can be safely neglected because the ion plasma frequency

$$\omega_{pi} = \left(\frac{n_i Z^2 e^2}{m_i \epsilon_0}\right)^{1/2} \tag{9.12}$$

is, for protons with $Z = 1$ and quasineutrality, $n_i \approx n_e$, by a factor of $(m_e/m_i)^{1/2} = 43$ smaller than ω_{pe}. At such low frequencies the electrons react almost without any inertia to the change in the electric field. Under this condition electron dynamics reduces to a simple balance between electron pressure and electric force (note that $\partial n_0/\partial x = 0$)

$$e\delta E = -\gamma_e k_B T_e \frac{\partial \ln n_e}{\partial x} \tag{9.13}$$

where $n_e = n_0 + \delta n_e$. When introducing the electric potential, $\delta E = -\partial \delta\phi/\partial x$, the above equation reduces to a Boltzmann-like dependence

$$n_e = n_0 \exp\left(\frac{e\delta\phi}{\gamma_e k_B T_e}\right) \tag{9.14}$$

of the electron density on the electric potential. The linearized version of this equation

$$\frac{\delta n_e}{n_0} = \frac{e\delta\phi}{\gamma_e k_B T_e} \tag{9.15}$$

describes the linear electron response to the low-frequency wave potential oscillation. Adding to it the linearized ion equations

$$\frac{\partial \delta n_i}{\partial t} = -n_i \frac{\partial \delta v_{i,x}}{\partial x}$$

$$\frac{\partial \delta v_{i,x}}{\partial t} = \frac{e}{m_i} \delta E \qquad\qquad (9.16)$$

where we have neglected the ion pressure term, supposing that the ions are much colder than the electrons, and assumed charge neutrality, $\delta n_e = \delta n_i = \delta n$, one arrives at

$$\frac{\partial^2 \delta n}{\partial t^2} - \frac{\gamma_e k_B T_e}{m_i} \frac{\partial^2 \delta n}{\partial x^2} = 0 \qquad\qquad (9.17)$$

as the ionic equivalent of Eq. (9.10). For plane waves its solution yields

$$\omega_{ia}^2 = \frac{\gamma_e k_B T_e}{m_i} k^2 \qquad\qquad (9.18)$$

which is the dispersion relation of *ion-acoustic waves*. These waves are called acoustic, since they have the same properties as sound waves in a gaseous medium. Both waves have linear dispersion, $\omega \propto k$, and are pure density fluctuations.

Dividing both sides of Eq. (9.18) by k^2, one finds the phase velocity of ion-acoustic waves as $v_{ph,ia} = c_{ia}$, where

$$c_{ia} = \left(\frac{\gamma_e k_B T_e}{m_i} \right)^{1/2} \qquad\qquad (9.19)$$

is the *ion-acoustic speed*. The latter is given by the square root of the ratio of electron temperature and ion mass and, for protons, is a factor of 43 smaller than the electron thermal velocity. The linear dispersion of ion-acoustic waves implies also that their group velocity is equal to the phase velocity, $v_{gr,ia} = c_{ia}$.

In the above derivation of ion-acoustic waves we have neglected the contribution of ion pressure. Correcting for this imprecision requires the replacement of $\gamma_e T_e$ in the above expressions by the sum of the electron and ion contributions, $\gamma_e T_e + \gamma_i T_i$. Hence, for high ion temperatures the ion sound speed becomes the ion thermal velocity, and the contribution of the electrons to sound waves is lost.

The second approximation used above was the assumption of quasineutrality even for the fluctuating quantities. At higher frequencies close to ω_{pi} this assumption is incorrect because the electron and ion motions in the wave field become uncorrelated. We

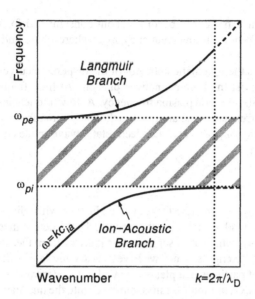

Fig. 9.3. Dispersion of Langmuir and ion-acoustic waves.

therefore replace the condition $\delta n_e = \delta n_i$ with Poisson's equation

$$\frac{\partial^2 \delta\phi}{\partial x^2} = \frac{en_0}{\epsilon_0}\left(\frac{\delta n_e}{n_0} - \frac{\delta n_i}{n_0}\right) \tag{9.20}$$

where we assumed quasineutrality of the undisturbed state, $n_e = n_i = n_0$. The electron and ion equations (9.15) and (9.16) can now be used to manipulate Poisson's equation (9.20) into the following form

$$\left(\frac{\partial^2}{\partial t^2} - c_{ia}^2 \frac{\partial^2}{\partial x^2}\right)\delta\phi = \frac{c_{ia}^2}{\omega_{pi}^2}\frac{\partial^4 \delta\phi}{\partial t^2 \partial x^2} \tag{9.21}$$

This is again a linear equation for the electric potential of the wave, $\delta\phi$, and one can apply the plane wave ansatz to obtain the more precise dispersion relation

$$\omega_{ia}^2 = \frac{k^2 c_{ia}^2}{1 + k^2 c_{ia}^2/\omega_{pi}^2} \tag{9.22}$$

This expression shows that ω becomes a linear function of k only for long wavelengths or small k. Here the wave has the character of a sound wave. But for short wavelengths comparable to the Debye length introduced in Eq. (1.3) the character of the sound wave is destroyed. The wave frequency becomes about constant and for very short wavelengths approaches the ion plasma frequency, ω_{pi}, where phase and group velocities both vanish.

Figure 9.3 shows the schematic behavior of the dispersion curves of the two electrostatic waves that exist in an unmagnetized plasma. At high frequencies the Langmuir branch starts at the electron plasma frequency. At low frequencies the ion-acoustic branch starts at zero frequency and approaches the ion plasma frequency. Between the two plasma frequencies, ω_{pi} and ω_{pe}, no electrostatic wave mode can propagate in an unmagnetized plasma.

Debye Length

To our surprise we have encountered an old acquaintance when discussing the dispersion relations for ion sound waves, the Debye length. That it determines the properties of the fundamental electrostatic waves suggests that it arises from thermal charge separation effects at short wavelengths. Since we have now accumulated sufficient knowledge about the dynamics of particles in a plasma we will give a derivation of this quantity.

Assume that a heavy, motionless ion is immersed into the quasineutral plasma. This ion will cause an electric charge separation field to arise in its vicinity that will attract electrons to charge-neutralize the ion. Because the electrons are highly mobile, they will be accelerated toward the ion, pass around the ion, and subsequently escape into the ambient plasma, but in the average there will be more electrons near the ion than outside at large distances. This poses the question of up to what distance the electron density will be slightly distorted due to the presence of the ion.

In the region where the density is distorted, charge neutrality becomes violated and a non-vanishing electric potential, $\phi(r)$, will arise which must satisfy Poisson's equation for an electron-proton plasma

$$\nabla^2 \phi = -\frac{e}{\epsilon_0}(n_i - n_e) \tag{9.23}$$

The ion density is the quasineutral density of the ambient plasma, $n_i = n_0$, while the electron density includes the distortion by the presence of the test ion. For an equilibrium between electron thermal motion and electric force the electrons are Maxwellian and their density in obeys Boltzmann's law

$$n_e(r) = n_0 \exp\left[\frac{e\phi(r)}{k_B T_e}\right] \tag{9.24}$$

For weak potentials, $|e\phi| \ll k_B T_e$, this expression can be expanded into a Taylor series and inserted into Poisson's equation to obtain

$$\nabla^2 \phi = \frac{e^2 n_0 \phi}{\epsilon_0 k_B T_e} \tag{9.25}$$

The problem will be spherically symmetric with radius r centered on the position of the ion, and the potential, $\phi(r)$, must diverge as $1/r$ for $r \to 0$. Dimensionally the left-hand side of the last equation is equal to the electrostatic potential divided by a characteristic length squared, ϕ/λ_D^2. Comparing the two sides of the equation yields for this length

$$\lambda_D = \left(\frac{\epsilon_0 k_B T_e}{n_0 e^2} \right)^{1/2} \tag{9.26}$$

which is the *Debye length* postulated in Sec. 1.1. The Debye length is thus the typical average screening distance of the electrostatic field of an ion charge in a quasineutral plasma by electrons of temperature T_e. It indicates that around each ion there is a cloud of excess electrons which screen the ion field from the plasma. The sphere of radius λ_D around the ion is called the *Debye sphere*, and the number of electrons within this sphere, roughly $\frac{4}{3}\pi n_0 \lambda_D^3$, is the *Debye number*, or approximately the plasma parameter. Inside the Debye sphere the potential of the ion is not screened. As a consequence, on distances λ_D in a plasma quasineutrality is distorted. Over these distances the electrostatic field experiences relatively strong fluctuations, which are the reason for the change in the dispersive properties of the waves at wavelengths comparable to the Debye length. Such short wavelength waves contain contributions from the unscreened electric fluctuations caused by the electron motions across the Debye sphere around each ion in the plasma.

The electron motions in the Debye sphere are thermal and do always exist. In thermal equilibrium between electrons and electrostatic fluctuations, the energy of the fluctuations is equal to the mean electron energy, $k_B T_e$, and the energy density of the fluctuations is this energy divided by the volume of the Debye sphere. Hence, the energy density contained in the thermal fluctuation is equivalent to the thermal energy of one electron per Debye sphere

$$W_{\text{tf}} \approx \frac{k_B T_e}{\lambda_D^3} \tag{9.27}$$

We have so far neglected the screening of electron charges by the ions. If one takes into account this often minor effect, the effective Debye length becomes

$$\lambda_{D,\text{eff}}^{-2} = \lambda_D^{-2} + \lambda_{Di}^{-2} \tag{9.28}$$

where

$$\lambda_{Di} = \left(\frac{\epsilon_0 k_B T_i}{n_0 e^2} \right)^{1/2} \tag{9.29}$$

is the ion Debye length. In an isothermal plasma with similar electron and ion temperatures both Debye lengths contribute equally to the effective Debye length.

The Debye length can be written as the ratio of electron thermal velocity to electron plasma frequency, $\lambda_D = v_{the}/\omega_{pe}$. This allows to write the dispersion relation of Langmuir waves in the form

$$\boxed{\omega_l^2 = \omega_{pe}^2 \left(1 + \gamma_e k^2 \lambda_D^2\right)} \tag{9.30}$$

Ordinary Electromagnetic Waves

The appearance of purely electrostatic disturbances which propagate as wave modes is a very particular property of plasmas, which is connected with the presence of free charges in a plasma. Moving charges can, however, contribute also to oscillating plasma currents which should become sources of electromagnetic waves. A large number of such electromagnetic wave modes may propagate in a magnetized plasma. In this introductory section, where we deal with an unmagnetized plasma, we will only consider the most familiar electromagnetic mode, the free-space electromagnetic wave.

An electromagnetic wave of frequency ω in the plasma will set the electrons into motion to generate a linear electron current

$$\delta \mathbf{j}_{em} = -en_0 \delta \mathbf{v}_e \tag{9.31}$$

Only the disturbance of the electron velocity contributes to the current, since the plasma was initially at rest. This disturbance can be calculated from the electron equation of motion in the electromagnetic plane wave field, $\delta \mathbf{E}$

$$\delta \mathbf{v}_e = -\frac{ie}{\omega m_e} \delta \mathbf{E} \tag{9.32}$$

Inserting this into the above expression for the current, one finds that the induced current $\delta \mathbf{j}_{em} = \sigma_{em} \delta \mathbf{E}$ is proportional to the wave electric field in direct analogy to Ohm's law. The constant of proportionality is the electromagnetic wave conductivity

$$\boxed{\sigma_{em} = \frac{i \epsilon_0 \omega_{pe}^2}{\omega}} \tag{9.33}$$

It depends on the wave frequency and on the electron plasma frequency and is an imaginary quantity. It is zero for very-high frequencies and vanishes in the absence of a plasma. In both cases the electromagnetic wave will become an ordinary electromagnetic free-space wave. The dispersion relation of a free-space electromagnetic wave is

$$N^2 = \frac{k^2 c^2}{\omega^2} \tag{9.34}$$

where N is the *refraction index*. In a vacuum $N^2 = 1$. In an unmagnetized plasma, one may replace it by the dielectric function $\epsilon(\omega, \mathbf{k})$ of the plasma to obtain the dispersion relation of the electromagnetic wave as

$$\frac{k^2 c^2}{\omega^2} = \epsilon(\omega, \mathbf{k}) \tag{9.35}$$

We will show in the next section that there is a unique relation between $\epsilon(\omega, \mathbf{k})$ and the wave conductivity, $\sigma(\omega, \mathbf{k})$, which in our special case can be written as

$$\epsilon(\omega) = 1 + \frac{i\sigma_{em}(\omega)}{\epsilon_0 \omega} = 1 - \frac{\omega_{pe}^2}{\omega^2} \tag{9.36}$$

With the help of this expression, the dispersion relation of the free-space electromagnetic wave in the presence of a plasma becomes

$$\boxed{\omega_{om}^2 = \omega_{pe}^2 + c^2 k^2} \tag{9.37}$$

The wave described by this dispersion relation is called the *ordinary mode*, because it has the same dispersion as the free-space wave when the plasma is discarded. The main difference to the free-space wave is that for frequencies below the electron plasma frequency there is no real solution for the wavenumber, and the wave ceases to exist. In other words, the wavenumber of the ordinary mode vanishes at the plasma frequency, ω_{pe}, which is a *cut-off* for the ordinary mode.

A cut-off is a point in the dispersion diagram as well as in real space, where the wavenumber turns zero and, hence, the direction of propagation of the wave reverses. Cut-offs are *wave reflection* points which are related simply to the refractive properties of the plasma and wave propagation across the plasma. The practical implication of the ordinary mode cut-off is that in an unmagnetized plasma electromagnetic waves cannot propagate below ω_{pe}.

This reflection has important practical consequences. It is responsible for the trapping of long wavelength radio waves with frequencies below the plasma frequency of the dense ionosphere between the Earth's surface and the ionosphere. Moreover, it is responsible for the screening of low-frequency radio waves generated in the topside auroral ionosphere from reaching the ground. This radiation, the *auroral kilometric radiation*, and the radio emission from the radiation belts, the *trapped radiation*, has frequencies well below the ionospheric plasma frequency and is reflected from the topside ionosphere so that it becomes unobservable from the ground while cosmic high-frequency radio waves, for instance solar radio bursts, can propagate without any serious attenuation down to the Earth's surface. On the other hand, launching low-frequency radio waves from the Earth into the ionosphere and recording the reflected wave provides a tool for sounding the ionospheric density profile. This technique is used in ionosondes. Figure 9.4 summarizes the reflection of radio waves at the ionosphere.

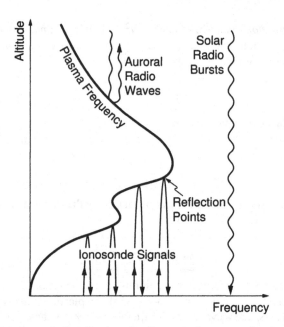

Fig. 9.4. Ionospheric reflection of radio waves from ground and space.

9.2. General Dispersion Relation

The examples of the previous section have shown that waves can propagate in a plasma. The three simplest wave modes in an unmagnetized plasma have been identified as the electrostatic Langmuir and ion-acoustic waves and the electromagnetic ordinary wave. However, the presence of a magnetic field introduces a large variety of other possible wave modes. On the other hand, the complexity of the interactions between the many particles and fields in the plasma allows only for a finite number of waves to propagate. These waves are the linear eigenmodes of the plasma. To determine their possible branches one can develop a general procedure based on the so-called *general dispersion relation* of a plasma.

General Wave Equation

Let us write Maxwell's equation taking into account a selfconsistent current, $\mathbf{j}$, which consists of the various contributions of the moving plasma particles. Similarly, the self-consistent plasma charge density in the plasma is ρ. In addition to these selfconsistent sources of the field, there may be external currents and charges, $\mathbf{j}_{ex}$, ρ_{ex}, which are the sources of electromagnetic fields applied externally to the plasma. Then

$$\nabla \times \mathbf{B} - \epsilon_0 \mu_0 \frac{\partial \mathbf{E}}{\partial t} + \mu_0 (\mathbf{j} + \mathbf{j}_{ex}) \tag{9.38}$$

$$\nabla \times \mathbf{E} = -\frac{\partial \mathbf{B}}{\partial t} \tag{9.39}$$

$$\nabla \cdot \mathbf{B} = 0 \tag{9.40}$$

$$\nabla \cdot \mathbf{E} = \frac{1}{\epsilon_0} (\rho + \rho_{ex}) \tag{9.41}$$

Taking the derivative of the first of these equations with respect to time and eliminating the magnetic field with the help of the second equation, one finds an inhomogeneous wave equation for the electric field

$$\nabla^2 \mathbf{E} - \nabla (\nabla \cdot \mathbf{E}) - \epsilon_0 \mu_0 \frac{\partial^2 \mathbf{E}}{\partial t^2} = \mu_0 \left(\frac{\partial \mathbf{j}}{\partial t} + \frac{\partial \mathbf{j}_{ex}}{\partial t} \right) \tag{9.42}$$

We have explicitly included external charges and currents which may be imposed into the plasma by outer means. But because Maxwell's equations are linear in the charges and fields, their contribution can be added subsequently. As the magnetic field can be eliminated from the internal current, $\mathbf{j}$, with the help of Maxwell's equations, it is sufficient to retain only the dependence of the current on the electric field as

$$\mathbf{j} = \int d^3 x' \int_{-\infty}^{t} dt' \sigma(\mathbf{x}, \mathbf{x}', t, t', \mathbf{E}) \cdot \mathbf{E} \tag{9.43}$$

This dependence is not necessarily linear, but can be quite complicated. However, for small perturbations of the fields and plasma properties it can be approximated by a linear relation between the current and the field corresponding to a time-varying Ohm's law

$$\mathbf{j} = \int d^3 x' \int_{-\infty}^{t} dt' \sigma(\mathbf{x}, \mathbf{x}', t, t') \cdot \mathbf{E} \tag{9.44}$$

The integration from $-\infty$ to t contains the concept of causality, where the history of the plasma contributes to its response at time t, while the future behavior is determined by the solution of Maxwell's equations.

The condition (9.44) closes the above system of equations, if the tensor σ is assumed to be known. This tensor contains all of the properties of the plasma. Since the wave current, $\mathbf{j}$, is the sum over all particle motions generated by the wave disturbance, it is clear that the process of finding the wave conductivity, σ, involves the plasma dynamics and thus depends on the choice of the plasma model. However, assuming that the plasma responds linearly to the presence of the wave disturbance, the wave conductivity becomes independent of the wave amplitude. Hence, the general linear dispersion relation can be derived by using a form of the wave conductivity that depends on relative position and time only, $\sigma(\mathbf{x} - \mathbf{x}_0, t - t_0)$. Under these conditions one may linearize the general wave equation (9.42) with $\mathbf{E}_0 = 0$ or $\mathbf{E}(t, \mathbf{x}) = \delta\mathbf{E}(t, \mathbf{x})$ to obtain

$$\nabla^2 \delta\mathbf{E} - \nabla(\nabla \cdot \delta\mathbf{E}) - \epsilon_0 \mu_0 \frac{\partial^2 \delta\mathbf{E}}{\partial t^2} = \mu_0 \frac{\partial \mathbf{j}}{\partial t} \qquad (9.45)$$

with Ohm's law given by

$$\mathbf{j}(t, \mathbf{x}) = \int d^3x' \int_{-\infty}^{t} dt' \sigma(\mathbf{x} - \mathbf{x}', t - t') \cdot \delta\mathbf{E} \qquad (9.46)$$

Equation 9.45 is the general linear wave equation. It is applicable to any medium with a linear response to an applied field fluctuation. Its left-hand side represents the purely electromagnetic part independent on the presence of any medium. The response of the medium is entirely included in the fluctuating current term on the right-hand side and, since this current is proportional to the fluctuating field, is contained only in the fluctuating conductivity tensor, σ.

General Wave Dispersion Relation

Interpreting the electric field fluctuations according to Eq. (9.1) as plane waves

$$\delta\mathbf{E}(\omega, \mathbf{k}) = \delta\mathbf{E}_0(\omega, \mathbf{k}) \exp(i\mathbf{k} \cdot \mathbf{x} - i\omega t) \qquad (9.47)$$

the dependence of the conductivity tensor on relative spatial and temporal distances turns the integral on the right-hand side of Eq. (9.46) into a folding integral. This fact considerably simplifies the solution of Eq. (9.45) and reduces it to

$$\left[\left(k^2 - \frac{\omega^2}{c^2} \right) \mathbf{I} - \mathbf{kk} - i\omega\mu_0 \sigma(\omega, \mathbf{k}) \right] \cdot \delta\mathbf{E}_0(\omega, \mathbf{k}) = 0 \qquad (9.48)$$

for the constant wave amplitude, $\delta\mathbf{E}_0(\omega, \mathbf{k})$. Hereby, the field and conductivity satisfy the symmetry conditions

$$\begin{aligned} \delta\mathbf{E}^*(\mathbf{k}, \omega) &= \delta\mathbf{E}(-\mathbf{k}, -\omega) \\ \sigma^*(\omega, \mathbf{k}) &= \sigma(-\omega, -\mathbf{k}) \end{aligned} \qquad (9.49)$$

which result from the requirement of real wave field amplitudes. Equation (9.48) is a dyadic (or tensor) equation. Its nontrivial solution requires that the determinant of the expression in brackets vanishes, thereby yielding the general dispersion relation

$$D(\omega, \mathbf{k}) = \text{Det}\left[\left(k^2 - \frac{\omega^2}{c^2}\right)\mathbf{I} - \mathbf{kk} - i\omega\mu_0\sigma(\omega, \mathbf{k})\right] = 0 \qquad (9.50)$$

Sometimes it is more convenient to include the current in Ampère's law into the electric field term on the left-hand side by defining the electric induction

$$\delta\mathbf{D} = \epsilon \cdot \delta\mathbf{E} \qquad (9.51)$$

where ϵ is the *dielectric tensor*. It satisfies the symmetry relations of Eq. (9.49)

$$\epsilon^*(\omega, \mathbf{k}) = \epsilon(-\omega, -\mathbf{k}) \qquad (9.52)$$

With its help the current density assumes the representation

$$\delta\mathbf{j}(\omega, \mathbf{k}) = -i\omega\epsilon_0\left[\epsilon(\omega, \mathbf{k}) - \mathbf{I}\right] \cdot \delta\mathbf{E}(\omega, \mathbf{k}) \qquad (9.53)$$

Using this version of Ohm's law to replace the current by the wave electric field, the dielectric tensor is defined as

$$\boxed{\epsilon(\omega, \mathbf{k}) = \mathbf{I} + \frac{i}{\omega\epsilon_0}\sigma(\omega, \mathbf{k})} \qquad (9.54)$$

This definition allows to rewrite the dispersion relation (9.50) as

$$\boxed{\text{Det}\left[\frac{k^2 c^2}{\omega^2}\left(\frac{\mathbf{kk}}{k^2} - \mathbf{I}\right) + \epsilon(\omega, \mathbf{k})\right] = 0} \qquad (9.55)$$

Equation (9.55) is the general dispersion relation of any active medium. Its solutions describe propagating linear waves of frequency $\omega = \omega(\mathbf{k})$. Since it is an eigenvalue equation, it has only a finite number of discrete solutions. To determine these solutions, one must first find the plasma dielectric tensor, which itself implies solving the linear dynamic plasma equations. This step depends on the choice of the plasma model.

Isotropic Plasma

The dielectric tensor, $\epsilon(\omega, \mathbf{k})$, contains all the relevant linear properties of the plasma. In general it is an anisotropic tensor. However, in the absence of an external magnetic field, when the plasma is unmagnetized, it becomes isotropic. In this case the dispersion relation can be simplified, since the only particular direction in a plasma is then given by the direction of the wave vector, **k**. It allows to construct a tensor

$$\mathsf{I}_\mathsf{L} = \frac{\mathbf{kk}}{k^2} \tag{9.56}$$

which is the *longitudinal unit tensor* prescribing the direction of the electrostatic fluctuations. Subtracting it from the unit tensor, **I**, we obtain the transverse unit tensor

$$\mathsf{I}_\mathsf{T} = \mathsf{I} - \frac{\mathbf{kk}}{k^2} \tag{9.57}$$

which in contrast accounts only for the electromagnetic directions of propagation. These two tensors permit to decompose the dielectric tensor into longitudinal and transverse components

$$\epsilon(\omega, \mathbf{k}) = \epsilon_\mathsf{L}(\omega, k)\mathsf{I}_\mathsf{L} + \epsilon_\mathsf{T}(\omega, k)\mathsf{I}_\mathsf{T} \tag{9.58}$$

In this representation the coefficients of the longitudinal and transverse unit tensors are scalar functions, which depend only on frequency and wavenumber, but not on the direction of the latter. Knowing the isotropic dielectric tensor of the plasma, these two dielectric functions can be calculated by multiplying ϵ from both sides with **k**

$$
\begin{aligned}
\epsilon_\mathsf{L}(\omega, k) &= \frac{\mathbf{k} \cdot \epsilon(\omega, \mathbf{k}) \cdot \mathbf{k}}{k^2} \\
\epsilon_\mathsf{T}(\omega, k) &= \frac{\mathrm{tr}\epsilon(\omega, \mathbf{k}) - \epsilon_\mathsf{L}(\omega, k)}{2}
\end{aligned}
\tag{9.59}
$$

The symbol 'tr' denotes the trace (of the dielectric tensor). In an isotropic plasma the dispersion tensor becomes very simple

$$\mathrm{Det}\left[\epsilon_\mathsf{L}(\omega, k)\mathsf{I}_\mathsf{L} + \left(\epsilon_\mathsf{T}(\omega, k) - \frac{k^2 c^2}{\omega^2}\right)\mathsf{I}_\mathsf{T}\right] = 0 \tag{9.60}$$

Obviously, the two tensors are linearly independent, and the two dispersion relations of an isotropic plasma reduce to the two decoupled scalar equations

$$
\begin{aligned}
\epsilon_\mathsf{L}(\omega, k) &= 0 \\
\epsilon_\mathsf{T}(\omega, k) - \frac{k^2 c^2}{\omega^2} &= 0
\end{aligned}
\tag{9.61}
$$

for the longitudinal electrostatic and transverse electromagnetic waves which can propagate in the plasma.

Dielectric Response Function

There is a close relation between the longitudinal dielectric function in Eq. (9.61) and the dielectric of an anisotropic plasma, the *dielectric response function*

$$\epsilon(\omega, \mathbf{k}) = \frac{\mathbf{k} \cdot \epsilon(\omega, \mathbf{k}) \cdot \mathbf{k}}{k^2} \qquad (9.62)$$

The former describes longitudinal waves in an unmagnetized plasma and is independent of the direction of the wave vector, while the latter governs the linear response of a plasma to a disturbance in density or electric field and depends on the full wave vector. In this sense the latter function is not restricted to an isotropic plasma but applies to a much wider range of plasma models.

To derive Eq. (9.62), we take advantage of the fact that external charges or currents can be added to the linear electromagnetic equations. Consider an external disturbance ρ_{ex} of the electric space charge in the plasma, for instance a number of test particles added to the plasma from outside. The total disturbance of the charge written in $(\omega, \mathbf{k})$ space is the sum of both linear disturbances

$$\rho_{tot}(\omega, \mathbf{k}) = \rho_{ex}(\omega, \mathbf{k}) + \rho(\omega, \mathbf{k}) \qquad (9.63)$$

where $\rho_{tot}(\omega, \mathbf{k})$ as the total disturbance of the charge density is defined to be related to the disturbance $\rho_{ex}(\omega, \mathbf{k})$ via the response function

$$\rho_{tot}(\omega, \mathbf{k}) = \frac{\rho_{ex}(\omega, \mathbf{k})}{\epsilon(\omega, \mathbf{k})} \qquad (9.64)$$

From Poisson's equation it then follows for the electric field disturbance, $\delta \mathbf{E}(\omega, \mathbf{k})$, that

$$\delta \mathbf{E}(\omega, \mathbf{k}) = -i \frac{\rho_{tot}(\omega, \mathbf{k}) \mathbf{k}}{k^2 \epsilon_0} \qquad (9.65)$$

Now, using the linear Ohm's law, $\mathbf{j} = \sigma \cdot \delta \mathbf{E}$, one arrives for an intermediate step at

$$\mathbf{j} = -i \frac{\rho_{tot} \sigma \cdot \mathbf{k}}{k^2 \epsilon_0} \qquad (9.66)$$

This current causes the induced space-charge disturbance, ρ. We can determine it from the charge-current continuity equation as

$$\rho(\omega, \mathbf{k}) = -\frac{i}{\omega \epsilon_0} \frac{\mathbf{k} \cdot \sigma \cdot \mathbf{k}}{k^2} \qquad (9.67)$$

Inserting this expression into Eq. (9.63) and comparing with Eq. (9.64) leads to

$$\epsilon(\omega, \mathbf{k}) = 1 + \frac{i}{\omega \epsilon_0} \frac{\mathbf{k} \cdot \sigma(\omega, \mathbf{k}) \cdot \mathbf{k}}{k^2} \qquad (9.68)$$

and with Eq. (9.54) to the expression for the dielectric response function in Eq. (9.62).

9.3.   Plasma Wave Energy

All waves contain energy even though the wave in the average of the amplitude taken over more than one wave train does not exist. This energy is the energy of the fluctuations of the electric and magnetic fields in the wave which is transported across the plasma at the wave group velocity.

Langmuir Plasmons

A simple example shows that waves are carrier of energy. Consider for instance the dispersion relation of Langmuir waves

$$\omega^2 = \omega_{pe}^2(1 + \gamma_e k^2 \lambda_D^2) \tag{9.69}$$

From quantum mechanics we know that any finite frequency implies the existence of a quantum of energy, $\hbar\omega$. Hence, multiplying the above relation by $\hbar^2$ yields the square of the energy contained in one single Langmuir wave packet, a Langmuir *plasmon*. Let us stress the analogy a bit further. The energy of any particle of rest mass m_0 can, for low velocities, be expressed as

$$W = m_0 c^2 + \frac{\mathbf{p}^2}{2m_0} \tag{9.70}$$

Because the momentum of the particle, $\mathbf{p} = -i\hbar\nabla$, is nothing else but the gradient, it can be written as $\mathbf{p} = \hbar\mathbf{k}$. The above dispersion relation can thus be rewritten as

$$\hbar\omega = \hbar\omega_{pe}\left(1 + \frac{\gamma_e \mathbf{p}_l^2 \lambda_D^2}{2\hbar^2}\right) \tag{9.71}$$

where we have expanded the square root and introduced the plasmon momentum, $\mathbf{p}_l$. Comparison with the expression for the energy of a particle shows that the second term in the bracket on the right-hand side of the Langmuir wave dispersion relation corresponds to the kinetic energy of the plasmon, $\hbar\gamma_e k^2 v_e^2/2$, while the first term corresponds to either a potential energy provided by the plasma or to the rest mass energy of the plasmon. Adopting the second interpretation, the mass is given by $m_{0l} = \hbar\omega_{pe}/c^2$ and is typically very small. For instance, in the solar wind where $f_{pe} = \omega_{pe}/2\pi \approx 10\,\mathrm{kHz}$, its value is $m_{0l} \approx 10^{-46}\,\mathrm{kg}$.

Average Energy

Plasma waves consist of many plasmons. By their transport of energy across the plasma they contribute to the redistribution of energy and information. It is of considerable interest to find a macroscopic expression for their average energy content as it is of interest to know how much energy is contained in any other plasma wave. In electrodynamics the flux of energy in an electromagnetic wave is given by the *Poynting vector*

$$\boxed{\mu_0 \mathbf{P} = \delta \mathbf{E} \times \delta \mathbf{B}} \tag{9.72}$$

Obviously, the Poynting vector, $\mathbf{P}$, is a nonlinear quantity. Being the energy flux, its divergence must balance the leakage of wave energy and the decrease of wave energy density, W_w, in a given volume. The latter is the sum of the changes in the electric and magnetic field energy densities. Calculating $\nabla \cdot \mathbf{P}$ from Eq. (9.72) results in

$$-\mu_0 \nabla \cdot \mathbf{P} = \delta \mathbf{E} \cdot (\nabla \times \delta \mathbf{B}) - \delta \mathbf{B} \cdot (\nabla \times \delta \mathbf{E}) \tag{9.73}$$

The curls in this formula can be replaced with the help of Maxwell's equations in order to obtain the conservation law of wave energy density

$$\frac{\partial W_w}{\partial t} + \delta \mathbf{j} \cdot \delta \mathbf{E} = -\nabla \cdot \mathbf{P} \tag{9.74}$$

where W_w is defined as

$$W_w = \epsilon_0 \delta \mathbf{E}^* \cdot \epsilon \cdot \delta \mathbf{E} + \frac{|\delta \mathbf{B}|^2}{2\mu_0} \tag{9.75}$$

The first term on the right-hand side of Eq. (9.75) is easily recognized as the electrostatic energy density stored in the wave, the second is its magnetic energy density. In a plasma the dielectric tensor contributes to storage of electric field energy. In other words, electric field energy can be stored in the interaction of the plasma particles which leads to the polarization of the plasma. However, the dielectric tensor contributes to the electric energy density only through its product from the right and from the left with the electric wave field. Thereby it reduces to the dielectric response function, $\epsilon(\omega, \mathbf{k})$, in Eq. (9.62). Finally, the term containing the electric current density in Eq. (9.74) is that part of the wave energy which is dissipated by Joule heating or Ohmic losses. This term vanishes if there is no dissipation which holds in general for collisionless plasmas.

We are interested in the two contributions of waves to the energy. Energy is a real quantity. Hence, for a complex wave only the products of wave amplitudes and their conjugates contribute to energy. Calculation of the magnetic energy becomes trivial. Assuming that the wave field is complex one simply obtains $|\delta \mathbf{B}|^2 / 2\mu_0$.

The calculation of the electrostatic part is more involved. The wave field changes in time and space with phase $\varphi = \omega t - \mathbf{k} \cdot \mathbf{x}$. For complex frequencies, $\omega = \omega_r + i\gamma$, the phase itself is a complex quantity. To simplify the procedure let us assume that the wave amplitude is small so that we are dealing with linear waves. Let us further temporarily abbreviate the product of the dielectric response function and the wave electric field to its right in Eq. (9.75) as the dielectric displacement, $\delta \mathbf{D} = \epsilon(\omega, \mathbf{k}) \delta \mathbf{E}(\omega, \mathbf{k})$, where we have implicitly represented the electric wave field by a Fourier integral

$$\delta \mathbf{E}(t, \mathbf{x}) = \frac{1}{16\pi^4} \int d^3k \int_{-\infty}^{\infty} d\omega \, \delta \mathbf{E}(\omega, \mathbf{k}) \exp(i\varphi)$$

$$\delta \mathbf{D}(t, \mathbf{x}) = \frac{1}{16\pi^4} \int d^3k \int_{-\infty}^{\infty} d\omega \, \epsilon(\omega, \mathbf{k}) \delta \mathbf{E}(\omega, \mathbf{k}) \exp(i\varphi)$$

(9.76)

We are not interested in the instantaneous energy carried by the waves at a particular time, but in their average energy. This average energy is defined as the change in electrostatic energy averaged over volume and time and over the whole ensemble of waves present in the wave mode

$$\langle W_w \rangle = \epsilon_0 \int_{-\infty}^{t} dt' \int d^3x \left\langle \delta \mathbf{E}(t', \mathbf{x}) \cdot \frac{\partial \delta \mathbf{D}(t', \mathbf{x})}{\partial t'} \right\rangle$$

(9.77)

When we now substitute from Eq. (9.76) into the last expression (remember that one now needs two different integration variables ω and ω'), replace the time derivative with $-i\omega$, and make the resulting expression symmetric, we find

$$\langle W_w \rangle = -\frac{i\epsilon_0}{64\pi^5} \int_{-\infty}^{t} dt' \int d^3k \int_{-\infty}^{\infty} d\omega \, d\omega'$$

$$[\omega \epsilon(\omega, \mathbf{k}) - \omega' \epsilon(-\omega', -\mathbf{k})] \langle \delta \mathbf{E}(\omega, \mathbf{k}) \cdot \delta \mathbf{E}(-\omega', -\mathbf{k}) \rangle \exp[i(\omega' - \omega)t'] \quad (9.78)$$

The terms with the dielectric response function can be simplified if the latter is expanded around the point where the two frequencies coincide ($\omega = \omega'$)

$$\omega' \epsilon(-\omega', -\mathbf{k}) = \omega \epsilon(-\omega, -\mathbf{k}) + (\omega' - \omega) \frac{\partial [\omega \epsilon(-\omega, -\mathbf{k})]}{\partial \omega}$$

(9.79)

Since we have assumed that dissipation is small, the symmetry of the response function requires simply that $\epsilon(-\omega, -\mathbf{k}) = \epsilon(\omega, \mathbf{k})$. Integrating over t' we obtain

$$\langle W_w \rangle = \frac{\epsilon_0}{64\pi^5} \int d^3k \int_{-\infty}^{\infty} d\omega \, d\omega'$$

$$\langle \delta \mathbf{E}(\omega, \mathbf{k}) \cdot \delta \mathbf{E}(-\omega', -\mathbf{k}) \rangle \frac{\partial [\omega \epsilon(\omega, \mathbf{k})]}{\partial \omega} \exp[i(\omega' - \omega)t]$$

(9.80)

The term in angular brackets under the integral sign is the spectral energy density function of the electric field

$$\langle \delta \mathbf{E}(\omega, \mathbf{k}) \cdot \delta \mathbf{E}(-\omega', -\mathbf{k}) \rangle = 4\pi^2 \langle |\delta \mathbf{E}(\omega, \mathbf{k})|^2 \rangle \delta(\omega' - \omega)$$

(9.81)

With its help we find ultimately for the total wave energy

$$\langle W_w \rangle = \frac{\epsilon_0}{16\pi^3} \int d^3k \int\limits_{-\infty}^{\infty} \langle |\delta E(\omega, \mathbf{k})|^2 \rangle \frac{\partial[\omega \epsilon(\omega, \mathbf{k})]}{\partial \omega} d\omega \qquad (9.82)$$

Since the energy and the spectral electric energy are real, only the real part of the dielectric response function enters into Eq. (9.82). However, we are not so much interested in the total energy of the electrostatic waves, but in their spectral energy density and its dependence on the wave frequency and the dispersive properties of the plasma waves. This can be directly read from Eq. (9.82) when we write

$$\langle W_w \rangle = \frac{1}{8\pi^3} \int d^3k \int\limits_{-\infty}^{\infty} W_w(\omega, \mathbf{k}) \, d\omega \qquad (9.83)$$

Comparing the latter two expressions, we obtain for the spectral energy density

$$\boxed{W_w(\omega, \mathbf{k}) = \frac{\epsilon_0}{2} \langle |\delta E(\omega, \mathbf{k})|^2 \rangle \frac{\partial[\omega \epsilon(\omega, \mathbf{k})]}{\partial \omega}} \qquad (9.84)$$

Again, only the real part of the dielectric response function enters the last expression.

It is very important to note once more that it is not the spectral function of the electric field alone (as it is in the case of the magnetic fluctuations) which contributes to the spectral energy density of the wave. The polarization of the plasma in response to the disturbance contributes as well. This contribution is contained in the derivative of the dielectric function with respect to frequency in Eq. (9.84). The actual electrostatic energy stored in the waves can therefore be considerably different from the measured spectral fluctuation of the electric field. To find the correct energy, knowledge of the relevant wave mode and its dispersive properties is required.

We can now use the above formulae to calculate the energy density of Langmuir waves and ion acoustic waves. The Langmuir wave dielectric function is

$$\epsilon_l(\omega, k) = 1 - \frac{\omega_{pe}^2}{\omega^2}(1 + \gamma_e k^2 \lambda_D^2) \qquad (9.85)$$

Multiplying by ω and differentiating with respect to ω yields a Langmuir wave energy density of $W_l = 2W_E$, just twice the electric field energy density

$$\boxed{W_E = \epsilon_0 |\delta E|^2 / 2} \qquad (9.86)$$

Hence, half of the wave energy is stored in the thermal electron motion providing the polarization of the plasma. The same holds for ion acoustic waves, both in the long- and short-wavelength domains.

9.4. Magnetohydrodynamic Waves

Waves in an ideal magnetohydrodynamic fluid can be treated on the basis of the previous section by calculating the wave conductivities and dielectric functions. But because of the relative simplicity of the ideal magnetohydrodynamic equations derived in Sec. 7.3 it is much simpler to return to these and linearize directly.

Magnetohydrodynamic Dispersion Relation

We assume stationary ideal homogeneous conditions as the initial state of the single-fluid plasma with vanishing average velocity and electric fields, overall pressure equilibrium, and vanishing magnetic stresses

$$
\begin{aligned}
\mathbf{v}_0 &= 0 \\
\mathbf{E}_0 &= 0 \\
\nabla \left(p_0 + B_0^2/2\mu_0 \right) &= 0 \\
(\mathbf{B}_0 \cdot \nabla)\mathbf{B}_0 &= 0
\end{aligned}
\tag{9.87}
$$

Plasma density, velocity, magnetic field and electric field are then decomposed as the sums of their initial values and a space and time dependent fluctuation according to

$$
\begin{aligned}
n &= n_0 + \delta n \\
\mathbf{v} &= \delta\mathbf{v} \\
\mathbf{E} &= \delta\mathbf{E} \\
\mathbf{B} &= \mathbf{B}_0 + \delta\mathbf{B}
\end{aligned}
\tag{9.88}
$$

Because the magnetohydrodynamic equations contain nonlinear terms, the fluctuations must be small. This assumption has to be justified, because some of the initial average quantities are zero. The idea is that we want to arrive at a homogeneous set of linear equations which will lead us to a dispersion relation for the eigenmodes of the plasma. In such a set all variables can be expressed through one single variable, whose value remains free. The assumption of smallness can thus be reduced to the assumption that this one remaining variable will be small compared with its initial value. Hence, if the ambient magnetic field is sufficiently strong, as is usually the case in magnetohydrodynamics, one can assume that the fluctuation amplitude of the magnetic field is much weaker than the stationary magnetic field

$$
|\delta\mathbf{B}| \ll B_0
\tag{9.89}
$$

With these assumptions the continuity equation becomes

$$
\frac{\partial \delta n}{\partial t} + n_0 \nabla \cdot \delta\mathbf{v} = 0
\tag{9.90}
$$

Similarly, the magnetohydrodynamic momentum conservation equation reduces to

$$m_i n_0 \frac{\partial \delta \mathbf{v}}{\partial t} = -\nabla \left(\delta p + \frac{1}{\mu_0} \mathbf{B}_0 \cdot \delta \mathbf{B} \right) + \frac{1}{\mu_0} (\mathbf{B}_0 \cdot \nabla) \delta \mathbf{B} \qquad (9.91)$$

Since the plasma is typically unable to extinguish the fast temperature variations caused by the fluctuations, one can use the adiabatic pressure law, and the variation of the pressure becomes

$$\frac{\partial \delta p}{\partial t} = m_i c_s^2 \frac{\partial \delta n}{\partial t} = -m_i n_0 c_s^2 \nabla \cdot \delta \mathbf{v} \qquad (9.92)$$

where in the second part of the equation we made use of Eq. (9.90). The quantity $c_s^2 = \gamma p_0 / m_i n_0$ is the square of the sound velocity introduced in Sec. 8.1. The only remaining equation is Faraday's induction law which after linearization becomes

$$\frac{\partial \delta \mathbf{B}}{\partial t} = (\mathbf{B}_0 \cdot \nabla) \delta \mathbf{B} - \mathbf{B}_0 (\nabla \cdot \delta \mathbf{v}) \qquad (9.93)$$

Equations (9.90) through (9.93) represent the desired linear and homogeneous system of equations for δn, $\delta \mathbf{v}$ and $\delta \mathbf{B}$. Because we assumed a uniform plasma with straight magnetic field lines, the direction of the ambient magnetic field is the only direction of symmetry. Hence, we can choose the magnetic field to be aligned with the z axis of our orthogonal system of coordinates, $\mathbf{B}_0 = B_0 \, \hat{\mathbf{e}}_\parallel$. With these conventions Eqs. (9.91) and (9.93) can be written as

$$\begin{aligned} \frac{\partial \delta \mathbf{v}}{\partial t} &= v_A^2 \nabla_\parallel \left(\frac{\delta \mathbf{B}_\perp}{B_0} \right) - \nabla \left(\frac{\delta p}{m_i n_0} \right) \\ \frac{\partial}{\partial t} \left(\frac{\delta \mathbf{B}}{B_0} \right) &= \nabla_\parallel \delta \mathbf{v}_\perp - \hat{\mathbf{e}}_\parallel (\nabla_\perp \cdot \delta \mathbf{v}_\perp) \end{aligned} \qquad (9.94)$$

where v_A is the Alfvén velocity, introduced in Eq. (8.11). These two equations together with Eq. (9.92) form a closed system of first-order differential equations for the fluctuating components of the magnetic field, pressure and velocity in the plasma. We have now the choice to either directly operate on it with the plane wave ansatz or to first derive one second-order wave equation for one of the field variables. In order to make the physics more transparent, we decide for the second choice. Taking the time derivative of the first equation and eliminating the variation of the magnetic field and pressure we obtain

$$\frac{\partial^2 \delta \mathbf{v}}{\partial t^2} = c_{ms}^2 \nabla (\nabla \cdot \delta \mathbf{v}) + v_A^2 \left(\nabla_\parallel^2 \delta \mathbf{v} - \nabla \nabla_\parallel \delta v_\parallel - \hat{\mathbf{e}}_\parallel \nabla_\parallel \nabla \cdot \delta \mathbf{v} \right) \qquad (9.95)$$

which is second-order in all terms and therefore describes travelling waves. Its solution is found by introducing $\delta \mathbf{v} = \delta \mathbf{v}_0 \exp(i \mathbf{k} \cdot \mathbf{x} - i \omega t)$ where $\delta \mathbf{v}_0$ is an arbitrary constant amplitude of the velocity field. This yields

$$\left[(\omega^2 - k_\parallel^2 v_A^2) \mathbf{I} - c_{ms}^2 \mathbf{k} \mathbf{k} + (k \hat{\mathbf{e}}_\parallel + \hat{\mathbf{e}}_\parallel \mathbf{k}) k_\parallel v_A^2 \right] \cdot \delta \mathbf{v}_0 = 0 \qquad (9.96)$$

A meaningful solution of this equation is obtained only if $\delta\mathbf{v}_0 \neq 0$, and thus the determinant of the tensor in square brackets must vanish. If we choose a right-handed system where the perpendicular component of the wave vector is parallel to the x axis so that $\mathbf{k} = k_\parallel \hat{\mathbf{e}}_\parallel + k_\perp \hat{\mathbf{e}}_x$, the last equation can be written in the form

$$\begin{bmatrix} \omega^2 - v_A^2 k_\parallel^2 - c_{ms}^2 k_\perp^2 & 0 & -c_s^2 k_\parallel k_\perp \\ 0 & \omega^2 - v_A^2 k_\parallel^2 & 0 \\ -c_s^2 k_\parallel k_\perp & 0 & \omega^2 - c_s^2 k_\parallel^2 \end{bmatrix} \begin{bmatrix} \delta v_{0x} \\ \delta v_{0y} \\ \delta v_{0\parallel} \end{bmatrix} = 0 \qquad (9.97)$$

where c_{ms} is the magnetosonic speed introduced in Eq. (8.73).

Shear Alfvén Wave

The above system shows that the velocity fluctuation in the y direction decouples from all other fields, representing a wave with linear dispersion relation

$$\boxed{\omega_A = \pm k_\parallel v_A} \qquad (9.98)$$

This wave propagates parallel to the ambient field and is purely transverse. It is an electromagnetic wave which is called *shear Alfvén wave*. According to Eq. (9.94), the magnetic component of this wave is parallel to the velocity component, $\delta B_y / B_0 = -\delta v_y / v_A$, and thus the wave has a zero electric fluctuation field in the direction of the ambient magnetic field, $\mathbf{B}_0$. The wave electric field $\delta E_x = \delta B_y / v_A$ points in the x direction. In addition, because $\delta B_y \ll B_0$, we find that the velocity fluctuation is small compared with the Alfvén speed, $\delta v_y \ll v_A$.

The frequency of the shear Alfvén wave depends linearly on the wavenumber. The shear Alfvén wave is thus non-dispersive and its wave energy flows along $\mathbf{B}_0$, as is recognized from its group velocity, $v_{gr,A\parallel} = v_A$, $v_{gr,A\perp} = 0$. It represents simple string-like oscillations of the magnetic field lines.

Magnetosonic Waves

The remaining four matrix elements couple the parallel velocity component, $\delta v_\parallel$, to the other transverse velocity fluctuation, δv_x. The dispersion relation of this wave is obtained from the vanishing of their determinant

$$\omega^4 - \omega^2 c_{ms}^2 k^2 + c_s^2 v_A^2 k^2 k_\parallel^2 = 0 \qquad (9.99)$$

The two roots of this relation are

$$\boxed{\omega_{ms}^2 = \frac{k^2}{2} \left\{ c_{ms}^2 \pm \left[\left(v_A^2 - c_s^2 \right)^2 + 4 v_A^2 c_s^2 \frac{k_\perp^2}{k^2} \right]^{1/2} \right\}} \qquad (9.100)$$

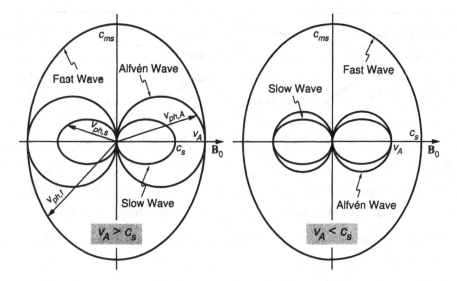

Fig. 9.5. Phase velocity diagrams of the three MHD wave modes.

The expressions in the curly brackets are the phase velocities of the two *magnetosonic wave* modes described by this dispersion relation. They depend only on the angle θ between the magnetic field and the wave vector through $k_\perp^2 / k^2 = \sin^2 \theta$. The root with the positive sign is called the *fast magnetosonic wave*, the root with the negative sign is the *slow magnetosonic wave*. We have encountered these waves already in Sec. 8.4, in connection with our discussion of fast and slow shocks. In fact, fast and slow shocks are the final states of fast and slow magnetosonic waves when evolving to large amplitudes.

Inspection of the dispersion relations of the two magnetosonic waves reveals that for $k = k_\perp$ the root on the right-hand side of the dispersion relation becomes trivial and the dispersion relation can be written as

$$\omega^2 = \tfrac{1}{2}k^2(c_{ms}^2 \pm c_{ms}^2) \qquad (9.101)$$

The fast mode (positive sign) propagates into the perpendicular direction with phase velocity $v_{ph,f\perp} = c_{ms}$, while the slow mode (negative sign) does not propagate. Since the shear Alfvén wave does not propagate into the perpendicular direction, too, we find that the only perpendicular magnetohydrodynamic wave is the fast magnetosonic mode.

In the parallel direction the dispersion relation reduces to

$$\omega^2 = \tfrac{1}{2}k^2 \left[c_s^2 + v_A^2 \pm (c_s^2 - v_A^2) \right] \qquad (9.102)$$

The character of the waves depends on whether v_A or c_s is higher. For $v_A > c_s$ the parallel phase velocity of the slow mode becomes $v_{ph,s\parallel} = c_s$. The parallel slow mag-

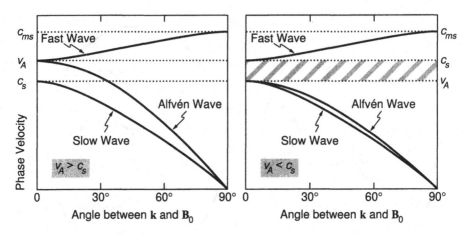

Fig. 9.6. Dependence of MHD wave phase velocities on the angle between $\mathbf{k}$ and $\mathbf{B}_0$.

netosonic mode is a simple sound wave, while the fast mode propagates at Alfvén speed, $v_{ph,f\parallel} = v_A$. In the opposite case, when $c_s > v_A$, the slow mode approaches the Alfvén speed, and the fast mode has sound velocity. This behavior of the phase velocities of the three magnetohydrodynamic modes is illustrated in Figs. 9.5 and 9.6.

Figure 9.5 is the phase velocity diagram of the three magnetohydrodynamic wave modes. In this diagram the ambient magnetic field, $\mathbf{B}_0$, points in the x direction. The arrows are the vectors of the phase velocities for the different modes drawn as functions of their angle with the magnetic field. Their absolute length is the value of the phase velocity at the particular angle. The figure can be interpreted as the instantaneous form of a wave front at unit time having started from the origin and propagating in all directions. This wave front will have become displaced farthest at the angle where the phase velocity is maximum for the mode under consideration.

Figure 9.6 shows the variation of the phase velocities of the three magnetohydrodynamic wave modes as function of the angle between their wave vectors and the direction of the ambient magnetic field for the two cases $v_A > c_s$ and $v_A < c_s$. One recognizes the increase in the fast wave speed from parallel to perpendicular direction and the corresponding decreases in the Alfvén and slow wave velocities.

For the Alfvén wave we found that the components δv_y, δB_y, and δE_x build up the shear Alfvén mode. Correspondingly δB_x, $\delta B_\parallel$, δv_x, $\delta v_\parallel$, δp, and δn belong to the two magnetosonic wave modes. In order to understand what happens physically, let us return to the equation of motion and write it in components

$$\omega \delta \mathbf{v} = \frac{\mathbf{k}}{m_i n_0} \left(\delta p + \frac{1}{\mu_0} \mathbf{B}_0 \cdot \delta \mathbf{B} \right) - \frac{\mathbf{k} \cdot \mathbf{B}_0}{\mu_0 m_i n_0} \delta \mathbf{B} \qquad (9.103)$$

In the parallel direction this equation reduces to the simple expression

$$\omega v_\parallel = \frac{k_\parallel \delta p}{m_i n_0} \tag{9.104}$$

showing that parallel pressure variations cause variations in the parallel flow but not in the magnetic field. Moreover, dotting the equation of motion with $\mathbf{k}$ and using $\mathbf{k} \cdot \delta \mathbf{B} = k_\parallel \delta B_\parallel + k_\perp \delta B_x$ yields

$$\omega(k_\parallel v_\parallel + k_\perp v_x) = \frac{k^2 \delta p_{\text{tot}}}{m_i n_0} \tag{9.105}$$

where $p_{\text{tot}} = p + B^2/2\mu_0$. We already know that $v_\parallel$ is generated by pressure fluctuations only. Hence, we conclude that perpendicular fluid motions are connected with variations in the total pressure and in particular with variations in the magnetic field strength. Because these variations can be in-phase or out-of-phase they either amplify the force on the plasma in which case the wave becomes the fast (accelerated) magnetosonic mode. In the opposite case the force is weakened by the out-of-phase total pressure variation, with the result that the magnetosonic mode is retarded and becomes the slow mode.

9.5. Cold Electron Plasma Waves

The one-fluid magnetohydrodynamic theory is valid only at very low frequencies $\omega \ll (\omega_{gi}, \omega_{pi})$, well below the ion-cyclotron and plasma frequencies where electron inertia can safely be neglected. Near both of these frequencies differences between electron and ion dynamics begin to become important and magnetohydrodynamic waves transform into modes which are not contained in the apparatus of one-fluid theory. When electron and ion inertia are to be taken into account in wave propagation one possibility is to relax the one-fluid assumption and to analyse wave propagation in a two-component plasma consisting of equal numbers of positive and negative charges, ions and electrons. In the following we proceed in two steps. First we consider a cold magnetized electron plasma where the ions constitute merely a neutralizing background. The waves derived will thus also have frequencies well above all ion frequencies, $\omega \gg (\omega_{gi}, \omega_{pi})$. In the next section we consider waves with frequencies intermediate between the magnetohydrodynamic and high-frequency electron waves in a magnetized plasma. Only in this case ion dynamics will explicitly be taken into account.

Cold Plasma Dispersion Relation

Cold electron dynamics is governed by the single-particle electron motion in a strong magnetic field as given by Eqs. (2.8) and (2.16). For our purposes where the electrons are 'cold', the magnetic field of any wave is not affected by the electron motion. In other words, in the Lorentz force we retain only the linear term $\delta \mathbf{v} \times \mathbf{B}_0$. Since $\mathbf{B}_0$ is a constant vector we can extract it from the product and include it into the electron gyrofrequency vector, $\omega_{ge} = e\mathbf{B}_0/m_e$. Then the parallel and perpendicular equations of motion become

$$\frac{d\delta v_{\parallel}}{dt} = -\frac{e}{m_e}\delta E_{\parallel}$$

$$\frac{d\delta \mathbf{v}_{\perp}}{dt} = -\frac{e}{m_e}\delta \mathbf{E}_{\perp} + \omega_{ge} \times \delta \mathbf{v}_{\perp} \tag{9.106}$$

Differentiating this equation with respect to time, substituting for the cross-product from the undifferentiated equation, and rearranging, we obtain the oscillator equation

$$\frac{\partial^2 \delta \mathbf{v}_{\perp}}{\partial t^2} + \omega_{ge}^2 \delta \mathbf{v}_{\perp} = -\frac{e}{m_e}\left(\frac{\partial \delta \mathbf{E}}{\partial t} + \omega_{ge} \times \delta \mathbf{E}_{\perp}\right) \tag{9.107}$$

It is our goal to obtain an expression for the linear wave conductivity, $\sigma(\omega, \mathbf{k})$, which we will use later to find the dielectric tensor and the cold plasma dielectric response function. Recall that we are dealing here with a cold plasma of zero pressure. As a consequence the thermal gradient force drops out from the momentum conservation equation and density fluctuations do not contribute to the wave. This is the reason why we are allowed to use the single-electron equation of motion and can ignore the continuity equation. The condition for the validity of the cold model is clearly that the thermal velocity of the plasma electrons is much less than the wave phase velocity.

The relation between the wave velocity and current is given by the sum over all particle momenta. But since in a cold plasma of zero temperature all electrons have the same velocity caused by the wave electric field fluctuation, the sum over all momenta reduces to the product of the undisturbed particle density, n_0, and the disturbance of the electron velocity, $\delta \mathbf{v}$. Hence, the current density becomes

$$\delta \mathbf{j} = -en_0\delta \mathbf{v} = \sigma \cdot \delta \mathbf{E} \tag{9.108}$$

From Eq. (9.107) one realizes that for $\delta \mathbf{E}_{\perp} = 0$ the electron motion becomes a pure gyration with velocity $\mathbf{v}_{\perp 0}$. The cyclotron motion is the homogeneous solution of Eq. (9.107). We are interested in the particular solution of the inhomogeneous equation

$$\delta \mathbf{v} - \mathbf{v}_{\perp 0} = -\frac{1}{en_0}\sigma \cdot \delta \mathbf{E} \tag{9.109}$$

which is a periodic oscillation of frequency ω according to $\delta \mathbf{v} \propto \exp(-i\omega t)$. Inserting into Eq. (9.107) yields (note that here and from now on we disregard the sign of the electron charge and take ω_{ge} as a positive number)

$$\sigma(\omega) = \epsilon_0 \omega_{pe}^2 \begin{bmatrix} \dfrac{i\omega}{\omega^2 - \omega_{ge}^2} & \dfrac{\omega_{ge}}{\omega^2 - \omega_{ge}^2} & 0 \\ -\dfrac{\omega_{ge}}{\omega^2 - \omega_{ge}^2} & \dfrac{i\omega}{\omega^2 - \omega_{ge}^2} & 0 \\ 0 & 0 & \dfrac{i}{\omega} \end{bmatrix} \tag{9.110}$$

The last line in this tensor arises from the parallel component of the electron equation of motion which is independent of the magnetic field. The conductivity tensor of cold plasma waves depends only on the wave frequency. Going back to the definition in Eq. (9.54) it is easy to find the dielectric tensor of the cold plasma

$$\epsilon_{cold}(\omega) = \begin{bmatrix} 1 + \dfrac{\omega_{pe}^2}{\omega_{ge}^2 - \omega^2} & -\dfrac{i\omega_{ge}}{\omega}\dfrac{\omega_{pe}^2}{\omega_{ge}^2 - \omega^2} & 0 \\ \dfrac{i\omega_{ge}}{\omega}\dfrac{\omega_{pe}^2}{\omega_{ge}^2 - \omega^2} & 1 + \dfrac{\omega_{pe}^2}{\omega_{ge}^2 - \omega^2} & 0 \\ 0 & 0 & 1 - \dfrac{\omega_{pe}^2}{\omega^2} \end{bmatrix} \tag{9.111}$$

Inserting this into Eq. (9.48) yields the cold electron plasma dispersion relation

$$\boxed{\text{Det}\left[\frac{k^2 c^2}{\omega^2}\left(\mathbf{1} - \frac{\mathbf{kk}}{k^2}\right) - \epsilon_{cold}\right] = 0} \tag{9.112}$$

The cold dielectric tensor is independent of the wave vector, which enters only through the electrodynamic equations. From this fact we conclude that all electrostatic variations in a cold electron plasma will be mere oscillations while the propagating waves are pure electromagnetic waves. The only oscillation in a cold electron plasma is the Langmuir oscillation with frequency $\omega = \pm\omega_{pe}$ and infinite wavenumber. We found this oscillation already in Sec. 9.1. Equation (9.112) therefore turns out to be the basic dispersion relation of high-frequency electromagnetic waves.

High-Frequency Electromagnetic Waves

We can write the cold plasma dielectric tensor in Eq. (9.111) in a shorthand version

$$\epsilon = \begin{bmatrix} \epsilon_1 & -i\epsilon_2 & 0 \\ i\epsilon_2 & \epsilon_1 & 0 \\ 0 & 0 & \epsilon_3 \end{bmatrix} \tag{9.113}$$

where the matrix components have been defined as

$$\epsilon_1 = 1 - \frac{\omega_{pe}^2}{\omega^2 - \omega_{ge}^2}$$

$$\epsilon_2 = -\frac{\omega_{ge}}{\omega} \frac{\omega_{pe}^2}{\omega^2 - \omega_{ge}^2} \qquad (9.114)$$

$$\epsilon_3 = 1 - \frac{\omega_{pe}^2}{\omega^2}$$

The electron-cyclotron frequency is taken to be independent of the sign of the electron charge. Then defining the vectorial refractive index, $\mathbf{N} = \mathbf{k}c/\omega$, with $N^2 = N_\perp^2 + N_\parallel^2$, and assuming without any restriction of generality that the wave vector, $\mathbf{k}$, is in the (x, z) plane ($k_y = 0$), the cold plasma dispersion relation (9.112) can be written as

$$\text{Det} \begin{bmatrix} N_\parallel^2 - \epsilon_1 & i\epsilon_2 & -N_\parallel N_\perp \\ -i\epsilon_2 & N^2 - \epsilon_1 & 0 \\ -N_\parallel N_\perp & 0 & N_\perp^2 - \epsilon_3 \end{bmatrix} = 0 \qquad (9.115)$$

This is the basic dispersion relation of a charge-compensated electron plasma of zero temperature which serves as the starting point of a discussion of high-frequency electromagnetic wave propagation in a plasma. One should, however, keep in mind that the cold plasma approximation is a very crude approximation which is valid only at wavelengths much larger than the gyroradii of the particles and for high phase velocities much larger than the thermal velocity of the electrons. In the following we distinguish between propagation parallel and perpendicular to the magnetic field.

Parallel Electromagnetic Propagation

For parallel propagation we have $N_\perp = 0$, $N_\parallel = N$. Since the wave vector is parallel to the magnetic field, $\mathbf{B}_0$, the dispersion relation splits into the parallel dispersion relation, $\epsilon_3 = 0$, and into a transverse dispersion relation. The former yields the already well-known plasma oscillations, $\omega^2 = \omega_{pe}^2$, belonging to an electrostatic parallel electric wave field, $\delta E_\parallel || k_\parallel$. The transverse dispersion relation contains the two perpendicular electric field components, δE_x, δE_y. The cylindrical geometry of this wave suggests the introduction of the following combination of electric wave field components

$$\sqrt{2}\,\delta E_{\mathrm{R,L}} = (\delta E_x \mp i\delta E_y) \qquad (9.116)$$

These new wave fields describe right-hand (R) and left-hand (L) circularly polarized waves. This is easily realized by taking the ratio of the Cartesian field components

$$(\delta E_y/\delta E_x)_{\mathrm{R,L}} = \pm i \qquad (9.117)$$

showing that the electric field vector of the R-wave rotates in positive y direction while that of the L-wave rotates in negative y direction.

The transformation from $\delta E_{x,y}$ to $\delta E_{R,L}$ does not change the electric field. It is a unitary matrix satisfying $\mathbf{U} \cdot \mathbf{U}^\dagger = \mathbf{U}^\dagger \cdot \mathbf{U} = 1$ and is defined as

$$\mathbf{U} = \begin{bmatrix} 1/\sqrt{2} & -i/\sqrt{2} & 0 \\ -i/\sqrt{2} & 1/\sqrt{2} & 0 \\ 0 & 0 & 1 \end{bmatrix} \tag{9.118}$$

Accordingly the dielectric tensor transforms as $\mathbf{U} \cdot \epsilon \cdot \mathbf{U}^\dagger$. The result of this procedure is the diagonal dielectric matrix

$$\mathbf{U} \cdot \epsilon \cdot \mathbf{U}^\dagger = \begin{bmatrix} \epsilon_R & 0 & 0 \\ 0 & \epsilon_L & 0 \\ 0 & 0 & \epsilon_3 \end{bmatrix} \tag{9.119}$$

Its transverse components are given by $\epsilon_{R,L} = \epsilon_1 \pm \epsilon_2$ or explicitly as

$$\epsilon_{R,L} = 1 - \frac{\omega_{pe}^2}{\omega(\omega \mp \omega_{ge})} \tag{9.120}$$

The diagonalized dielectric tensor separates the transverse dispersion relation into two independent dispersion relations for the right- and left-circular polarized parallel modes

$$N^2 = \frac{k^2 c^2}{\omega^2} = \epsilon_{R,L} \tag{9.121}$$

The right-hand side of these dispersion relations is independent of the wavenumber, indicating that the dispersion of the transverse modes is entirely due to the refraction index, N. Hence, Eq. (9.121) is already in its explicit form for the parallel wavenumber.

The right-circular polarized electromagnetic wave has refraction index

$$\boxed{\frac{k^2 c^2}{\omega^2} = 1 - \frac{\omega_{pe}^2}{\omega(\omega - \omega_{ge})}} \tag{9.122}$$

For $\omega \to 0$ as well as for $\omega \to \omega_{ge}$ the refraction index diverges. In the latter case this implies that the wavenumber diverges, $k \to \infty$. Hence, $\omega - \omega_{ge} \to -0$ and

$$\omega_{R,res} = \omega_{ge} \tag{9.123}$$

is the *electron-cyclotron resonance frequency* for the right-circular polarized mode.

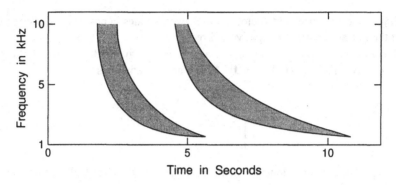

Fig. 9.7. Schematic sonogram of two whistlers recorded on the ground.

Resonances are quite complex physical phenomena requiring deep insight into the physical processes of the interaction of the wave with the plasma itself. The observation that $k \to \infty$ at a resonance implies that the resonant wavelength becomes very short, $\lambda = 2\pi/k \to 0$, microscopically small indeed. In addition, because the frequency of the wave is constant, the phase velocity of the wave in the plasma frame becomes zero, while the wave plasmon momentum, $\hbar \mathbf{k}$, increases. In such a case the interaction between the cold plasma particles and the wave must necessarily become very strong, while over the length of one wavelength the wave starts 'resolving' single particles and violently affects their orbits. During this strong interaction the wave will either give energy to the particles and will become damped, or it will extract energy and momentum from the particles in resonance and will grow. Either case is possible so that one can distinguish between *resonant absorption* of wave energy or *resonant amplification* of the wave.

At sufficiently low frequencies one can neglect the one in the dispersion relation of the low-frequency slow mode and obtain

$$\omega = \frac{\omega_{ge}}{1 + \omega_{pe}^2/k^2 c^2} \tag{9.124}$$

In the long-wavelength limit of small k this dispersion relation simplifies to

$$\boxed{\omega_{\mathrm{w}} = k^2 c^2 \omega_{ge} \omega_{pe}^{-2}} \tag{9.125}$$

This is the dispersion relation of the *electron whistlers* frequently observed on the ground and in the Earth's magnetosphere and often excited by lightnings. They propagate nearly strictly along the magnetic field lines from one hemisphere to the other. Their phase and group velocities are both proportional to k and thus to $\omega_{\mathrm{w}}^{1/2}$, implying that higher-frequency waves have higher group and phase velocities. Hence, the high-frequency

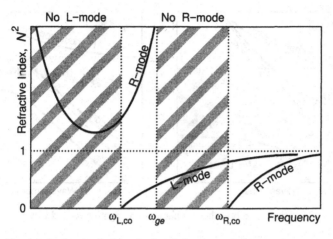

Fig. 9.8. Refractive indices for parallel propagating R- and L-waves.

part of a whistler injected from, say, lightning in the southern hemisphere will reach the northern hemispheric footpoint of the magnetic field line earlier than its low-frequency part. The whistler will appear as a falling tone in a frequency-time sonogram like the one shown in Fig. 9.7, a property which gave it its name.

For large $\omega > \omega_{ge}$ the refraction index, and thus the wavenumber, vanishes at the *right-hand cut-off* frequency

$$\omega_{R,co} = \tfrac{1}{2}\left[\omega_{ge} + (\omega_{ge}^2 + 4\omega_{pe}^2)^{1/2}\right] \tag{9.126}$$

which is found by setting the right-hand side of Eq. (9.122) to zero.

Let us now turn to the left-circular wave mode. Its dispersion relation reads

$$\boxed{\frac{k^2 c^2}{\omega^2} = 1 - \frac{\omega_{pe}^2}{\omega(\omega + \omega_{ge})}} \tag{9.127}$$

One immediately recognizes that this dispersion relation does not have any resonances. Moreover, since $N^2 < 1$, left-circular waves have phase velocities larger than c. They are high-frequency waves and have a low-frequency cut-off at

$$\omega_{L,co} = \tfrac{1}{2}\left[(\omega_{ge}^2 + 4\omega_{pe})^{1/2} - \omega_{ge}\right] \tag{9.128}$$

This cut-off is below the R-wave cut-off. In our approximation, where the effects of the ions are neglected, the left-circular wave exists only at frequencies above $\omega_{L,co}$.

Figure 9.8 shows the dependence of the refraction index on frequency. There is no wave propagation for negative values of N^2. The dispersion branches of the R and L

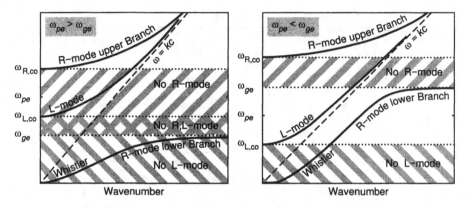

Fig. 9.9. Dispersion branches for parallel propagating R- and L-waves.

modes for a dense, $\omega_{pe} > \omega_{ge}$, and a dilute, $\omega_{pe} < \omega_{ge}$, plasma are shown in Fig. 9.9. At high frequencies the L-mode and R-mode upper branches become free-space modes of light velocity. It is easily recognized from the tangents to all the curves that the group velocities of the waves are smaller than c. One also observes that there is no connection between the two branches of the R-mode. Both branches are separated by a stop band. High-frequency R-modes cannot penetrate into a plasma beyond the point where their frequency matches the R-mode cut-off. Similarly L-modes cannot penetrate beyond the L-mode cut-off. In a dilute plasma the low-frequency R- and L-mode branches overlap in the interval between the cyclotron frequency and the L-mode cut-off.

Perpendicular Electromagnetic Propagation

The other limiting case are waves propagating perpendicular to the magnetic field with a wave vector $\mathbf{k} = \mathbf{k}_\perp$. In a homogeneous plasma the direction of the wave vector in the perpendicular plane is arbitrary and we can chose it to be parallel to the x axis. For $k_y = k_\parallel = 0$ the general cold plasma dispersion relation (9.115) can be reduced to

$$\text{Det} \begin{bmatrix} -\epsilon_1 & i\epsilon_2 & 0 \\ -i\epsilon_2 & N_\perp^2 - \epsilon_1 & 0 \\ 0 & 0 & N_\perp^2 - \epsilon_3 \end{bmatrix} = 0 \qquad (9.129)$$

Let us first consider the third line of the matrix in Eq. (9.129). It tells us that the wave with $\delta E_\parallel$ decouples from the two transverse field components. Its dispersion relation is $N_\perp^2 = \epsilon_3$ or explicitly using the definition of ϵ_3 in Eq. (9.114)

$$\boxed{\omega_{om}^2 = \omega_{pe}^2 + k_\perp^2 c^2} \qquad (9.130)$$

Because the wave vector and wave electric field of this mode are perpendicular to each other, this wave is a transverse electromagnetic wave. For $\omega_{pe} \to 0$ it becomes the usual free-space electromagnetic wave. We have recovered the *ordinary wave*, abbreviated as *O-mode*, in a magnetized plasma which we already met in Eq. (9.37) for an unmagnetized plasma. As shown there, it is easy to recognize that this mode has a low-frequency cut-off at the local plasma frequency. The ordinary wave smoothly connects to the high-frequency L-mode when its wave vector turns parallel to the magnetic field.

The dispersion relation of the remaining wave is obtained by solving the determinant of the perpendicular part of Eq. (9.129)

$$\epsilon_2^2 + \epsilon_1(N_\perp^2 - \epsilon_1) = 0 \qquad (9.131)$$

which, when using Eq. (9.114), is rewritten as

$$k_\perp^2 c^2 = \frac{(\omega^2 - \omega_{R,co}^2)(\omega^2 - \omega_{L,co}^2)}{\omega^2 - \omega_{ge}^2 - \omega_{pe}^2} \qquad (9.132)$$

We have factorized the right-hand side of this expression in a convenient way, making use of the previously defined cut-off frequencies of the R- and L-modes, which turn out to be cut-off frequencies of the new perpendicular mode, too. Since this is topologically impossible with one single branch, the mode described by the above dispersion relation will have two different branches, one being cut off at the R-mode cut-off, the other at the L-mode cut-off. We call this mode the *extraordinary mode*, abbreviated as *X-mode*, to distinguish it from the O-mode.

Before discussing its dispersive properties, we note that its electric field components are in the (x, y) plane. Since $k_\perp = k_x$ by definition, the extraordinary mode has an electric field component parallel to the wave vector. This wave is thus of mixed electrostatic and electromagnetic polarization. It propagates perpendicular to the magnetic field, but has no electric field component parallel to the magnetic field. On the other hand, because its wave vector must be perpendicular to the wave magnetic field, it has magnetic components in the y direction as well as parallel to the magnetic field.

The X-mode has a resonance at the *upper-hybrid frequency*

$$\omega_{uh}^2 = \omega_{ge}^2 + \omega_{pe}^2 \qquad (9.133)$$

At this frequency the plasma and cyclotron properties of the electrons mix. Since the upper-hybrid frequency is higher than the L-mode cut-off, the lower-frequency branch of the X-mode is resonant at the upper-hybrid frequency. This branch can thus propagate between its cut-off at $\omega_{L,co}$ and the resonance at ω_{uh}. On the other hand, the R-mode cut-off is higher than the upper-hybrid frequency and the upper branch of the X-mode,

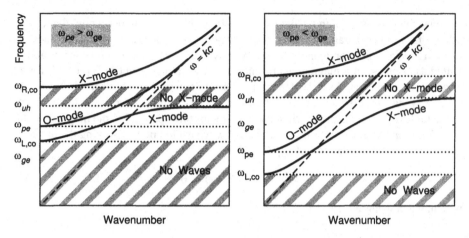

Fig. 9.10. Dispersion branches for perpendicular propagating O- and X-modes.

which for high frequencies becomes a free-space mode propagating above $\omega_{R,co}$. Thus there is a stop-band for the X-mode between $\omega_{R,co} > \omega > \omega_{uh}$.

Figure 9.10 shows the transverse mode branches for perpendicular propagation in dense and dilute plasmas. A comparison with Fig. 9.9 shows that the two branches of the perpendicular extraordinary wave become the two branches of the right-circular R-mode when the wave vector turns into parallel propagation.

9.6. Two-Fluid Plasma Waves

At very low frequencies comparable to the ion-cyclotron frequency, $\omega \approx \omega_{gi}$, one cannot anymore neglect the ion dynamics. In linear theory it is easy to include the ion terms into the dispersion relation, observing that the contribution of the plasma is contained only in the linear conductivity and thus in the dielectric tensor in an additive way. This is clear when one remembers that both the space charges and linear currents are additive quantities containing the sum of the electron and ion contributions.

Effect of Ion Dynamics

The total linear conductivity is the sum of the electron and ion conductivities. The dielectric does, in addition, contain a unit tensor which results from the field equations and is independent of the plasma. Hence, one must subtract the unit terms in the diagonal terms of the dielectric tensor to obtain the plasma contribution. Doing this and adding the ion terms to the components of the cold dielectric tensor components in Eq. (9.114) one obtains the following tensor elements for low-frequency waves

$$\epsilon_1 = 1 - \frac{\omega_{pe}^2}{\omega^2 - \omega_{ge}^2} - \frac{\omega_{pi}^2}{\omega^2 - \omega_{gi}^2}$$

$$\epsilon_2 = -\frac{\omega_{ge}}{\omega}\frac{\omega_{pe}^2}{\omega^2 - \omega_{ge}^2} + \frac{\omega_{gi}}{\omega}\frac{\omega_{pi}^2}{\omega^2 - \omega_{gi}^2} \qquad (9.134)$$

$$\epsilon_3 = 1 - \frac{\omega_{pe}^2}{\omega^2} - \frac{\omega_{pi}^2}{\omega^2}$$

Parallel Wave Propagation

For parallel propagation, $N_\perp = 0$, the dispersion relation including electron and ion dynamics becomes

$$\boxed{N_{\parallel R,L}^2 = 1 - \frac{\omega_{pe}^2}{\omega(\omega \mp \omega_{ge})} - \frac{\omega_{pi}^2}{\omega(\omega \pm \omega_{gi})}} \qquad (9.135)$$

where as before the upper sign applies to the R-mode, the lower to the L-mode. The R-mode has the well-known resonance $N_{\parallel R} \to \infty$ at the electron-cyclotron frequency

$$\omega_{R,res} = \omega_{ge} \qquad (9.136)$$

while now, in contrast to a pure electron plasma, the L-mode has also a resonance at the ion-cyclotron frequency, called the left-hand *ion-cyclotron resonance frequency*

$$\omega_{L,res} = \omega_{gi} \qquad (9.137)$$

The above dispersion relation (9.135) has cut-offs for $N_\parallel \to 0$. These cut-off frequencies are nearly the same as those derived before and given in Eqs. (9.126) and (9.128), with slight corrections due to the ion dynamics. These corrections can be included by replacing $\omega_{pe}^2 \to \omega_{pe}^2 + \omega_{pi}^2$ and $\omega_{ge} \to \omega_{ge} \pm \omega_{gi}$, where the upper sign applies inside the square roots, the lower sign outside the square roots. Because of the small mass ratio these corrections are of little importance.

Solving for the low-frequency R-mode one also finds that the whistler dispersion relation (9.125) is changed by the inclusion of the ion correction

$$\omega = \frac{\omega_{ge}}{2}\left(1 + \frac{\omega_{pe}^2}{k^2 c^2}\right)^{-1}\left[\left(1 + \frac{4\omega_{pi}^2}{k^2 c^2}\right)^{1/2} + 1\right] \qquad (9.138)$$

For long wavelengths, $k^2 \ll \omega_{pi}^2/c^2$, where the ion term in the square root dominates, this dispersion relation reduces to the Alfvén wave, $\omega_A = \pm k_\parallel v_A$, so that in a quasineutral electron-ion plasma the long-wavelength, low-frequency limit of the R- or whistler mode is a right-handed Alfvén wave with vanishing frequency at infinitely long wavelengths. The Alfvén wave at low frequencies therefore has a right-handed and a left-handed component, which both are dominated by ion inertia. In general, Alfvén waves will have an elliptical polarization, but they can become linearly polarized when the two oppositely polarized modes have about equal amplitudes.

At shorter wavelengths, where the ion term can be neglected, the R-mode with the dispersion relation (9.138) becomes a whistler, whose parallel wavenumber satisfies

$$4\omega_{pi}^2 \ll k^2 c^2 \ll \omega_{ge}^2 \qquad (9.139)$$

showing that the whistler wavelength is much longer than the *electron inertial length*, c/ω_{pe}, but still much shorter than the *ion inertial length*, c/ω_{pi}.

The dispersion relation of the L-mode at low frequencies can be written down in full analogy to that of the R-mode as

$$\omega = \frac{\omega_{ge}}{2}\left(1+\frac{\omega_{pe}^2}{k^2 c^2}\right)^{-1}\left[\left(1+\frac{4\omega_{pi}^2}{k^2 c^2}\right)^{1/2}-1\right] \qquad (9.140)$$

The long-wavelength limit is again an Alfvén wave, this time the left-handed mode of Eq. (9.98). At slightly shorter wavelengths the L-mode wave has a dispersion similar to the electron whistler and becomes an *ion whistler* which runs into the L-mode ion cyclotron resonance, $\omega_{L,res}$, when the wavenumber increases further.

Perpendicular Wave Propagation

For perpendicular propagation, the dispersion relation (9.131) can be written as

$$N_\perp^2 = \epsilon_1 - \epsilon_2^2/\epsilon_1 = 0 \qquad (9.141)$$

In the extremely-low frequency limit, $\omega \to 0$, we find for $\omega_{ge} \gg \omega_{gi}$

$$\lim_{\omega\to 0}\epsilon_1 = \lim_{\omega\to 0}\epsilon_2 = 1+\frac{c^2}{v_A^2} \qquad (9.142)$$

as the low-frequency dielectric constant for the extraordinary wave. This is the relevant low-frequency dielectric constant of a plasma, because at such low frequencies no ordinary wave mode can propagate. Instead the dispersion relation becomes that of the two-fluid Alfvén wave, $k^2 c^2/\omega^2 = \epsilon_1$, or

$$\omega = \pm k v_A\left(1+\frac{v_A^2}{c^2}\right)^{-1/2} \qquad (9.143)$$

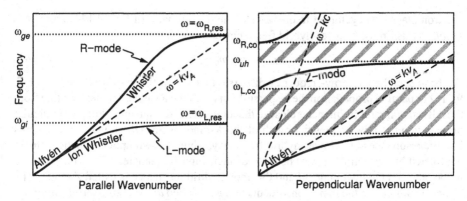

Fig. 9.11. Low-frequency dispersion branches for parallel and perpendicular propagation.

As claimed earlier, the X-mode at very low frequencies close to zero frequency becomes an Alfvén wave. Since its wavenumber vanishes, too, it turns out to be a long wavelength mode. Its actual phase velocity is slightly slower than the Alfvén velocity, however. Hence, the L-X-mode very-low frequency branch is an Alfvén wave which propagates parallel as well as perpendicular to the magnetic field. This two-fluid Alfvén mode is nearly isotropic.

Using Eq. (9.134) and assuming $\omega \ll \omega_{ge}$, one finds from Eq. (9.141) that for $\epsilon_1 \to 0$ another resonance appears at

$$\omega_{lh}^2 = \frac{\omega_{pi}^2 + \omega_{gi}^2}{1 + \omega_{pe}^2/\omega_{ge}^2} \qquad (9.144)$$

It is called the *lower-hybrid resonance*. The numerator of the lower-hybrid resonance frequency resembles the upper-hybrid resonance frequency in Eq. (9.133) in that the electron plasma and cyclotron frequencies in the latter are replaced here with the corresponding ion frequencies. But it also contains an electron contribution in its denominator which results from the electron response to the low-frequency electric wave field, $\delta \mathbf{E}_\perp$. The cold electrons respond to this transverse electric field by performing an electric drift motion perpendicular to the magnetic field.

In most cases the ion-cyclotron frequency can be neglected in the numerator of Eq. (9.144), if the Alfvén velocity is slower than the speed of light, $v_A^2/c^2 = \omega_{gi}^2/\omega_{pi}^2 \ll 1$. This condition may be violated in diluted plasmas or in very strong magnetic fields in which case the ion-cyclotron frequency contributes to the lower-hybrid resonance. It is easy to show that under the condition $m_e/m_i \ll v_A^2/c^2 \ll 1$ the lower-hybrid frequency is equal to the ion plasma frequency, $\omega_{lh} = \omega_{pi}$, while for $v_A^2/c^2 > 1$ it is equal to the ion-cyclotron frequency, $\omega_{lh} = \omega_{gi}$. In the opposite case of dense plasmas the ratio of electron plasma to cyclotron frequencies in the denominator of Eq. (9.144) dominates unity. In this limit the lower-hybrid frequency approaches

$$\omega_{lh} = (\omega_{ge}\omega_{gi})^{1/2} \tag{9.145}$$

We have sketched the low-frequency dispersion curves of the two-fluid waves in Fig. 9.11. For parallel propagation the dispersion curves of the R- and L-modes separate from the Alfvén branch at frequencies close to the ion-cyclotron resonance. For perpendicular propagation the low-frequency branch of the X-mode is an Alfvén wave which turns into the resonance at the lower-hybrid frequency, while the two upper branches remain unaffected by the ion dynamics because of their high frequencies. The non-escaping mode between the lower cut-off and the upper hybrid resonance is called *Z-mode*. The Z- and X-modes are the main constituents of the so-called *auroral kilometric radiation*.

Oblique Propagation

The above theory can be extended to arbitrary directions of propagation. In this case we return to the general dispersion relation of a cold plasma in Eq. (9.112), but to include the ion contribution, so that the components of the dielectric tensor are given by Eq. (9.134). The right-hand and left-hand components of the dielectric tensor in Eq. (9.120) are similarly redefined. These redefinitions permit us to rewrite the cold plasma dispersion relation (9.115) as function of k^2 and the angle θ between $\mathbf{B_0}$ and $\mathbf{k}$

$$\text{Det} \begin{bmatrix} \left(\dfrac{kc}{\omega}\right)^2 \cos^2\theta - \epsilon_1 & i\epsilon_2 & -\left(\dfrac{kc}{\omega}\right)^2 \sin\theta\cos\theta \\ -i\epsilon_2 & \left(\dfrac{kc}{\omega}\right)^2 - \epsilon_1 & 0 \\ -\left(\dfrac{kc}{\omega}\right)^2 \sin\theta\cos\theta & 0 & \left(\dfrac{kc}{\omega}\right)^2 \sin^2\theta - \epsilon_3 \end{bmatrix} = 0 \tag{9.146}$$

This equation can be solved either for the wavenumber, k^2, or for the angle, θ. Writing it as an equation for k^2, it becomes a biquadratic equation in $N^2 = k^2 c^2/\omega^2$

$$AN^4 - BN^2 + C = 0 \tag{9.147}$$

The coefficients of this equation are

$$\begin{aligned} A &= \epsilon_1 \sin^2\theta + \epsilon_3 \cos^2\theta \\ B &= \epsilon_R\epsilon_L \sin^2\theta + \epsilon_1\epsilon_3(1 + \cos^2\theta) \\ C &= \epsilon_3\epsilon_R\epsilon_L \end{aligned} \tag{9.148}$$

The solution of Eq. (9.147) is the *Appleton-Hartree equation*

$$N^2 = \frac{B \pm \sqrt{B^2 - 4AC}}{2A} \tag{9.149}$$

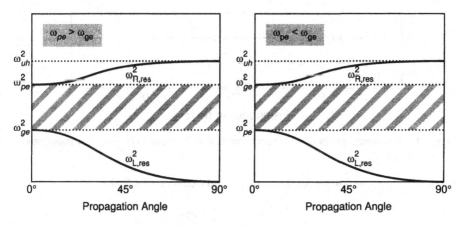

Fig. 9.12. Angular variation of cut-offs and resonances.

which shows that there are two principal propagating wave solutions, which may have different branches depending on the sign of the expression under the square root. For parallel and perpendicular propagation we have identified these modes as the (R,L)- and (O,X)-modes, respectively. For oblique angles these modes are smoothly connected.

Solving for the angular function, one finds a rearrangement of Eq. (9.149)

$$\tan^2 \theta = -\frac{\epsilon_3 (N^2 - \epsilon_R)(N^2 - \epsilon_L)}{(\epsilon_1 N^2 - \epsilon_R \epsilon_L)(N^2 - \epsilon_3)} \tag{9.150}$$

From which the angular dependence of the resonances, $k \to \infty$, may be determined as

$$\tan^2 \theta_{\text{res}} = -\frac{\epsilon_3}{\epsilon_1} \tag{9.151}$$

For a pure electron plasma this condition can be written as a biquadratic equation

$$\omega_{\text{res}}^4 - \omega_{uh}^2 \omega_{\text{res}}^2 + \omega_{ge}^2 \omega_{pe}^2 \cos^2 \theta = 0 \tag{9.152}$$

There are two resonance frequencies, whose variation with θ is plotted in Fig. 9.12 for a dense, $\omega_{pe} > \omega_{ge}$, and a dilute, $\omega_{ge} > \omega_{pe}$, plasma. In the first case the upper-hybrid resonance at perpendicular propagation moves to a resonance at the plasma frequency for parallel propagation, while the cyclotron resonance decreases from the ω_{ge} at 0° to zero at 90°. In the second case the cyclotron resonance increases from the cyclotron frequency at parallel propagation to the upper-hybrid frequency for perpendicular propagation, while the other resonance moves from zero at 90° to ω_{pe} at 0°.

Including the ion contribution leads only to minor changes in the angular dependence of the higher frequency resonances. However, the lower-hybrid resonance becomes weakly sensitive on angle. If we approximate Eq. (9.151) for low frequencies,

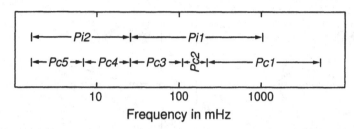

Fig. 9.13. Frequency ranges of ULF continuous (Pc) and irregular (Pi) magnetic pulsations.

$\omega_{gi}^2 \ll \omega^2 \ll \omega_{ge}^2$, and solve for the lower-hybrid frequency we find

$$\omega_{lh}^2 = \frac{\omega_{pi}^2 + \omega_{pe}^2 \cot^2 \theta}{1 + \omega_{pe}^2/\omega_{ge}^2} \tag{9.153}$$

For perpendicular propagation we have $\cot\theta = 0$ and Eq. (9.153) becomes identical with the corresponding approximation to Eq. (9.144), while for parallel propagation this expression breaks down due to our approximation. Writing (9.151) as

$$\epsilon_3 + \epsilon_1 \tan^2 \theta = 0 \tag{9.154}$$

we find that for $\theta = 0$ this condition implies that $\epsilon_3 = 0$ which means that for parallel propagation the resonance approaches the value $\omega_{\text{res},\parallel} = \omega_{pe}(1+m_e/m_i)^{1/2}$. The lower-hybrid resonance thus smoothly approaches a frequency close to the plasma frequency. The general expression for the cut-offs follows from setting $C = 0$ in Eq. (9.149). This is equivalent to putting $\epsilon_3 \epsilon_R \epsilon_L = 0$. The latter does not depend on angle and, hence, the cut-off frequencies do not change with the angle of propagation.

As a final example, we consider the change in the whistler dispersion relation when including the angle of propagation. In this case Eq. (9.122) becomes

$$\frac{k^2 c^2}{\omega^2} = 1 + \frac{\omega_{pe}^2}{\omega(\omega_{ge} \cos \theta - \omega)} \tag{9.155}$$

The whistler resonance frequency decreases from the cyclotron frequency at parallel propagation to a certain angle, $\cos\theta_{\text{res}} = \omega/\omega_{ge}$, beyond which the whistler does not propagate anymore. This is the *whistler resonance angle*. A whistler with a given frequency is confined to a certain cone of propagation around the field line.

Investigation of wave propagation in a cold plasma has turned out to be a complicated play with the dispersive properties of the different wave modes. An exhaustive encyclopedia is beyond the scope of this text, but we note that App. B.5 describes a useful tool for plasma wave analysis, the *CMA-diagram*.

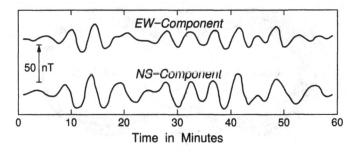

Fig. 9.14. Ground magnetic disturbance of a Pc5 pulsation.

9.7. Geomagnetic Pulsations

As an example of the application of the above theory to the Earth's environment we turn to the discussion of a phenomenon which is known in geomagnetism since about one century. This is the existence of fast fluctuations of the Earth's surface magnetic field in the frequency range from a few Millihertz up to a few Hertz, corresponding to oscillation periods from several hundred seconds to a fraction of a second. Here we are in the ultra-low frequency, range which is conventionally divided into five intervals, Pc1-Pc5, for *continuous pulsations* and into two intervals, Pi1 and Pi2, for *irregular pulsations* (see Fig. 9.13). In many cases, the pulsating disturbance fields observed are associated with Alfvén waves

The class of continuous pulsations covers quasi-sinusoidal oscillations of narrow spectral bandwidth, like shown in Fig. 9.14. They may have a comparably long duration from several minutes up to hours. Pc pulsations can generally be observed over a wide latitudinal and longitudinal range on the Earth's surface and in the magnetosphere, but their frequencies and amplitudes often exhibit a latitudinal variation.

The irregular pulsations, in contrast, are shorter-lived, each pulsation sometimes being composed of only a few oscillations decaying in time and neither being of sinusoidal form nor having a well-expressed spectral peak. Rather they have a broad spectrum. Pi pulsations are usually also more localized, both in latitude and longitude and have similar spectrum over the region where they are observed.

Field Line Resonance

The Pc5 pulsations shown in Fig. 9.14 are caused by oscillations of magnetospheric field lines. The simplest wave mode which can be called for the explanation of standing magnetic field line oscillations is the single-fluid shear Alfvén wave. For a standing wave the length of the field line between the two reflection points, ℓ, must be a multiple of half the parallel wavelength or $\nu_h \lambda_\parallel = 2\ell$, where $\nu_h = 1, 2, 3, \ldots$ From the dispersion relation

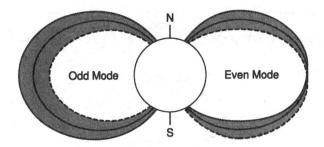

Fig. 9.15. Fundamental poloidal field line resonances.

of the shear Alfvén wave, $\omega = k_\parallel v_A$, one finds for the possible oscillation frequencies of standing Pc oscillations

$$\omega_h = \frac{v_h \pi \langle v_A \rangle}{\ell} \qquad (9.156)$$

where $\langle v_A \rangle$ is the average Alfvén velocity along the field line. This approximate formula shows that each particular field line has a number of distinct Alfvénic resonances. Since the length of the field lines increases with latitude, the resonance frequency decreases with latitude. For an average Alfvén velocity of 1000 km/s, the fundamental resonance frequency on closed field lines, outside the polar cap, ranges between 1 and 100 mHz and thus falls into the Pc3 to Pc5 range.

Figure 9.15 shows schematically how the dipolar field configuration changes for two fundamental types of field line resonances. The footpoints of the field lines are fixed on the Earth while the dipole field lines may either perform a breathing motion (fundamental odd mode) or a wobbling motion (even mode). In addition to these poloidal modes, the field lines can also exert toroidal oscillations, in which case the elongation of the field line and the plasma bulk flow are purely azimuthal. Pc pulsations often are a mixture of poloidal and toroidal oscillations.

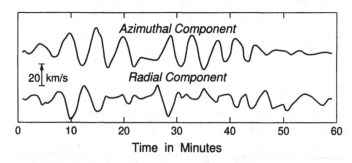

Fig. 9.16. Magnetospheric plasma drift during a Pc5 pulsation event.

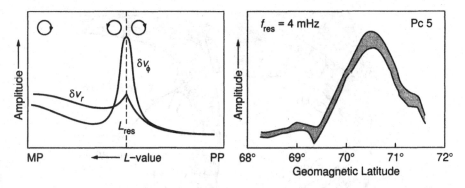

Fig. 9.17. L-dependence of the amplitude of Pc5 field line resonance.

The Alfvénic oscillations of the magnetospheric field lines are, of course, associated with oscillations of the plasma bulk flow velocity. Figure 9.16 shows a spacecraft measurement of the perpendicular magnetospheric plasma drift at geostationary altitude, decomposed into azimuthal, v_ϕ, and radial, v_r, components for the same Pc5 pulsation event shown in Fig. 9.14. Both, on the ground and in space the phase difference between the components indicates elliptic polarization. In particular the plasma drift velocity exhibits a large phase shift of about 180° with v_ϕ being ahead of v_r.

Sources of Pc Pulsations

One can imagine the field line resonances to occur as modes which are excited on the entire magnetosphere. Some periodic disturbance of frequency ω_{ex} arriving at the magnetopause may set the field line with resonance frequency $\omega_{res} = \omega_{ex}$ into oscillation while all other field lines, whose resonance frequencies do not match ω_{ex}, are only marginally excited and do not contribute to the pulsation. Since ω_{res} is a function of space, one particular field line or L-shell is excited.

The left-hand panel of Fig. 9.17 provides a schematic view how the magnetospheric flux tubes, i.e., the field lines and their plasma content, respond to the an excitation at a particular frequency, ω_{ex}. Since the field line eigenperiod varies with L, the external frequency picks out the resonant field line, where the amplitude of the flux tube motion maximizes. Interestingly, the polarization of the oscillation changes when crossing the resonance, as indicated in the figure. It is nearly linear at the resonant shell, L_{res}.

The second part of Fig. 9.17 shows a ground-based radar measurement of the latitudinal dependence of a Pc5 field line resonance region in high latitudes. This measurement shows a resonance region of about 1° latitudinal width, indicating that the source wave may not have been monochromatic but has resonantly excited a whole geomagnetic flux tube of about 1000 km radial width in the magnetosphere.

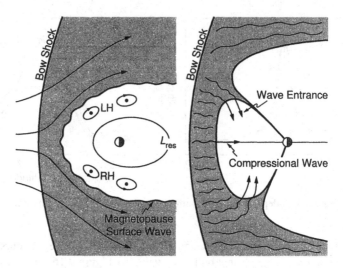

Fig. 9.18. Magnetopause surface waves and magnetosheath turbulence as sources of pulsations.

Concerning the initial wave source which excites the field line resonance, there are two possible scenarios. Surface waves excited via the Kelvin-Helmholtz instability by the flow of the solar wind around the magnetopause (left part of Fig. 9.18) can set the inner magnetosphere into oscillation and become resonant at the resonant L-shell, L_{res}. For Pc5 pulsations this mechanism seems to work and the polarization pattern is in rough agreement with the polarization pattern of the surface wave showing left-handed (LH) oscillations at dawn and right-handed (RH) oscillations at dusk. Other possibilities are compressional waves which enter the magnetosphere at its nose or turbulence from the magnetosheath which may enter directly through the cusps and leak into the magnetosphere (right-hand panel of Fig. 9.18).

Pi2 Pulsations

The short-period irregular Pi2 pulsations are associated with the development of the substorm current wedge described in Sec. 5.7. Whenever field-aligned currents are suddenly switched on somewhere in the magnetosphere, they must be transported to the ionosphere via shear Alfvén waves. Only this transverse magnetohydrodynamic wave mode can carry field-aligned current. Launched in the magnetosphere, the shear Alfvén waves are then reflected back and forth between the ionosphere and the current generator in the tail until a stationary equilibrium is reached.

Figure 9.19 shows qualitatively the development of the magnetic disturbance field and thus the field-aligned current flow after switch-on of a current generator in the mag-

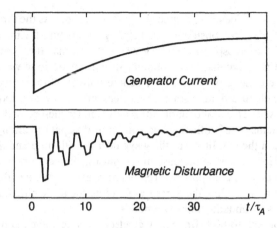

Fig. 9.19. Magnetic disturbance due to switch-on of a current generator.

netotail. At $t = 0$ an Alfvén wave is launched which carries a current corresponding to the generator current and thus has a magnetic disturbance field, ΔB_1, corresponding to the magnetic disturbance caused by the generator current, $\Delta B_1 = \Delta B_g(0)$. This wave reaches the ionosphere at $t = \tau_A$, the Alfvén wave travel time between magnetosphere and ionosphere of some 30 s. Here about 80% of its amplitude is reflected with the magnetic field of the reflected wave, ΔB_2, adding to the primary disturbance field, ΔB_1.

At $t = 2\tau_A$ the reflected wave comes back to the generator and launches a third wave, whose magnetic disturbance, ΔB_3, is determined by the requirement that the total disturbance matches that caused by the current of the generator at time $t = 2\tau_A$, namely $\Delta B_g(2\tau_A) = \Delta B_1 + \Delta B_2 + \Delta B_3$. Since typically $\Delta B_1 + \Delta B_2 > \Delta B_g(0) > \Delta B_g(2\tau_A)$, the magnetic disturbance of the third wave must be of opposite polarity and thus decrease the total wave magnetic field. Multiple bounces of the wave lead to magnetic field disturbances which oscillate with a period of $4\tau_A$ until they finally converge to match the generator current. In particular, the magnetic oscillations with periods of some 100 s are readily observable as Pi2 pulsations. They are often used as a good indicator for the onset of substorms.

Concluding Remarks

The fluid theory of plasma waves is only the first step in a more sophisticated approach to plasma wave generation and propagation. Since it is based on the fluid model, it neglects all microscopic particle correlations. Such an approach is justified on relatively large scales, if the waves have sufficiently long wavelengths and sufficiently low frequencies. Moreover, dissipative effects must be negligible to justify the ideal plasma model.

The first restriction does not provide any problem as long as the three basic magnetohydrodynamic modes are considered. Neglecting dissipation for these waves is also justified, except in the ionosphere and the lower solar corona. It is, however, not a big problem to include resistive dissipation into the discussion of these waves. Since in the linear treatment the conductivity does not become infinite in a resistive plasma and is independent of the electric field, a simple replacement of the wave frequency by $\omega \rightarrow \omega - i\nu_c$, where ν_c is the classical or anomalous collision frequency, will account for dissipative effects. Inclusion of collision frequencies or dissipation simply leads to *wave damping*. As a result the amplitude of the wave decays with time and the wave energy turns into heat, which causes an increase of the plasma temperature and ultimately violates the assumption of a cold plasma. Temperature effects can be taken into account in a fluid approach by including the pressure term into the basic equations, but leads to further splitting of the fundamental wave modes.

The fluid approach to high-frequency electromagnetic waves is justified because for such frequencies the plasma reacts passively like a dielectric medium. Nevertheless, in the following chapters we will encounter some important microscopic effects on electromagnetic radiation as well.

Further Reading

An exhaustive elementary description of fluid plasma waves is given in [1]. A compendium of wave propagation in the magnetosphere and ionosphere can be found in [2]. For more elaborate theories of wave phenomena consult [5, 6]. In particular, monograph [6] is a classic among the plasma wave books. However, as its starting point it takes the kinetic theory which will we will develop in the next chapter. If one is interested in whistlers, the old monograph [5] is still very useful. Further information about pulsations is contained in [3] and the references therein.

[1] J. A. Bittencourt, *Fundamentals of Plasma Physics* (Pergamon Press, Oxford, 1986).

[2] K. G. Budden, *Radio Waves in the Ionosphere* (Cambridge University Press, Cambridge, 1961).

[3] K. H. Glaßmeier, in *Handbook of Atmospheric Electrodynamics, Vol. 2*, ed. H. Volland (CRC Press, Boca Raton, 1995), p. 463.

[4] R. A. Helliwell, *Whistlers and Related Ionospheric Phenomena* (Stanford University Press, Stanford, 1965).

[5] D. G. Swanson, *Plasma Waves* (Academic Press, Boston, 1989).

[6] T. H. Stix, *Waves in Plasma* (American Institute of Physics, New York, 1992).

10. Wave Kinetic Theory

The most general theory of plasma waves makes use of the kinetic theory of a plasma. As has been demonstrated in Chap. 6, the only approximations underlying the kinetic theory are the statistical assumption and the assumption that higher order particle correlations can be neglected in a statistical theory based on a perturbation approach. But in contrast to the fluid theory of plasma waves, the wave kinetic theory takes explicitly care of the properties of the particle distribution function and its variations and of the correlations between particles and fields. Hence, entirely new effects will appear in this theory which cannot be covered by the fluid approach to a plasma.

Because in wave kinetic theory we are dealing with distribution functions and their evolution, the set of mass, momentum and energy conservation equations of fluid theory is replaced by the set of Vlasov equations for the different components of the plasma while the field equations remain the same. This implies that for the investigation of linear waves the formal structure of the general linear wave dispersion relation in Eq. (9.55) remains unchanged. The only quantity to be replaced is the dielectric tensor, $\epsilon(\omega, \mathbf{k})$, because it contains the particle dynamics. Its calculation is instrumental to the subsequent investigation of the kinetic properties of the plasma waves.

10.1. Landau-Laplace Procedure

In kinetic plasma wave theory the microscopic electric charge separation fields and particle currents become important. The dispersive properties of the plasma will therefore look rather complicated. Because of this reason, we start by neglecting any magnetic fields and thus considering the plasma as unmagnetized and the waves as purely electrostatic. One should, however, be aware that these assumptions are valid only when there are no oscillating microscopic currents (or fast particle motions) in the plasma, which is a poorly satisfied assumption when considering high-frequency electron oscillations against ions at rest. But in a plasma where the electron temperature is sufficiently small compared to the electron rest energy, i.e., in a nonrelativistic plasma with $v_e \ll c$, the electromagnetic part of the waves is usually weak.

Let us initially consider only the one-dimensional case. Under these assumptions the Vlasov equation Eq. (6.20) for electrons and ions simplifies considerably

$$\frac{\partial f_{e,i}(v, x, t)}{\partial t} + v\frac{\partial f_{e,i}(v, x, t)}{\partial x} \pm \frac{e}{m_{e,i}}E(x, t)\frac{\partial f_{e,i}(v, x, t)}{\partial v} = 0 \qquad (10.1)$$

The electric field is purely electrostatic and satisfies the Poisson equation

$$\frac{\partial E(x, t)}{\partial x} = \frac{e}{\epsilon_0}\int\limits_{-\infty}^{\infty} dv\,[f_i(v, x, t) - f_e(v, x, t)] \qquad (10.2)$$

where on the right-hand side we have expressed the particle densities through the moments of the corresponding distribution functions. Although the Poisson equation is a linear equation, the system of Eqs. (10.1) and (10.2) is nonlinear, because eliminating the electric field from the Vlasov equation with the help of Poisson's equation produces a product between the partial derivative of the distribution function with respect to the velocity in the last term of the Vlasov equation and an integral over the distribution function. In addition, the Poisson equation couples the two particle species together.

Considering waves on an otherwise quiet background requires a separation of the distribution functions and the electric field into undisturbed parts, $f_{e,i0}$, E_0, and perturbations, $\delta f_{e,i}$, δE, according to

$$f_{e,i} = f_{e,i0} + \delta f_{e,i} \qquad (10.3)$$
$$E = E_0 + \delta E \qquad (10.4)$$

We now assume that the perturbations of the distribution function are linear, i.e., they are, for all times and at all positions in the plasma, much smaller then the local undisturbed value of the distribution, $|\delta f| \ll f_0$. The physical content of this assumption is that the probability to find a given number of particles at any time and position in the equilibrium state is much larger than to find them in an excited and perturbed state. Hence, the unperturbed particle density will be much larger than the density of particles participating in the oscillation. This assumption permits us to linearize the Vlasov-Poisson system of equations in order to obtain the linear set

$$\frac{\partial \delta f_{e,i}}{\partial t} + v\frac{\partial \delta f_{e,i}}{\partial x} \pm \frac{eE_0}{m_{e,i}}\frac{\partial \delta f_{e,i}}{\partial v} = \mp\frac{e\delta E}{m_{e,i}}\frac{\partial f_{e,i0}}{\partial v} \qquad (10.5)$$

$$\frac{\partial \delta E}{\partial x} = \frac{e}{\epsilon_0}\int\limits_{-\infty}^{\infty} dv\,[\delta f_i - \delta f_e] \qquad (10.6)$$

for $\delta f_{e,i}(v, x, t)$ and $\delta E(x, t)$. Yet these equations are still very general, but we can simplify them if we suppose that the undisturbed distribution function varies very slowly in time and space compared to the perturbation. In such a case the equilibrium distribution, $f_0(v)$, becomes a function of velocity only. In addition the undisturbed plasma should be in a quasineutral state. Then $E_0 = 0$ and the third term on the left-hand side of the linearized Vlasov equation (10.5) vanishes. The remaining three coupled equations constitute a linear but inhomogeneous system of partial integro-differential equations in the three variables v, x, t with constant coefficients for the perturbations of the ion and electron distribution functions. After solving them, the electric perturbation field can be calculated separately.

Before proceeding, remember that we wanted to calculate the dielectric function of the plasma. Referring to its definition in Eq. (9.54) this requires to find the plasma wave conductivity, $\sigma(\omega, \mathbf{k})$, which in our one-dimensional and isotropic case is a scalar and is determined from the perturbed current density

$$\delta j = e n_0 (\delta v_i - \delta v_e) = \sigma \delta E \tag{10.7}$$

by comparing the coefficients of δE, after having expressed the perturbations of the particle velocities calculated from the perturbed distribution functions through the electric perturbation field.

Langmuir Waves

In order to demonstrate the method of how to solve the Vlasov-Poisson system we restrict ourselves to the case of high-frequency perturbations applied to the plasma. For frequencies far above the ion plasma frequency the ions can then be considered as an immobile stationary charge-neutralizing background. The Vlasov-Poisson system reduces to the two equations

$$\frac{\partial \delta f(v, x, t)}{\partial t} + v \frac{\partial \delta f(v, x, t)}{\partial x} = \frac{e}{m_e} \delta E(x, t) \frac{\partial f_0(v, x, t)}{\partial v} \tag{10.8}$$

$$\frac{\partial \delta E(x, t)}{\partial x} = -\frac{e}{\epsilon_0} \int_{-\infty}^{\infty} dv\, \delta f(v, x, t) \tag{10.9}$$

We have dropped the index e on the electron distribution function here in order to simplify notation. We also note that for the purely electrostatic longitudinal perturbations the electric field can be written as the gradient of an electrostatic potential

$$\delta E = -\frac{\partial \phi}{\partial x} \tag{10.10}$$

There are two ways to solve the above linear system of Vlasov-Poisson equations. Because the system is linear, one may use the plane wave ansatz which corresponds to a solution by Fourier transformation both in space and time. This method implicitly assumes that the plasma is periodic in space and time. However, time evolves on a linear scale and assuming periodicity is physically not entirely justified. To make physics more transparent we therefore choose the more involved method of Fourier transforming in space, while solving the time-dependent equations in a different way.

Fourier transformation in space yields the following representation of the perturbations of the distribution function and electric field

$$[\delta f(v, x, t), \delta E(x, t)] = (2\pi)^{-1/2} \int_{-\infty}^{\infty} dk\, [\delta f(v, k, t), \delta E(k, t)]\, e^{ikx} \qquad (10.11)$$

where the Fourier transform of the electric field can also be expressed by the Fourier transformed potential as $\delta E(k, t) = -ik\phi(k, t)$, and the linearized Poisson equation (10.9) yields

$$\delta E(k, t) = \frac{ie}{\epsilon_0 k} \int_{-\infty}^{\infty} dv\, \delta f(v, x, t) \qquad (10.12)$$

while from the Vlasov equation (10.8) we obtain the differential equation

$$\frac{\partial \delta f(k, v, t)}{\partial t} + ikv\, \delta f(k, v, t) - \frac{e}{m_e} \delta E(k, t) \frac{\partial f_0(k, v, t)}{\partial v} = 0 \qquad (10.13)$$

for the Fourier transformed perturbation of the distribution function.

Landau-Laplace Procedure

We solve this equation by taking the Laplace transform of the two unknown functions

$$[\delta f(k, v, p), \delta E(k, p)] = \int_{0}^{\infty} dt\, [\delta f(k, v, t), \delta E(k, t)]\, e^{-pt} \qquad (10.14)$$

under the assumption that this integral converges. This assumption requires that the real part of p is sufficiently large and positive, larger than any possible negative real part of any time-dependent exponential factor contained in $\delta E, \delta f$. The Laplace transform explicitly takes into account that the perturbation evolves in time. The Laplace transformed Eqs. (10.12) and (10.13) are

$$(p + ikv)\delta f(k, v, p) - \frac{e}{m_e}\delta E(k, p)\frac{\partial f_0(v)}{\partial v} = g(k, v) \qquad (10.15)$$

$$\delta E(k, p) = \frac{ie}{\epsilon_0 k} \int_{-\infty}^{\infty} dv\, \delta f(k, v, p) \qquad (10.16)$$

The inhomogeneity appearing on the right-hand side of the upper equation is the initial value of the perturbed distribution, $g(k, v) = \delta f(k, v, t = 0)$. It contains the asymmetric time-behavior of the perturbation. The inhomogeneous set of algebraic equations (10.15) and (10.16) has a well-defined unique solution for the evolution of the disturbances of the electron distribution function and the electric field. The Fourier-Laplace transform of this solution is

$$\delta f(k, v, p) = (p + ikv)^{-1} \left[\frac{e}{m_e} \delta E(k, p) \frac{\partial f_0(v)}{\partial v} + g(k, v) \right] \qquad (10.17)$$

$$\delta E(k, p) = \frac{ie}{\epsilon_0 k \epsilon(k, p)} \int_{-\infty}^{\infty} dv \frac{g(k, v)}{p + ikv} \qquad (10.18)$$

The new term $\epsilon(k, p)$ appearing in the denominator of the solution for $\delta E(k, p)$ and, as a consequence, also in the first term on the right-hand side of the solution for the perturbed distribution function is given by

$$\epsilon(k, p) = 1 - \frac{i\omega_{pe}^2}{n_0 k} \int_{-\infty}^{\infty} dv \frac{\partial f_0(k, v, p)/\partial v}{p + ikv} \qquad (10.19)$$

where we used the definition of the electron plasma frequency, $\omega_{pe}^2 = e^2 n_0/\epsilon_0 m_e$. To find the physically relevant solution one must transform back to real time. Inversion of the Laplace transform of the electric perturbation field gives

$$\delta E(k, t) = \frac{1}{2\pi i} \int_{a-i\infty}^{a+i\infty} dp \, e^{pt} \delta E(k, p) \qquad (10.20)$$

Here a is a real 'large enough' constant, and the contour of integration is a line which is parallel to the imaginary axis in the complex p plane to the right of all singularities of the integrand in Eq. (10.18) in order to warrant convergence

$$\delta E(k, t) = \frac{e}{2\pi \epsilon_0 k} \int_{a-i\infty}^{a+i\infty} dp \, e^{pt} \left[\frac{\displaystyle\int_{-\infty}^{\infty} dv \frac{g(k, v)}{p + ikv}}{1 - \frac{i\omega_{pe}^2}{n_0 k} \displaystyle\int_{-\infty}^{\infty} dv \frac{\partial f_0(k, v, p)/\partial v}{p + ikv}} \right] \qquad (10.21)$$

In this expression the integrals in the numerator and denominator contain poles at $v = ip/k$ even for the analytic functions $g(k, v)$, $\partial f_0(v)/\partial v$. In addition the Laplace integral

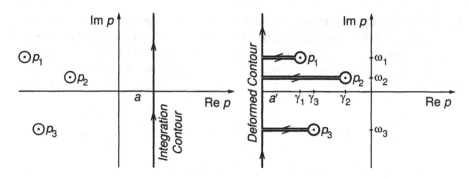

Fig. 10.1. Integration contours and poles of Eq. (10.21).

may has singularities at all points where $\epsilon(k, p) = 0$. Let us therefore assume that at least in a physically acceptable solution, $g(k, v)$ and $\partial f_0(v)/\partial v$ are analytic functions, so that the integrals over v can be analytically continued. In this case we can apply the simple analytic theory of App. A.7. Let us further assume without restriction of generality that $k > 0$. Then the two integrals over v become entire functions. The only possible singularities of the integral in Eq. (10.21) are the poles of its denominator

$$\epsilon(k, p) = 0 \qquad (10.22)$$

Equation (10.22) may have a finite number of solutions, $p_i(k)$. These are the poles of the integral in Eq. (10.21). Let us choose the position, a, of the path of integration in Eq. (10.21) so that it is larger than the largest of the real parts of the $p_i(k) = \gamma_i(k) - i\omega_i(k)$, where we split p_i into real and imaginary parts. Integrating along a line at $a = $ const parallel to the imaginary axis (as shown on the left-hand side of Fig. 10.1) and assuming that for physical solutions $E(k, p) \rightarrow 0$ for $|p| \rightarrow \infty$, the integral will exist, and we can deform the contour of integration by pulling it into the negative direction of the real p axis to a position $a = a'$, far beyond all the poles. It will consist of small circles around the poles and a piecewise continuous path parallel to the imaginary p axis with horizontal connections to the poles along which the contributions to the integral cancel (see right-hand side of Fig. 10.1). The integral becomes the sum of all the residua, $r_i(p_i)$, at the poles, $p_i(k)$, plus the integral along the piecewise continuous path parallel to the imaginary axis

$$\delta E(k, t) = \sum_i r_i(p_i) \exp[p_i(k)t] + (2\pi i)^{-1} \int_{a'-i\infty}^{a'+i\infty} dp\, e^{pt} \delta E(k, p) \qquad (10.23)$$

The integral taken along $a' = $ const vanishes in the long-time limit, $t \rightarrow \infty$, as

$$\lim_{t \rightarrow \infty} \exp(-|a'|t) \rightarrow 0 \qquad (10.24)$$

so that only the sum of the residua survives. Moreover, of these residua only the one with the smallest negative real part is, in the long-time limit, of interest. For this particular pole, $p_l(k) = \gamma_l(k) - i\omega_l(k)$, the electric perturbation field assumes the asymptotic form

$$\delta E(k, t) \propto \exp[\gamma_l(k)t - i\omega_l(k)t] \tag{10.25}$$

Hence, the linear solution of the Vlasov-Poisson set of equations is a damped oscillation of frequency, $\omega_l(k)$, and negative damping decrement, $\gamma_l(k) < 0$. As we will demonstrate below, this frequency is the Langmuir plasma oscillation frequency found in Chap. 9. But the damping decrement is entirely new. It is the so-called *Landau damping* of Langmuir waves, which does not come from collisions between the particles, but from decorrelations between particles and waves.

10.2. Landau Damping

Considering Langmuir wave propagation in a pure charge-compensated electron plasma from the point of view of kinetic theory thus leads to a number of surprises. The first surprise is that kinetic theory introduces a larger number of wave modes and in addition shows that harmonic waves do exist only in the time-asymptotic limit when the contribution of the integral at a' = const vanishes fast enough. The second surprise is the appearance of a *collisionless damping*, which cannot be avoided and which is a purely kinetic effect. Actually, it appears as a damping only as long as $\gamma_l(k) < 0$. When this condition is violated, the asymptotic time-limit breaks down and one speaks of an *instability*. But under equilibrium conditions all waves will tend to damp out. In the final state of the equilibrium only thermal fluctuations survive, where the thermal particle motion causes small-scale electric fluctuations which are immediately damped out by Landau damping. It is the equilibrium between damping and the thermal motions which causes the thermal fluctuation level given in Eq. (9.27).

Thermal Plasma

Having clarified the general properties of the solution, let us proceed to one special equilibrium distribution function. Assuming that the disturbance is introduced into a plasma in thermal equilibrium, we choose the undisturbed distribution

$$f_0(v) = n_0 \left(\frac{m_e}{2\pi k_B T_e}\right) \exp\left(-\frac{mv^2}{2k_B T_e}\right) \tag{10.26}$$

to be a Maxwellian of electron temperature T_e. Its derivative with respect to v is always negative, $\partial f_0 / \partial v = -(v/2k_B T_e) f_0$. Therefore, for $p > 0$ and $k \neq 0$, the function $\epsilon(k, p)$ has no zeros, as is obvious from its definition Eq. (10.19). All the poles of $\epsilon(k, p)$ have negative real parts of p, and the wave perturbation will be damped.

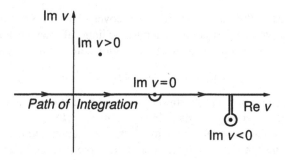

Fig. 10.2. Integration contours for three positions of the pole.

In order to determine p, one must calculate $\epsilon(k, p)$ explicitly and solve the dispersion relation in Eq. (10.22) which gives the real and imaginary parts of p at its poles. We write Eq. (10.19) in the two equivalent forms

$$\epsilon(k, p) = 1 - \frac{\omega_{pe}^2}{n_0 k^2} \int_{-\infty}^{\infty} dv \frac{\partial f_0(v)/\partial v}{v - ip/k}$$

$$= 1 + \frac{\omega_{pe}^2}{n_0} \int_{-\infty}^{\infty} dv \frac{f_0(v)}{(p + ikv)^2} \qquad (10.27)$$

where the second form arises from partial integration. The real and imaginary parts of the pole, $p = \gamma - i\omega$, are found by performing the v-integration in Eq. (10.27). Now let us assume that the real part of p at the relevant pole is small, so that in agreement with our former discussion of the surviving disturbance the damping will be small. The pole will be close to the imaginary p axis. The integration in $\epsilon(k, p)$ is along the real v axis with v having a pole at $v = (\omega + i\gamma)/k$. Remember that $\gamma < 0$ and small. The integral in Eq. (10.27) then consists of the contribution of the pole and the principal value taken along the real axis (see App. A.7). The paths of integration for the three possible cases are shown in Fig. 10.2 and the integral in Eq. (10.27) can be represented as

$$\int_{-\infty}^{\infty} dv \frac{\partial f_0(v)/\partial v}{(v - ip/k)} = \begin{cases} \int_{-\infty}^{\infty} dv_r \frac{\partial f_0(v)/\partial v}{(v_r - ip/k)} & \gamma > 0 \\[3mm] \int_{-\infty}^{\infty} dv_r \frac{\partial f_0(v)/\partial v}{(v_r - ip/k)} + 2\pi i \left. \frac{\partial f_0(v)}{\partial v} \right|_{v=ip/k} & \gamma < 0 \end{cases} \qquad (10.28)$$

The limiting case for $\gamma = 0$ is given by the Plemelj formula (see App. A.7). The contribution of the negative pole is therefore

$$-2\pi i \frac{\omega_{pe}^2}{n_0 k^2} \frac{\partial f_0(v)}{\partial v}\bigg|_{v=ip/k} \tag{10.29}$$

The main contribution to $\epsilon(k, p)$ comes from the integral along the real axis, v_r, outside the pole. It can be obtained by taking v as real and expanding the denominator in powers of ikv/p. Using the second version in Eq. (10.27) for $\epsilon(k, p)$. Expanding $(p + ikv)^{-2}$ up to second order, one obtains

$$\epsilon(k, p) = 1 + \frac{\omega_{pe}^2}{n_0 p^2} \int\limits_{-\infty}^{\infty} dv_r f_0(v_r) \left(1 - \frac{2ikv_r}{p} - \frac{3k^2 v_r^2}{p^2}\right) - 2\pi i \frac{\omega_{pe}^2}{n_0 k^2} \frac{\partial f_0(v)}{\partial v}\bigg|_{v=ip/k} \tag{10.30}$$

The real and imaginary parts of this equation can be used to find $\gamma(k, \omega)$ and $\omega(k)$ For large γ this is quite involved, but for weakly damped disturbances one can treat γ as a small correction to p. The integral of the first term in Eq. (10.30) is the unperturbed density, n_0. The integral over the second term vanishes because the plasma is at rest, $\langle v \rangle = 0$. The integral over the third term is $n_0 k_B T_e / 2m_e$. Keeping only these three terms and dropping the small contribution of the pole, the real part of the dispersion relation takes the form

$$1 - \frac{\omega_{pe}^2}{\omega^2} - \frac{3\omega_{pe}^2}{\omega^2} \frac{k^2 v_{the}^2}{\omega^2} = 0 \tag{10.31}$$

This is the dispersion relation of Langmuir waves given in Eq. (9.30) for the one-dimensional case, $\gamma_e = 3$, if we replace one of the ω^2 in the denominator of the third term by ω_{pe}^2. This term turns out to be the thermal correction to the Langmuir waves and the dispersion relation, including the damping term, becomes

$$\boxed{\omega_l = \pm\left(1 + \frac{3}{2}k^2\lambda_D^2\right) + i\gamma_l(k)} \tag{10.32}$$

The collisionless damping decrement can now be calculated by inserting $p = i\omega_l$ from here into the imaginary correction of Eq. (10.30) and calculating the derivative of the Maxwellian. Under the assumption that the damping is weak, $\gamma_l \ll \omega_l$, it is reasonable to expect that the imaginary part of $\epsilon(k, \omega)$ is also small.

Damping Rate

Under this condition it is possible to develop a simple prescription of how the damping rate can be determined from the knowledge of the dispersion relation. Let us split

$$\epsilon(k, \omega, \gamma) = \epsilon_r(k, \omega, \gamma) + i\epsilon_i(k, \omega, \gamma) \tag{10.33}$$

into real and imaginary parts, $\epsilon_r(k, \omega, \gamma)$, $\epsilon_i(k, \omega, \gamma)$. Since $\gamma_l \ll \omega_l$, one can expand with respect to the real frequency at vanishing damping rate and obtain

$$\epsilon(k, \omega, \gamma) = \epsilon_r(k, \omega, 0) + i\gamma \left.\frac{\partial \epsilon_r(k, \omega, \gamma)}{\partial \omega}\right|_{\gamma=0} + i\epsilon_i(k, \omega, 0) = 0 \qquad (10.34)$$

Setting the real and imaginary parts of this expression separately equal to zero gives

$$\epsilon_r(k, \omega, 0) = 0 \qquad (10.35)$$

$$\gamma(k, \omega) = -\frac{\epsilon_i(k, \omega, 0)}{\partial \epsilon_r(k, \omega, \gamma)/\partial \omega|_{\gamma=0}} \qquad (10.36)$$

The first of these equations is the usual dispersion relation depending only on real quantities. But the second equation is a very useful expression for calculating the damping rate of any weakly damped wave. In the following we will make extensive use of this expression. Applying it to the Langmuir wave dispersion relation (10.30) yields

$$\boxed{\gamma_l(k) = -\left(\frac{\pi}{8}\right)^{1/2} \frac{\omega_{pe}}{k^3 \lambda_D^3} \exp\left(-\frac{1}{2k^2\lambda_D^2} - \frac{3}{2}\right)} \qquad (10.37)$$

as an expression for the Landau damping of high-frequency Langmuir waves in a collisionless unmagnetized plasma. This damping is not caused by particle collisions but is entirely due to particle decorrelation effects.

Physics of Landau-Damping

The appearance of collisionless dissipation in the dispersion of electron plasma waves is somewhat disturbing, since dissipation implies a preferred direction in time, while the Vlasov equation is reversible in time, as can be seen by substituting $t \rightarrow -t$ and $v \rightarrow -v$ into Eq. (10.1). However, Landau damping only affects a very small part of the distribution function. Only particles with velocities close to the phase velocity of the wave, $v_{ph} = \omega/k$ become resonant and are redistributed in phase space during their interaction with the wave. Hence, the directivity does not affect most of the distribution function and has thus no effect on the time symmetry of the Vlasov equation.

In order to understand the physics of Landau damping, let us consider a plasma wave propagating across a plasma in thermal equilibrium with a Maxwellian equilibrium distribution function, $f_0(v)$. The situation is depicted in the left-hand part of Fig. 10.3. Particles at position $v = v_{ph} = \omega/k$ are in resonance with the wave, since they are moving at the same speed as the wave in the plasma. Clearly, these particles interact strongest with the wave electric field. Depending on the direction of this field the particles will either be accelerated or decelerated. On the other hand, any particles moving slightly faster or slower than the wave will experience a different kind of interaction.

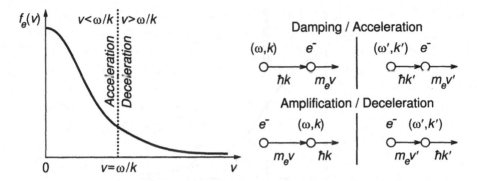

Fig. 10.3. The mechanism of Landau damping.

We can investigate this kind of interaction by exploiting the simple analogy of a collision between two particles (right-hand side of Fig. 10.3), taking the wave as an uncharged particle of energy $\hbar\omega$ and momentum $\hbar k$. In a collision between the two particles (electron and wave) the one with the higher momentum will always speed up the lower momentum particle, whereby itself will loose energy.

This kind of interaction is of elastic nature and thus dissipationless. It exactly resembles the process of Landau damping as an elastic interaction between particles and waves with no preferred direction. Fast electrons will speed up the waves, while slow electrons are pushed by the wave and gain energy. But why then an effective damping of the wave? The reason is the asymmetry of the Maxwellian distribution function with respect to the plasma wave phase velocity. There are more low than high velocity particles, and, hence, the wave looses more momentum and energy in the interaction with low momentum particles than it gains back from interaction with higher momentum particles. Clearly, during this process the distribution function must necessarily become slightly distorted as shown in Fig. 10.4. The retarded and accelerated particles right and left of the resonance are attracted by the resonance and accumulate there.

We can make this argument a little more quantitative by estimating the change in energy the which electron distribution experiences during the interaction with a Langmuir wave. This change is given by the integral

$$\delta W_e = m_e \int\limits_{-\infty}^{\infty} dv\, v \langle \Delta v \rangle f(v) \qquad (10.38)$$

where $\langle \Delta v \rangle$ is the electron velocity change averaged over one Langmuir wavelength. It is convenient to transform to a system moving with phase velocity ω/k, choosing $v' = v - \omega/k$. The change in velocity, $\langle \Delta v \rangle$, is independent of such a transformation, and

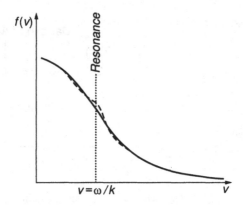

Fig. 10.4. Attraction of particles by a resonance.

the above integral becomes

$$\delta W_e = m_e \int_{-\infty}^{\infty} dv' \left(v' + \frac{\omega}{k} \right) \langle \Delta v \rangle f \left(v' + \frac{\omega}{k} \right) \qquad (10.39)$$

We now expand the distribution function around $v' = 0$ to obtain

$$\delta W_e = m_e \int_{-\infty}^{\infty} dv' \left(v' + \frac{\omega}{k} \right) \langle \Delta v \rangle \left[f \left(\frac{\omega}{k} \right) + v' \left. \frac{\partial f}{\partial v} \right|_{v'=\omega/k} \right] \qquad (10.40)$$

It can be shown by using the electron equation of motion in an oscillating electric wave field of amplitude δE_0

$$m_e \frac{dv_e}{dt} = -e\delta E_0 \sin\left[kx(t) - \omega t \right] \qquad (10.41)$$

that the average velocity variation, $\langle \Delta v \rangle$, is an odd function of v' (calculate $v_e(t)$, $x(t)$ from the above equation of motion and substitute $x(t) = x_0 + v't$ back to determine Δv). Hence, of the four product terms in the above integral only the two terms containing the product $v' \langle \Delta v \rangle$ survive the integration from $-\infty$ to $+\infty$. After integrating Δv over one oscillation period, one finds

$$\langle \Delta v(t) \rangle = \frac{A}{k^2 v'^3} \left[\cos(kv't) - 1 + \frac{1}{2}(kv't)\sin(kv't) \right] \qquad (10.42)$$

Between the constant of proportionality, A, and the wave energy density averaged over one wave oscillation, $W_w = \epsilon_0 \, \delta E_0^2 / 2$, there is the relation

$$A = W_w \frac{\omega_{pe}^2}{n_0 m_e} \qquad (10.43)$$

The change in particle energy is thus determined from the integral

$$\Delta W_e = m_e \int_{-\infty}^{\infty} dv' v' \langle \Delta \rangle \left[f\left(\frac{\omega}{k}\right) + \frac{\omega}{k} \frac{\partial f}{\partial v'} \bigg|_{v'=\omega/k} \right] \qquad (10.44)$$

The first term in the brackets can be neglected because it adds only a small contribution which is independent of the shape of the distribution function

$$\Delta W_e \approx \frac{m_e \omega}{k} \frac{\partial f}{\partial v'} \bigg|_{v'=\omega/k} \int_{-\infty}^{\infty} dv' v' \langle \Delta \rangle \qquad (10.45)$$

After integration and replacing $\omega \approx \omega_{pe}$, we get

$$\Delta W_e(t) = -A\pi t \frac{m_e \omega_{pe}}{k^2} \frac{\partial f}{\partial v} \bigg|_{v=\omega_{pe}/k} \qquad (10.46)$$

Hence, in the average over one wave oscillation period the electrons gain energy, if the derivative of the equilibrium distribution function in the vicinity of the resonance is negative. This energy is provided by the wave and leads to acceleration of the small number of resonant particles with velocities just below the wave phase speed. In equilibrium the energy transferred to the electrons per unit time equals the loss of wave energy

$$\frac{\Delta W_e(t)}{t} = -\frac{dW_w(t)}{dt} = 2\gamma_l W_w(0) \exp(-2\gamma_l t) \qquad (10.47)$$

The second part of this equation results from the definition of the average wave energy, $W_w = \epsilon_0 \delta E(t) \cdot \delta E^*(t)/2$, where after multiplication of the wave electric field with its conjugate complex part only twice the real part of the exponent survives. Inserting Eq. (10.46) for the gain in electron energy and Eq. (10.43) for the amplitude factor, A, one finds that the wave energy, W_w, appears on both sides of the equation and obtains another expression for the Landau damping

$$\gamma_l = \omega_{pe} \frac{\pi \omega_{pe}^2}{2n_0 k^2} \frac{\partial f}{\partial v} \bigg|_{v=\omega_{pe}/k} \qquad (10.48)$$

Inserting the Maxwellian distribution and remembering that it is normalized to the unperturbed density, n_0, one just recovers Eq. (10.37).

Equation (10.48) confirms that the derivative of the equilibrium distribution function in the vicinity of the resonance decides about the sign of the real part of p. Typically the derivative is negative, we have $\gamma_l < 0$ and the wave is damped. However, if distribution function has a positive derivative at the resonance, we have $\gamma_l > 0$ and the wave extracts energy from the resonant particles and grows. Such *inverse Landau damping* implies instability and will be discussed in our companion book, *Advanced Space Plasma Physics*.

10.3. Unmagnetized Plasma Waves

Landau Damping does not only affect the Langmuir waves, but also the other wave modes that propagate in a warm unmagnetized plasma. In addition, a new wave mode appears in the kinetic treatment.

Ion-Acoustic Waves

So far we have suppressed the contribution of ion inertia. From the derivation of the dispersion relation, $\epsilon(k, p) = 0$, it has become clear that the contributions of different species (electrons, ions, etc.) can be accounted for by adding a singular integral over the distribution function of the corresponding species of the same kind as in Eq. (10.27) to $\epsilon(k, p)$. Hence, including the ion contribution requires solving the following dispersion relation

$$\epsilon(k, p) = 1 + \frac{\omega_{pe}^2}{n_{0e}k^2} \int\limits_{-\infty}^{\infty} \frac{dv f_{0e}(v)}{(v - ip/k)^2} + \frac{\omega_{pi}^2}{n_{0i}k^2} \int\limits_{-\infty}^{\infty} \frac{dv f_{0i}(v)}{(v - ip/k)^2} = 0 \qquad (10.49)$$

At the high frequencies corresponding to electron plasma oscillations we can use the same expansion for the two integrals as before and find for the real part

$$1 - \frac{\omega_{pe}^2 + \omega_{pi}^2}{\omega^2} - \frac{3k^2}{\omega^4} \left(\omega_{pe}^2 v_{the}^2 + \omega_{pi}^2 v_{thi}^2\right) = 0 \qquad (10.50)$$

which leads to a slightly modified dispersion relation of Langmuir waves, corrected for the effect of ions

$$\boxed{\omega_l^2(k) = \omega_{pe}^2 \left(1 + \frac{m_e}{m_i}\right) \left[1 + \frac{3k^2 \lambda_D^2}{1 + m_e/m_i} \left(1 + \frac{m_e^2}{m_i^2} \frac{T_i}{T_e}\right)\right]} \qquad (10.51)$$

The difference between the simple Langmuir dispersion relation and this corrected version is small. The plasma frequency is corrected by a term of the order of the electron-to-ion mass ratio. The correction of the Debye length turns out even smaller and becomes important only for extremely high ion temperatures.

Similarly, the Landau damping now contains an ion contribution. Denoting the complete Landau damping as $\gamma_l'(k)$, one obtains

$$\gamma_l'(k, \omega_l) = \gamma_l(k, \omega_l) \left[1 + \left(\frac{T_e}{T_i} \right)^{3/2} \exp \left(\frac{\omega_l^2}{2k^2\lambda_D^2} \frac{T_e - T_i}{T_i} \right) \right] \tag{10.52}$$

Ion damping at high frequencies is small compared to electronic Landau damping. The important contribution of ions to wave propagation is met at frequencies well below the electron plasma frequency and for wave phase velocities intermediate between the electron and ion thermal velocities, under the assumption that the ion temperature is considerably less than the electron temperature (this condition might not be satisfied in many astrophysical and space plasmas)

$$\frac{k_B T_i}{m_i} \ll \frac{\omega^2}{k^2} \ll \frac{k_B T_e}{m_e} \tag{10.53}$$

The expansion of the electron integral must, under these conditions, be taken in the low phase velocity limit, $\omega/k \ll v$, while the ion integral is expanded in the large velocity limit, $\omega/k \gg v$. Such a double expansion leads to the following expression for the real part of the dielectric function

$$\epsilon(k, \omega) = 1 + \frac{1}{k^2\lambda_D^2} - \frac{\omega_{pi}^2}{\omega^2} \left(1 + \frac{3k^2}{\omega^2} \frac{k_B T_i}{m_i} \right) \tag{10.54}$$

Of course, because of the same kind of expansion, the ion term in this formula is of the same kind as the corresponding electronic term in the Langmuir dispersion relation. However, the resulting waves turn out to be very different from ion Langmuir waves. This difference is due to the second term on the right-hand side of the above formula, which is the low-frequency electronic contribution. Solving and iterating for the frequency yields

$$\omega_{ia}^2 = \frac{\omega_{pi}^2}{1 + 1/k^2\lambda_D^2} \left[1 + \frac{3T_i}{T_e} \left(1 + k^2\lambda_D^2 \right) \right] \tag{10.55}$$

Up to the correction factor in parentheses this is the familiar ion-acoustic dispersion relation. In the long-wavelength limit, $k^2\lambda_D^2 \ll 1$, it reproduces the sound-wave branch of the ion-acoustic wave mode

$$\omega = \pm k c_{ia}' \tag{10.56}$$

where $c_{ia}^2 = k_B T_e/m_i$ is the ion sound velocity and $c_{ia}' = c_{ia}(1 + 3T_i/T_e)^{1/2}$ is a modified ion-acoustic speed. At short wavelengths it becomes a modified *ion plasma wave* with a dispersion like the electron Langmuir wave

$$\omega^2 = \omega_{pi}^2 (1 + 3k^2\lambda_{Di}^2) \tag{10.57}$$

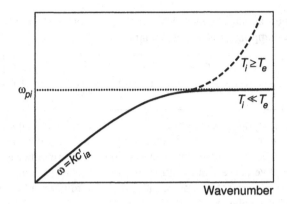

Fig. 10.5. Ion-acoustic dispersion branches.

Here the ion Debye length has taken the position of the electron Debye length in the Langmuir wave. If the ion temperature is low, this wave becomes a simple ion plasma oscillation. Figure 10.5 shows a schematic plot of the ion-acoustic wave dispersion.

Following the procedure of the previous sections to calculate the Landau damping of the ion-acoustic wave from the imaginary part of the dielectric function

$$\epsilon_i(k, \omega) = -\left(\frac{\pi}{2}\right)^{1/2} \frac{1}{k^3 \lambda_D^3} \frac{\omega}{\omega_{pe}} \left[1 + \left(\frac{T_e}{T_i}\right)^{3/2} \exp\left(-\frac{T_e}{T_i} \frac{1}{2k^2 \lambda_D^2} \frac{\omega^2}{\omega_{pe}^2}\right)\right] \quad (10.58)$$

Assuming that the damping is small, one obtains in the long-wavelength domain

$$\gamma_{ia} = -\left(\frac{\pi}{8}\right)^{1/2} \left[\left(\frac{m_e}{m_i}\right)^{1/2} + \left(\frac{T_e}{T_i}\right)^{3/2} \exp\left(-\frac{T_e}{2T_i} - \frac{3}{2}\right)\right] \quad (10.59)$$

for the damping of the ion-sound wave. The first term is the ion damping term. If the electron temperature is very large, $T_e \gg T_i$, the damping reduces to only this term and is very weak

$$\gamma_{ia} \approx \omega_{ia} (\pi/8)^{1/2} (m_e/m_i)^{1/2} \quad (10.60)$$

Because of this reason ion-acoustic waves at long wavelengths and in large electron temperature plasmas are practically undamped modes. Being one of the eigenmodes of an unmagnetized plasma, they play an important role in the dynamics of a collisionless plasma.

We can now calculate the ion-acoustic wave spectral density. Using Eq. (9.84) and multiplying the response function of ion-acoustic waves by frequency and differentiating we get

$$W_{ia} = 2\left(1 + \frac{1}{k^2\lambda_D^2}\right)W_E \qquad (10.61)$$

with $W_E = \epsilon_0|\delta E|^2/2$.

Electron-Acoustic Waves

Another interesting wave mode can be extracted from a dispersion relation similar to the ion-acoustic dispersion relation (10.49), if one assumes that the electron plasma consists of two independent components with different densities and temperatures. Designating the colder electron population by an index c, the hotter one by an index h, and splitting the electron distribution function into cold and hot distributions, $f_{0e}(v) = f_{0c}(v) + f_{0h}(v)$, the dielectric function (10.49) becomes

$$\epsilon(k, p) = 1 + \frac{\omega_{pc}^2}{n_{0c}k^2}\int_{-\infty}^{\infty}\frac{dv f_{0c}(v)}{(v - ip/k)^2} + \frac{\omega_{ph}^2}{n_{0h}k^2}\int_{-\infty}^{\infty}\frac{dv f_{0h}(v)}{(v - ip/k)^2}$$

$$+ \frac{\omega_{pi}^2}{n_{0i}k^2}\int_{-\infty}^{\infty}\frac{dv f_{0i}(v)}{(v - ip/k)^2} \qquad (10.62)$$

Due to the requirement of quasineutrality, the undisturbed densities satisfy the relation

$$n_{0c} + n_{0h} = n_{0i} = n_0 \qquad (10.63)$$

The interesting range of phase velocities is

$$\frac{k_B T_i}{m_i} \sim \frac{k_B T_c}{m_e} \ll \frac{\omega^2}{k^2} \ll \frac{k_B T_h}{m_e} \qquad (10.64)$$

This approximation permits to expand the hot electron integral in the small phase velocity limit, while the ion and cold electron integrals are expanded in the large phase velocity limit. This procedure which yields a result equivalent to Eq. (10.54)

$$\epsilon(k, \omega) = 1 + \frac{1}{k^2\lambda_{Dh}^2} - \frac{\omega_{pc}^2}{\omega^2}\left(1 + \frac{3k^2}{\omega^2}\frac{k_B T_c}{m_e}\right) - \frac{\omega_{pi}^2}{\omega^2}\left(1 + \frac{3k^2}{\omega^2}\frac{k_B T_i}{m_i}\right) \qquad (10.65)$$

Multiplying this equation by ω^2 and iterating ω^2 on the right-hand side, one obtains the dispersion relation of electron-acoustic waves

$$\omega_{ea}^2 = \frac{\omega_{pc}^2\left(1 + \frac{m_e n_0}{m_i n_{0c}}\right)}{1 + \frac{1}{k^2\lambda_{Dh}^2}}\left[1 + 3\left(k^2\lambda_{Dc}^2 + \frac{n_{0h}}{n_{0c}}\frac{T_c}{T_h}\right)\frac{1 + \frac{m_e n_0 T_i}{m_i n_{0c} T_c}}{\left(1 + \frac{m_e n_0}{m_i n_{0c}}\right)^2}\right] \qquad (10.66)$$

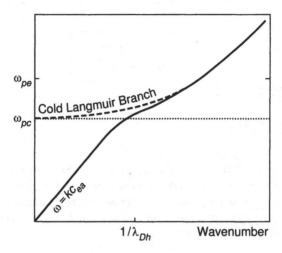

Fig. 10.6. Electron-acoustic dispersion branches.

This relation can be simplified by neglecting all electron-to-ion mass ratios, yielding

$$\omega_{ea}^2 = \frac{\omega_{pc}^2}{1 + 1/k^2\lambda_{Dh}^2}\left(1 + 3k^2\lambda_{Dc}^2 + 3\frac{n_{0h}}{n_{0c}}\frac{T_c}{T_h}\right) \qquad (10.67)$$

as the dispersion relation of *electron-acoustic waves*. In this last form all ion contributions have been neglected, and the mode is purely electronic. Ion corrections become important only for high ion temperatures and low cold electron densities.

The term in front of the parentheses is of the same kind as the ion-acoustic wave dispersion relation. Hence, in the long-wavelength limit, $k^2\lambda_{Dh}^2 \ll 1$, the dispersion relation becomes that of an acoustic wave, with an acoustic wave speed of the order of the *electron-acoustic velocity*

$$c_{ea} = (n_{0c}/n_{0h})^{1/2}v_{thh} \qquad (10.68)$$

This long-wavelength electron-acoustic wave satisfies

$$\omega = \pm kc_{ea} \qquad (10.69)$$

In the short-wavelength limit, this wave becomes a modified Langmuir wave

$$\omega^2 = \omega_{pc}^2\left(1 + 3k^2\lambda_{Dc}^2 + 3\frac{n_{0h}}{n_{0c}}\frac{T_c}{T_h}\right) \qquad (10.70)$$

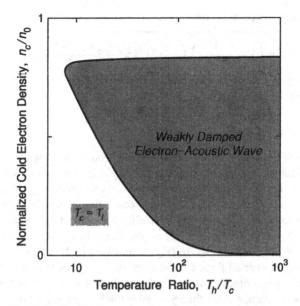

Fig. 10.7. Weakly damped region of electron-acoustic waves.

Hence, electron-acoustic waves at long wavelengths propagate like sound waves, while at short wavelengths they become cold electron Langmuir waves. Figure 10.6 gives an impression of the electron-acoustic wave dispersion.

These waves are heavily damped by electron Landau damping, as can be seen by rewriting the Landau damping of ion-acoustic waves for the electron-acoustic mode. In particular, the long-wavelength branch with linear dispersion is strongly damped. Because of this strong damping, electron-acoustic waves are typically ignored. However, the intermediate branch corresponding to Eq. (10.67) and

$$\frac{n_h}{n_c}\frac{T_c}{T_h} < k^2\lambda_{Dc}^2 < 1 \tag{10.71}$$

is only weakly damped. Here the cold electrons play the role of the heavy ions in the ion-acoustic mode, while charge neutralization is provided by the mobile hot electrons. In this regime the frequency is relatively close to the cold plasma frequency.

Figure 10.7 shows, for $T_c = T_i$, where in parameter space the electron-acoustic mode is weakly damped. Typically, one must have $T_h/T_c > 10$ for weak damping, but at high cold electron densities no weakly damped regime exists. For shorter wavelengths the electron-acoustic waves become strongly damped again.

Calculating the wave spectral density by using Eq. (9.84), we find an expression similar to that for ion-acoustic waves

$$W_{ea}(k) = \left(1 + \frac{1}{k^2\lambda_{Dh}^2}\right)\left(2 - 3k^2\lambda_{Dc}^2 - 3\frac{n_{0h}}{n_{0c}}\frac{T_c}{T_h}\right)W_E \qquad (10.72)$$

where W_E is given by Eq. (9.86).

Electromagnetic Waves

The calculations of Langmuir, ion-acoustic, and electron-acoustic waves have been performed in one dimension only. In an unmagnetized plasma this is no restriction because the only direction of importance is the direction of wave propagation, $\mathbf{k}$. For electrostatic waves, $\delta E(\omega, \mathbf{k}) = -i\mathbf{k}\,\delta\phi(\omega, \mathbf{k})$, the wave electric field is parallel to the wavenumber, and the situation is one-dimensional by itself. Hence, for the general case where the equilibrium distribution function depends on the full velocity vector, $f_0(\mathbf{v})$, it is sufficient to replace $k\partial f_0(v)/\partial v$ with $\mathbf{k} \cdot \partial f_0(\mathbf{v})/\partial\mathbf{v}$, and kv with $\mathbf{k} \cdot \mathbf{v}$. The integration over velocity is then taken over all three components, v_x, v_y, v_z. But since only the component parallel to k, say v_z, comes into play, the integrations over the two other components can trivially be performed, and the theory remains unchanged, with $f_0(v)$ understood as the distribution function integrated over v_x, v_y, $f_0(v_z) = \int dv_x dv_y f_0(\mathbf{v})$. This conclusion is of course violated if any anisotropy is introduced by, say, an anisotropy of the distribution function or by an external magnetic field.

Let us consider the case of electromagnetic waves propagating in an isotropic unmagnetized plasma with vanishing external electric and magnetic fields, $\mathbf{E}_0 = \mathbf{B}_0 = 0$. We must then use the full Vlasov equation, but the fields appearing are the magnetic field of the wave, $\delta\mathbf{B}$, and the transverse wave electric field, satisfying $\mathbf{k} \cdot \delta\mathbf{E} = 0$ or

$$\mathbf{k} \times \mathbf{k} \times \delta\mathbf{E} = -k^2\delta\mathbf{E} \qquad (10.73)$$

Since only oscillating electromagnetic fields exist in this case, we immediately write down the linearized Vlasov equation for particles of charge q and mass m

$$\frac{\partial\delta f(\mathbf{v})}{\partial t} + \mathbf{v}\cdot\nabla\delta f(\mathbf{v}) + \frac{q}{m}(\delta\mathbf{E} + \mathbf{v}\times\delta\mathbf{B})\cdot\frac{\partial f_0(\mathbf{v})}{\partial\mathbf{v}} = 0 \qquad (10.74)$$

The isotropy of the plasma has been taken into account by letting the undisturbed equilibrium distribution function depend only on the modulus, v, of the velocity. Calculating $\partial f_0(v)/\partial\mathbf{v} = (\partial f_0/\partial v)(\partial v/\partial\mathbf{v}) = 3(\partial f_0/\partial v)\mathbf{v}$ and dot-multiplying with $\mathbf{v}\times\delta\mathbf{B}$, it becomes apparent that the last term in the parentheses vanishes, when dotted with the velocity derivative of the equilibrium distribution. Hence, the magnetic field contribution drops out of the Vlasov equation, leaving us with

$$\frac{\partial\delta f(\mathbf{v})}{\partial t} + \mathbf{v}\cdot\nabla\delta f(\mathbf{v}) + \frac{q}{m}\delta\mathbf{E}\cdot\frac{\partial f_0(\mathbf{v})}{\partial\mathbf{v}} = 0 \qquad (10.75)$$

This equation can be understood as a linear inhomogeneous equation for $\delta f(\mathbf{v})$, which can be solved by superposition of plane waves or, in other words, by Fourier transformation both in space and time. The choice of this method, which is much simpler than the one applied in the previous electrostatic case, can be justified by arguing that we expect all physically reasonable distribution functions and fields to be analytic functions. In this case the dispersion relation will just depend on wavenumber, $\mathbf{k}$, and the complex variable $p = \gamma - i\omega$. Thus, interpreting the plane wave ansatz, $\exp[i(\mathbf{k} \cdot \mathbf{x} - \omega t) + \gamma t]$, as a complex Laplace factor, $\exp(pt)$, will yield the desired result.

With this philosophy in mind, we Fourier transform the linear equation, solve for the variation of the distribution function, and find

$$\delta f(\mathbf{v}, \mathbf{k}, \omega) = -\frac{q}{m} \frac{\partial f_0(v)/\partial \mathbf{v}}{\omega - \mathbf{k} \cdot \mathbf{v}} \cdot \delta \mathbf{E}(\mathbf{k}, \omega) \tag{10.76}$$

as a general expression for the disturbed distribution function in an isotropic plasma. It contains the velocity gradient of the equilibrium distribution and is proportional to the wave electric field.

In order to determine the dielectric function we need to know the plasma conductivity, which is calculated from the plasma current

$$\mathbf{j}(\mathbf{k}, \omega) = -\sum_{s=e,i} \frac{q_s}{m_s} \int d^3 v \, \mathbf{v} \frac{\partial f_{s0}(v)/\partial \mathbf{v}}{\omega - \mathbf{k} \cdot \mathbf{v}} \cdot \delta \mathbf{E}(\mathbf{k}, \omega) \tag{10.77}$$

The factor in front of $\delta \mathbf{E}(\mathbf{k}, \omega)$ is the expression for the conductivity. When using it in the general dispersion relation derived in Chap. 9, we find the dispersion relation for electromagnetic waves in an isotropic plasma

$$(\omega^2 - k^2 c^2)\mathbf{I} = -\sum_{s=e,i} \frac{\omega_{ps}^2}{n_0} \int \frac{d^3 v \, \omega}{\omega - \mathbf{k} \cdot \mathbf{v}} \mathbf{v} \frac{\partial f_{s0}(v)}{\partial \mathbf{v}} \tag{10.78}$$

From this expression we immediately find that for sufficiently high phase velocities, $\omega \gg \mathbf{k} \cdot \mathbf{v}$, the dispersion relation of the ordinary wave mode is recovered, since

$$(\omega^2 - k^2 c^2)\mathbf{I} = -\sum_{s=e,i} \frac{\omega_{ps}^2}{n_0} \int d^3 v \, \mathbf{v} \frac{\partial f_{s0}(v)}{\partial \mathbf{v}} \tag{10.79}$$

and the integral just has the value $-n_0 \mathbf{I}$. Hence, this expression finally yields

$$\boxed{\omega_{om}^2 = k^2 c^2 + \omega_{pe}^2} \tag{10.80}$$

where we have neglected the small ion plasma frequency correction. Such waves are undamped as is obvious from the disappearance of the resonance in the above expression. This mode, as we already know, is the only electromagnetic wave propagating in an isotropic unmagnetized plasma. Even a more precise calculation would show that it is practically undamped as long as relativistic particle effects are not included.

Dispersion Function

In the calculation of dispersion relations we have continuously encountered singular integrals of the kind

$$Z(\zeta) = \int_{-\infty}^{\infty} \frac{dx \; f_0(x)}{x - \zeta}$$ (10.81)

where $f_0(x)$ is some function related to the equilibrium distribution function which usually is an analytic function if its argument, x, is interpreted as the real part of a complex variable, z. The above integral is then taken along the real axis of the complex z plane. Integrals of this kind have been calculated in the previous sections. Clearly the value of the integral depends crucially on the choice of the distribution function. There is, however, a certain number of canonical distribution functions for equilibrium plasmas for which these integrals have been calculated. These functions are called *dispersion functions*. The best know dispersion function is the *plasma dispersion function* or *Fried-Conte function*. It is based on a Maxwellian equilibrium distribution. Therefore the plasma dispersion function is usually defined as

$$Z(\zeta) = \pi^{-1/2} \int_{-\infty}^{\infty} \frac{dx \; \exp(-x^2)}{x - \zeta}$$ (10.82)

The plasma dispersion function naturally plays an important role in most of the calculations of the linear properties of plasma waves propagating in an equilibrium background plasma. Its properties are listed in App. A.7.

In order to give an example of the application of the plasma dispersion function, we rewrite the dispersion relation of ion-acoustic waves in terms of the Z-function

$$\epsilon(\omega, k) = 1 - \frac{1}{k^2 \lambda_D^2} Z'(\zeta_e) - \frac{1}{k^2 \lambda_{Di}^2} Z'(\zeta_i) = 0$$ (10.83)

The integrals containing the velocity derivative of the electron and ion Maxwellian equilibrium distribution functions have turned into the total derivatives of the plasma dispersion functions of the electron and ion arguments, $\zeta_e = \omega/k v_{the}$ and $\zeta_i = \omega/k v_{thi}$. These arguments contain the frequency dependence. It is now simple to apply the small, $\zeta_e \ll 1$, and large, $\zeta_i \gg 1$, argument expansions given in App. A.7, to find the ion-acoustic dispersion relation and the associated Landau damping.

10.4. Magnetized Dispersion Relation

An external magnetic field introduces an anisotropy into the plasma. It affects the perpendicular particle motion while the parallel motion remains undisturbed. This must necessarily transform the scalar dielectric function into a tensor. In order to keep the mathematical difficulties at a minimum, this section does not make use of the Laplace method but returns to the easier though less precise procedure of Fourier transformation. We start from the linearized Vlasov equation of a magnetized plasma

$$\left(\frac{\partial}{\partial t} + \mathbf{v} \cdot \nabla + \frac{q}{m}\mathbf{v} \times \mathbf{B}_0 \cdot \frac{\partial}{\partial \mathbf{v}}\right) \delta f(\mathbf{v}, \mathbf{x}, t) = -\frac{q}{m}(\delta\mathbf{E} + \mathbf{v} \times \delta\mathbf{B}) \cdot \frac{\partial f_0(\mathbf{v})}{\partial \mathbf{v}} \quad (10.84)$$

where we have suppressed the species index, s. As before we have assumed that the undisturbed electric field vanishes, the ambient magnetic field and plasma densities are homogeneous. The equilibrium distribution function now depends on the velocity vector, in order to account for the anisotropy. The linearized current and charge densities are given by

$$\delta\mathbf{j} = \sum_s q_s \int d^3v\, \mathbf{v}\, \delta f_s$$

$$\delta\rho_e = \sum_s q_s \int d^3v\, \delta f_s \quad (10.85)$$

Because δf depends on time and the six phase space coordinates, the left-hand side of Eq. (10.84) is the total time derivative of δf along a particle phase space orbit, while the right-hand side describes the change of the distribution function along this orbit under the action of the wave field. Hence, this equation can be rewritten as

$$\frac{d\,\delta f[\mathbf{v}(t), \mathbf{x}(t), t]}{dt} = -\frac{q}{m}\{\delta\mathbf{E}[\mathbf{x}(t), t] + \mathbf{v}(t) \times \delta\mathbf{B}[\mathbf{x}(t), t]\} \cdot \frac{\partial f_0[\mathbf{v}(t)]}{\partial \mathbf{v}(t)} \quad (10.86)$$

Formally, one may calculate δf from this equation by integrating it over its entire history with respect to the time

$$\delta f[\mathbf{v}(t), \mathbf{x}(t), t] = -\frac{q}{m}\int_{-\infty}^{t} dt' \{\delta\mathbf{E}[\mathbf{x}(t'), t'] + \mathbf{v}(t') \times \delta\mathbf{B}[\mathbf{x}(t'), t']\} \cdot \frac{\partial f_0[\mathbf{v}(t')]}{\partial \mathbf{v}(t')} \quad (10.87)$$

But this procedure requires precise knowledge of the phase space orbit of all particles for all times $t' < t$, which is not available. But in a linearized theory, where the disturbance of the distribution function and the wave amplitudes remain small for all times, one can approximate the particle orbit by the orbit a particle would perform in a homogeneous and uniform external magnetic field, $\mathbf{B}_0 = B_0\hat{\mathbf{e}}_z$. This motion has been discussed in

Chap. 2. It consists of a uniform motion along $\mathbf{B}_0$ and a gyration of frequency, ω_g, and gyroradius, $r_g = v_\perp / \omega_g$. Hence, the velocity components at any time can be represented by

$$\mathbf{v}(t'-t) = \{v_\perp \cos[\omega_g(t'-t) + \psi], v_\perp \sin[\omega_g(t'-t) + \psi], v_\parallel\} \qquad (10.88)$$

where ψ is the initial phase angle. Correspondingly the position of the particle is given by the time integral of this expression

$$\mathbf{x}(t'-t) - \mathbf{x} = \omega_g^{-1} \{v_\perp \sin[\omega_g(t'-t) + \psi], -v_\perp \cos[\omega_g(t'-t) + \psi], v_\parallel(t'-t)\} \qquad (10.89)$$

We now transform the time integration in Eq. (10.87) into an integration with respect to $\tau = t' - t$ in order to obtain

$$\delta f(\mathbf{v}) = -\frac{q}{m} \int_0^\infty d\tau \, [\delta\mathbf{E}(\tau) + \mathbf{v}(\tau) \times \delta\mathbf{B}(\tau)] \cdot \frac{\partial f_0[\mathbf{v}(\tau)]}{\partial \mathbf{v}(\tau)} \qquad (10.90)$$

and introduce the plane wave ansatz, $\exp[-i\varphi(\tau)]$, for the electric and magnetic wave fields, with $\varphi(\tau) = -\omega\tau + \mathbf{k} \cdot \mathbf{x} - \mathbf{x}(\tau)$. From Maxwell's equations we further deduce that these wave fields are related through Faraday's law, yielding $\mathbf{k} \times \delta\mathbf{E} = \omega\,\delta\mathbf{B}$, so that the magnetic wave field amplitude can be eliminated from Eq. (10.90). This expression then transforms into

$$\delta f(\mathbf{v}) = -\frac{q\,\delta\mathbf{E}(\mathbf{k}, \omega)}{m\omega} \cdot \int_0^\infty d\tau \, e^{-i\varphi(\tau)} \left[\mathbf{v}\,\mathbf{k}(\tau) + \mathbf{I}\,(\omega - \mathbf{k} \cdot \mathbf{v}(\tau))\right] \cdot \frac{\partial f_0[\mathbf{v}(\tau)]}{\partial \mathbf{v}(\tau)} \qquad (10.91)$$

This is the expression which has to be used in calculating the linear current in order to find the linear conductivity of the magnetized plasma. The expression for the current density contains an integral over velocity. This integral transforms into another integral over the new phase space volume element, $v_\perp dv_\perp dv_\parallel d\psi$, a transformation which completes our linear approach. What is left is to explicitly calculate the current integral

$$\delta\mathbf{j}(\mathbf{k}, \omega) = -\sum_s \frac{\epsilon_0 \omega_{ps}^2}{n_0 \omega} \int_0^\infty \int_{-\infty}^\infty \int_0^{2\pi} v_\perp dv_\perp dv_\parallel d\psi$$

$$\delta\mathbf{E}(\mathbf{k}, \omega) \cdot \int_0^\infty d\tau \, e^{-i\varphi(\tau)} \left[\mathbf{v}\,\mathbf{k}(\tau) + \mathbf{I}\,(\omega - \mathbf{k} \cdot \mathbf{v}(\tau))\right] \cdot \frac{\partial f_0[\mathbf{v}(\tau)]}{\partial \mathbf{v}(\tau)} \qquad (10.92)$$

This calculation is performed in detail in App. B.6 for a gyrotropic equilibrium distribution function. With the help of Eq. (B.43) and splitting the wave vector into parallel and perpendicular components according to

$$\mathbf{k} = k_\perp \hat{\mathbf{e}}_\perp + k_\parallel \hat{\mathbf{e}}_\parallel \qquad (10.93)$$

the magnetized dielectric function of the plasma takes the following form

$$
\epsilon(\omega, \mathbf{k}) = \left(1 - \sum_s \frac{\omega_{ps}^2}{\omega^2}\right)\mathbf{1} - \sum_s \sum_{l=-\infty}^{l=\infty} \frac{2\pi\omega_{ps}^2}{n_{0s}\omega^2}
$$

$$
\int\limits_0^\infty \int\limits_{-\infty}^\infty v_\perp dv_\perp dv_\parallel \left(k_\parallel \frac{\partial f_{0s}}{\partial v_\parallel} + \frac{l\omega_{gs}}{v_\perp}\frac{\partial f_{0s}}{\partial v_\perp}\right)\frac{\mathbf{S}_{ls}(v_\parallel, v_\perp)}{k_\parallel v_\parallel + l\omega_{gs} - \omega} \quad (10.94)
$$

The tensor, $\mathbf{S}_{ls}$, appearing in the integrand is of the form

$$
\mathbf{S}_{ls}(v_\parallel, v_\perp) = \begin{bmatrix} \dfrac{l^2\omega_{gs}^2}{k_\perp^2}J_l^2 & \dfrac{ilv_\perp\omega_{gs}}{k_\perp}J_lJ_l' & \dfrac{lv_\parallel\omega_{gs}}{k_\perp}J_l^2 \\[3mm] -\dfrac{ilv_\perp\omega_{gs}}{k_\perp}J_lJ_l' & v_\perp^2 J_l'^2 & -iv_\parallel v_\perp J_lJ_l' \\[3mm] \dfrac{lv_\parallel\omega_{gs}}{k_\perp}J_l^2 & iv_\parallel v_\perp J_lJ_l' & v_\parallel^2 J_l^2 \end{bmatrix} \quad (10.95)
$$

and the Bessel functions, J_l, $J_l' = dJ_l/d\xi_s$, depend on the argument $\xi_s = k_\perp v_\perp/\omega_{gs}$.

Equation (10.94) is the most general expression for the linear dielectric function of a homogeneous nonrelativistic plasma immersed into a uniform magnetic field. It contains, when inserted into the general dispersion relation in Eq. (9.55), all electromagnetic and electrostatic wave eigenmodes, which can exist in such a plasma.

In the case of purely electrostatic (or longitudinal) modes with $\delta\mathbf{B} = 0$, the dielectric function simplifies considerably, since for such waves it is sufficient to consider the dielectric response function in Eq. (9.62). It is obtained by taking the dot-product of the dielectric tensor with the wave vector, $\mathbf{k}$, from both sides. Since $\mathbf{k}$ is in the (x, z) plane, one can show that after multiplication the term ω_{ps}^2/ω^2 in Eq. (10.94) is canceled by the $l = 0$ term of the sum. Furthermore, only two diagonal terms of the tensor $\mathbf{S}_{ls}$ survive, and their sum gives just J_l^2 (cf. App. A.7). Hence, the final result is

$$
\epsilon(\omega, \mathbf{k}) = 1 - \sum_s \sum_{l=-\infty}^{l=\infty} \frac{2\pi\omega_{ps}^2}{n_{0s}k^2}\int\limits_0^\infty \int\limits_{-\infty}^\infty v_\perp dv_\perp dv_\parallel
$$

$$
\left(k_\parallel \frac{\partial f_{0s}}{\partial v_\parallel} + \frac{l\omega_{gs}}{v_\perp}\frac{\partial f_{0s}}{\partial v_\perp}\right)\frac{J_l^2(\xi_s)}{k_\parallel v_\parallel + l\omega_{gs} - \omega} \quad (10.96)
$$

Setting this function equal to zero, $\epsilon(\omega, \mathbf{k}) = 0$, yields the dispersion relation for all the electrostatic waves propagating in a homogeneous uniformly magnetized plasma. In the following sections we will use it identify the dominant electrostatic modes.

Particle Resonance

As in the discussion of the Landau method for electrostatic electron waves the eigen-modes of a magnetized plasma are determined by the poles of the integrand of the di-electric tensor (10.94). These poles appear at the positions where

$$\omega - k_\| v_\| - l\omega_{gs} = 0 \tag{10.97}$$

which is the particle *resonance condition*. In the case of an unmagnetized plasma it re-duces to the *Landau resonance*

$$\omega = k_\| v_\| \tag{10.98}$$

These two conditions are conditions on a specific group of particles in the plasma and are therefore particle resonances. They do not affect the whole plasma. This becomes obvious for instance from the Landau resonance, $l = 0$, which indicates that parti-cles with velocities equal to the phase velocity of the wave are the only particles which contribute to the pole. These particles are in phase with the wave and the wave fre-quency seen by them is zero. Hence, these particles have a well defined parallel energy, $W_\| = m(\omega^2/2k_\|^2)$, the Landau resonant energy. In a magnetized plasma the selection of resonant particles becomes more complicated. Particles which move along the magnetic field case see the frequency of the wave Doppler-shifted to $\omega' = \omega - k_\| v_\| = l\omega_{gs}$. For $l = \pm 1$ this is just the gyrofrequency of species s; for $l \neq 1$ it is its l-th harmonic.

Hence, the group of particles whose parallel velocity just matches the parallel wave phase speed does not only see a constant parallel electric field of the wave, but if it is the right kind of particle also gyrates together with the perpendicular electric field com-ponent so that this component is also constant for them, or, for $|l| > 1$ sees a higher har-monic of its own gyration. Such particles interact strongly with the wave electric field because they become either accelerated or decelerated. It is these resonant particles who are responsible for the kinetic wave effects in magnetized and unmagnetized plasmas.

10.5. Electrostatic Plasma Waves

In this section we investigate the longitudinal eigenmodes of a warm magnetized plasma. These modes are the solutions of the dispersion relation

$$\epsilon(\omega, \mathbf{k}) = 0 \tag{10.99}$$

where the response function in a magnetic field has been defined in Eq. (10.96). This function still holds for an arbitrary distribution function, but in the remainder of this chapter we will use an equilibrium Maxwellian

$$f_{0s}(v_\perp, v_\|) = \frac{n_{0s}}{\pi^{3/2} v_{ths\|} v_{ths\perp}^2} \exp\left(-\frac{v_\|^2}{v_{ths\|}^2} - \frac{v_\perp^2}{v_{ths\perp}^2}\right) \tag{10.100}$$

where for greater generality we permitted for anisotropy in the thermal velocities. The perpendicular velocity integral over the Bessel function can be substantially simplified. Making use of the Weber integrals given in App. A.7 and inserting f_{0s} from Eq. (10.100), the response function reduces to an integral over the parallel velocities

$$\epsilon(\omega, \mathbf{k}) = 1 + \sum_s \sum_{l=-\infty}^{l=\infty} \frac{2\omega_{ps}^2 \Lambda_l(\eta_s)}{\pi^{1/2} k^2 v_{ths\perp}^2} \int_{-\infty}^{\infty} \frac{dv_\parallel}{v_{ths\parallel}} \left(\frac{T_{s\perp}}{T_{s\parallel}} k_\parallel v_\parallel + l\omega_{gs} \right) \frac{\exp(-v_\parallel^2/v_{ths\parallel}^2)}{k_\parallel v_\parallel + l\omega_{gs} - \omega}$$

(10.101)

where we defined a new function

$$\Lambda_l(\eta_s) = I_l(\eta_s) \exp(-\eta_s)$$

(10.102)

and $I_l(\eta_s)$ is the modified Bessel function with the argument

$$\eta_s = \frac{k_\perp^2 v_{ths\perp}^2}{2\omega_{gs}^2} = \frac{k_\perp^2 T_{s\perp}}{m_s \omega_{gs}^2} = \frac{k_\perp^2 r_{gs}^2}{2}$$

(10.103)

The $v_\parallel$-integration can be performed with the help of the plasma dispersion function (App. A.7), yielding

$$\epsilon(\omega, \mathbf{k}) = 1 - \sum_s \sum_{l=-\infty}^{l=\infty} \frac{\omega_{ps}^2 \Lambda_l(\eta_s)}{k^2 v_{ths\perp}^2} \left\{ \frac{T_{s\perp}}{T_{s\parallel}} Z'(\zeta_s) - \frac{2l\omega_{gs}}{k_\parallel v_{ths\parallel}} Z(\zeta_s) \right\}$$

(10.104)

where $Z'(\zeta_s) = dZ/d\zeta_s$. The argument of the $Z(\zeta_s)$ function is $\zeta_s = (\omega - l\omega_{gs})/k_\parallel v_{ths\parallel}$. It depends on the species and on the running index of the sum, l. The zeros of this function are the electrostatic eigenmodes of the magnetized plasma.

Magnetized Langmuir and Ion-Acoustic Waves

Let us consider nearly parallel propagation first. At high frequencies the only wave is the Langmuir mode. Because for $k_\perp \to 0$ and $l \neq 0$ all $\Lambda_l \to 0$, the sum in Eq. (10.104) reduces to the term with index $l=0$. Neglecting the ion contribution, we obtain

$$\epsilon(\omega, \mathbf{k}) = 1 - \frac{\omega_{pe}^2}{k^2 v_{the\parallel}^2} \Lambda_0(\eta_e) Z'(\zeta_e) = 0$$

(10.105)

where $\zeta_e = \omega/k_\parallel v_{the\parallel}$, $\eta_e = k_\perp^2 v_{the\perp}^2/2\omega_{ge}^2$, and we allow for an anisotropy in the parallel and perpendicular thermal velocities. Now the plasma dispersion function is expanded in the large argument limit (see App. A.7), valid for Langmuir oscillations

$$1 - \frac{\omega_{pe}^2}{k^2 v_{the\parallel}^2} \frac{\Lambda_0}{\zeta_e^2} \left[1 + \frac{3}{\zeta_e^2} - \frac{2i\pi^{1/2}}{\zeta_e} \exp(-\zeta_e^2) \right] = 0$$

(10.106)

Solving for the real part of the frequency under the assumption of weak damping we find

$$\omega_l^2(k, \theta) = \omega_{pe}^2 \Lambda_0(\eta_e) \cos^2\theta \left(1 + 3k^2\lambda_{D\parallel}^2 \cos^2\theta\right) \qquad (10.107)$$

Here $\lambda_{D\parallel}$ is the Debye length with respect to the parallel thermal velocity. Under the assumption that damping is weak, $\gamma_l \ll \omega_l$, one finds with the help of the general expression for the weak damping rate in Eq. (10.36)

$$\gamma_l(\omega_l, k, \theta) \approx -\left(\frac{\pi}{8}\right)^{1/2} \frac{\omega_l(k, \theta)\Lambda_0^{3/2}(\eta_e)}{k^3\lambda_{D\parallel}^3} \exp\left[-\frac{\Lambda_0(\eta_e)}{2k^2\lambda_{D\parallel}^2} - \frac{3}{2}\right] \qquad (10.108)$$

for the Landau damping in a magnetized plasma. The requirement that the waves are weakly damped is thus restricted to the region where Λ_0 is sufficiently large. It depends on the angle of propagation. But because Λ_0 decreases monotonically for increasing argument, which implies that it decreases with increasing angle of propagation, oblique waves become ever stronger damped. In order to estimate up to what angle the damping is weak, we put $\Lambda_0(\eta_e) \approx 2k^2\lambda_{D\parallel}^2$. This yields

$$\exp(-\eta_e) \approx 2k^2\lambda_{D\parallel}^2/I_0(\eta_e) \qquad (10.109)$$

which can be rewritten as

$$\eta_e \approx \ln[I_0(\eta_e)/2k^2\lambda_{D\parallel}^2] \approx \ln(1/2k^2\lambda_{D\parallel}^2) \qquad (10.110)$$

For not too large η_e we have $I_0(\eta_e) \approx 1$, which allows to write for the maximum weakly damped angle of propagation

$$\sin\theta < (kr_{ge})^{-1} \left|\ln(1/2k^2\lambda_{D\parallel}^2)\right|^{1/2} \qquad (10.111)$$

For nearly parallel propagation η_e is small, and we can expand $\Lambda_0(\eta_e) \approx 1 - \eta_e$. This yields the magnetized Langmuir wave dispersion relation in compact form

$$\omega_l^2(k, \theta) = \omega_{pe}^2 \cos^2\theta \left[1 + 3k^2\lambda_{D\parallel}^2 \cos^2\theta \left(1 - \frac{T_\perp}{6T_\parallel} \frac{\omega_{pe}^2}{\omega_{ge}^2} \tan^2\theta\right)\right] \qquad (10.112)$$

This dispersion relation is similar to that of ordinary Langmuir waves, but it shows that Langmuir waves in a magnetized plasma propagate essentially parallel to the magnetic field and depend only weakly on the ratio of plasma to gyrofrequency. The frequency of the wave decreases to zero with the wave vector turning to perpendicular propagation, but one must remember that the approximations made in the derivation of Eq. (10.112) break down for large angles of propagation. Nevertheless, in magnetized plasmas the Langmuir wave frequency at small k decreases to $\omega_l(\theta) < \omega_{pe}$ for oblique propagation.

Magnetized ion-acoustic waves can be treated along the same lines as Langmuir waves. The only difference is that instead of expanding the electron plasma dispersion function in the small argument limit, one expands the ion dispersion function in the large argument limit. As a final result one finds

$$
\omega_{ia}^2 = \frac{\omega_{pi}^2 \Lambda_0(\eta_i) \cos^2 \theta}{1 + \Lambda_0(\eta_e)/k^2\lambda_D^2} \left(1 + 3k^2\lambda_{Di}^2 \cos^2 \theta\right) \tag{10.113}
$$

Up to the Λ_0-correction this dispersion relation is similar to that of the unmagnetized ion-acoustic wave. This correction contains the effect of the magnetic field on wave propagation. But ion-acoustic waves in magnetized plasma do also propagate roughly parallel to the magnetic field, as is the case with the Langmuir wave.

Electron Bernstein Waves

We now turn to perpendicular propagation. If we put $k_\perp = 0$ in Eq. (10.96), the argument of the Bessel functions vanishes and only the term $l = 0$ in the sum survives. This implies that the influence of the magnetic field on the waves disappears and we recover the unmagnetized plasma case. Thus electrostatic waves in a plasma, which are susceptible to the presence of magnetic fields, propagate always oblique to the magnetic field. The most extreme case is to look for transverse propagation $k_\parallel = 0$, $k = k_\perp$. Inserting this into the response function, the dispersion relation can be written as

$$
1 - \sum_s \frac{\omega_{ps}^2}{k_\perp^2} \frac{m_s^2}{k_B^2 T_s} \sum_{l=-\infty}^{\infty} \frac{l\omega_{gs}}{\omega - l\omega_{gs}} \int_0^\infty v_\perp dv_\perp J_l^2(\eta_s) \exp\left(-\frac{v_\perp^2}{v_{ths\perp}^2}\right) = 0 \tag{10.114}
$$

or after performing the integration, again with the help of the Weber integral

$$
1 - \sum_s \frac{\omega_{ps}^2}{\omega_{gs}} \sum_{l=-\infty}^{\infty} \frac{l I_l(\eta_s)}{\omega - l\omega_{gs}} \frac{\exp(-\eta_s)}{\eta_s} = 0 \tag{10.115}
$$

This dispersion relation is free of any singular integrals. The only singularities are the resonances at the harmonics of the cyclotron frequencies of the various species of particles in the plasma. In particular, the $l = 0$ term is zero so that there is no wave below the first cyclotron harmonic. In the high-frequency range, $\omega \sim l\omega_{ge}$, the ion contribution to the dispersion relation can be neglected, and we find the dispersion relation of strictly perpendicular electrostatic electron-cyclotron waves or *Bernstein waves*

$$
1 - \frac{\omega_{pe}^2}{\omega_{ge}^2} \sum_{l=-\infty}^{\infty} \frac{\omega_{ge} l \Lambda_l(\eta_e)}{\eta_e(\omega - l\omega_{ge})} = 0 \tag{10.116}
$$

which shows that there is an infinite number of resonances or wave modes possible with frequencies $\omega \geq l\omega_{gs}$. Because of the symmetry of the modified Bessel functions with respect to their index, $I_{-l} = I_l$, Eq. (10.116) can be written as

$$1 - \frac{\omega_{pe}^2}{\omega_{ge}^2} \sum_{l=1}^{\infty} \frac{2l^2 \Lambda_l(\eta_e)}{\eta_e(v_{he}^2 - l^2)} = 0 \qquad (10.117)$$

where $v_{he} = \omega/\omega_{ge}$. The different dispersion branches are concentrated near the harmonics. The behavior of the dispersion branches, $v_{he} = v_{he}(k_\perp)$, at large wavenumbers $k_\perp \to \infty$, is given by the asymptotic expansion of $\Lambda_l(\eta_e)$

$$\Lambda_l(\eta_e) \to (2\pi\eta_e)^{-1/2} \exp(-l^2/2\eta_e) \qquad (10.118)$$

Substituting into Eq. (10.117), we find that with $\eta_e \to \infty$ the exponential vanishes and the dispersion relation can only be satisfied by $v_{he} = l$, which implies that all short wavelength waves are exact harmonics of the electron-cyclotron frequency, $\omega(k_\perp \to \infty) = l\omega_{ge}$. Moreover, the dispersion branches are bound to their harmonic bands, l, and cannot cross over into one of the adjacent bands, $l' = l \pm 1$. For small wavenumbers, $k_\perp \approx 0$, we have $\Lambda_l(\eta_e) \approx \eta_e^l/2l!$, and Eq. (10.117) turns into

$$1 - \frac{\omega_{pe}^2}{\omega_{ge}^2} \sum_{l=1}^{\infty} \frac{l\eta_e^{l-1}}{(l-1)!(v_{he}^2 - l^2)} = 0 \qquad (10.119)$$

Two limiting cases are of interest. For $l \neq 1$ and $\eta_e \to 0$ this equation can be satisfied only if $v_{he} \approx l$. On the other hand, for $v_{he} \neq l$ only the term $l = 1$ survives and yields

$$\omega^2(k_\perp = 0) = \omega_{pe}^2 + \omega_{ge}^2 \qquad (10.120)$$

This frequency is just the upper-hybrid frequency, ω_{uh}. The perpendicular electrostatic modes thus start at small wavenumber as a cyclotron harmonic oscillation with the exception of the band which contains the upper-hybrid frequency. Here the oscillation branch starts at ω_{uh}. For large perpendicular wavenumbers all dispersion branches approach their respective cyclotron harmonics. This behavior is sketched in Fig. 10.8.

To demonstrate this behavior explicitly for the lower harmonic bands, we neglect the ion contribution and write Eq. (10.116) up to the $l = 2$ term, using the small argument expansion for $\Lambda(\eta_e)$ together with the definition of η_e

$$\frac{\omega_{ge}^2}{\omega_{pe}^2} = \frac{1 - \eta_e}{v_{he}^2 - 1} + \frac{\eta_e(1 - \eta_e)}{v_{he}^2 - 4} \qquad (10.121)$$

Let us now take $v_{he} = 1 + \delta_{e,1}$ with $\delta \ll 1$ as the resonance of the first term on the right-hand side. Then, we find to first order

$$\delta_{e,1} = \frac{3}{2(1 - \eta_e)}\left(1 + \frac{6}{1 - 2\eta_e}\frac{\omega_{ge}^2}{\omega_{pe}^2}\right) > 0 \qquad (10.122)$$

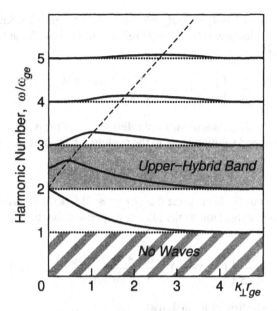

Fig. 10.8. Dispersion branches of Bernstein modes.

a correction which is always positive, confirming that there is no dispersion branch below the first harmonic resonance. The next resonance, $\nu_{he} = 2 + \delta_{e,2}$, yields

$$\delta_{e,2} = -\frac{\eta_e}{4} \left(-\frac{3}{1 - \eta_e} \frac{\omega_{ge}^2}{\omega_{pe}^2} \right) < 0 \qquad (10.123)$$

This dispersion branch starts at $\nu_{he} = 2$ and lies below the second harmonic. Only in the band containing the upper-hybrid frequency the dispersion branch starts at ω_{uh}. In the higher bands the branches all lie above the harmonic resonances.

Because the dispersion branches are all trapped into cyclotron harmonic bands, the general dispersion relation (10.117) of Bernstein waves can be written for each of the modes separately. Temporarily re-introducing ω instead of ν_{he} and solving for the real frequency, ω, one finds as dispersion relation of each electron-cyclotron Bernstein mode

$$\boxed{\omega_{ec}^2 = l^2 \omega_{ge}^2 \left[1 + \frac{2\omega_{pe}^2}{\omega_{ge}^2} \frac{\Lambda_l(\eta_e)}{\eta_e} \right]} \qquad (10.124)$$

where the wavenumber dependence enters only through the Λ_l function. This dispersion relation holds for strictly perpendicular propagation. We can now look for a solution of the dispersion relation (10.104) at nearly perpendicular propagation. Knowing that only modes near $\nu_{he} = l$ play any role, the sum reduces to the $l = 0$ component plus one single l-component

$$\epsilon(\omega, \mathbf{k}) = 1 + \frac{1}{k_\perp^2 \lambda_D^2} \left[1 - \Lambda_0(\eta_e) + \frac{\nu_{he} \Lambda_l(\eta_e)}{|k_\parallel| r_{ge}|} Z \left(\frac{\nu_{he} - l}{|k_\parallel r_{ge}|} \right) \right] = 0 \qquad (10.125)$$

The index, $l = \pm 1, \pm 2, \ldots$, is arbitrary and refers to any of the harmonic bands. Let us ask for a weakly damped solution close to the cyclotron harmonic number $\nu_{he} = l$

$$\nu_{he}(\mathbf{k}) = l[1 + \delta_{e,l}(\mathbf{k})] + i\gamma_l(\mathbf{k}) \qquad (10.126)$$

with $\delta_{e,l}(\mathbf{k}) \ll 1$. Let us further assume that $|k_\parallel r_{ge}| \ll l\delta_{e,l}(\mathbf{k})$. In this case we can make use of the asymptotic expansion of the plasma dispersion function for large argument and find

$$\epsilon(\omega, \mathbf{k}) \approx 1 + \frac{1}{k^2 \lambda_D^2} \left[1 - \Lambda_0(\eta_e) + \frac{l(1 + \delta_{e,l}) \Lambda_l(\eta_e)}{|k_\parallel r_{ge}|} \left(i\pi^{1/2} e^{-\zeta_e^2} - \frac{k_\parallel r_{ge}}{l\delta_{e,l}} \right) \right]$$
$$(10.127)$$

From here we obtain to first order in $\delta_{e,l}(\mathbf{k})$

$$\delta_{e,l}(\mathbf{k}) = \frac{\Lambda_l(\eta_e)}{k^2 \lambda_D^2 + 1 - \Lambda_0(\eta_e)} \qquad (10.128)$$

showing that the excursion of the dispersion curves for increasing harmonic number decreases both because of the inverse proportionality to l and because of the asymptotic behavior (10.118) of $\Lambda_l(\eta_e)$. The maximum of $\delta_{e,l}(\mathbf{k})$ decreases approximately as l^{-3}.

Investigating the behavior of the damping rate, $\gamma(\omega, \mathbf{k})$, we remind ourselves that γ is proportional to the imaginary part of the response function (10.127), which is proportional to $\exp(-l^2 \delta_{e,l}^2 / k_\parallel^2 r_{ge}^2)$. Hence, strictly perpendicular propagating Bernstein waves with $k_\parallel = 0$ have vanishing Landau damping and can propagate without any dissipation. However, damping becomes stronger with increasing angle of propagation so that Bernstein modes propagate nearly perpendicular to the magnetic field.

Upper-Hybrid Waves

The two most important branches of electron-cyclotron waves are the $l = 1$ harmonic branch, which is the electron-cyclotron resonance leading to the fundamental electron-cyclotron wave $\omega = l\omega_{ge}$, and the branch which contains the upper-hybrid frequency, ω_{uh}, because this is a natural plasma resonance frequency. The dispersion relation of upper-hybrid waves is found from the magnetized dielectric response function if we assume $\omega \approx \omega_{uh}$ and expand in the vicinity of perpendicular propagation, $k_\parallel \ll k_\perp$. Ne-

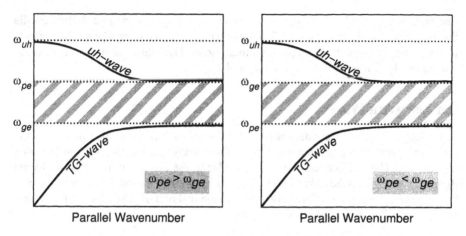

Parallel Wavenumber   Parallel Wavenumber

Fig. 10.9. Dispersion branches of upper-hybrid and Trivelpiece-Gould waves.

glecting the minor contribution of the ions at such frequencies, we find that

$$1 - \frac{\omega_{pe}^2}{\omega_2} \frac{k_\parallel^2}{k_\perp^2} - \frac{\omega_{pe}^2}{\omega^2 - \omega_{ge}^2} \frac{k_\perp^2}{k^2} = 0 \qquad (10.129)$$

In the limit of small gyrofrequency this equation describes oblique Langmuir waves (not including the thermal effects). The solution of this equation is

$$\omega^2 = \frac{1}{2}\omega_{uh}^2 \left[1 \pm \left(1 - \frac{4\omega_{pe}^2 \omega_{ge}^2}{\omega_{uh}^4} \right)^{1/2} \right] \qquad (10.130)$$

which for strictly perpendicular propagation becomes the upper-hybrid resonance. For oblique propagation the wave becomes strongly damped for

$$\cos^2 \theta > \omega_{uh}^4 / 4\omega_{ge}^2 \omega_{pe}^2 \qquad (10.131)$$

Expanding the root in the above dispersion relation one obtains two different modes

$$\omega^2(\mathbf{k}) = \begin{cases} \omega_{uh}^2 - \dfrac{\omega_{pe}^2 \omega_{ge}^2}{\omega_{uh}^2} \dfrac{k_\parallel^2}{k^2} \\[3mm] \dfrac{\omega_{pe}^2}{1 + \omega_{pe}^2/\omega_{ge}^2} \dfrac{k_\parallel^2}{k_\perp^2} \end{cases} \qquad (10.132)$$

The first of these modes is the *upper-hybrid wave*. It becomes an upper-hybrid oscillation for perpendicular propagation. The second is known as the *Trivelpiece-Gould mode* which is simply a modified oblique Langmuir mode. Their dispersion curves are shown schematically in Fig. 10.9 for $\omega_{pe} > \omega_{ge}$ and $\omega_{pe} < \omega_{ge}$.

Ion Bernstein Waves

In analogy to electron Bernstein waves, *ion Bernstein waves* can also propagate in a plasma perpendicular to the magnetic field. These waves are electrostatic ion-cyclotron resonances. Their dispersion relation is similar to the dispersion relation of electron Bernstein waves with the only difference that in addition to the ion terms the usual electron screening term $(k\lambda_D)^{-2}$ must be added to the real part of the dielectric function so that their general dispersion relation becomes

$$1 + \frac{1}{k_\perp^2 \lambda_D^2} - \frac{1}{k_\perp^2 \lambda_{Di}^2} \sum_{l=1}^{\infty} \frac{2l^2 \Lambda_l(\eta_i)}{v_{hi}^2 - l^2} = 0 \qquad (10.133)$$

where $v_{hi} = \omega/\omega_{gi}$. Hence, one expects *electrostatic ion-cyclotron waves* to be ordered into ion-cyclotron harmonic bands in a similar way as electron Bernstein waves are ordered into electron-cyclotron harmonic bands. Their dispersion will be slightly modified due to the electron term appearing in the dispersion relation

$$\omega = l\omega_{gi}[1 + \delta_{i,l}(\mathbf{k})] \qquad (10.134)$$

where

$$\delta_{i,l}(\mathbf{k}) = \frac{\Lambda_l(\eta_i)}{k^2 \lambda_{Di}^2 + 1 + (T_i/T_e) - \Lambda_0(\eta_i)} \qquad (10.135)$$

There are, however, several differences in the behavior of ion-cyclotron waves and electron Bernstein modes. The first is that in the former the lower-hybrid frequency plays the role the upper-hybrid frequency plays in the latter. In the harmonic band containing the lower-hybrid frequency, ω_{lh}, putting $k_\perp = 0$ (which corresponds to putting $\eta_i = 0$) the dispersion relation (10.133) reduces to

$$\frac{\omega_{pe}^2}{\omega^2 - \omega_{ge}^2} + \frac{\omega_{pi}^2}{\omega^2 - \omega_{gi}^2} = 1 \qquad (10.136)$$

which we are familiar with from earlier considerations and which has as solution for $\omega_{gi} \ll \omega \ll \omega_{ge}$, the lower-hybrid resonance frequency. Hence, in this band the dispersion curve starts at ω_{lh}. The second difference is the possibility that the ions may behave unmagnetized. This leads to waves propagating at the lower-hybrid frequency.

For long wavelength ion-cyclotron waves the dominant mode is the $l = 1$ mode. This becomes obvious from Eq. (10.135) in the small argument expansion, $\eta_i \ll 1$. Using the expansions of $\Lambda_l(\eta_i)$, this mode satisfies, with $v_{hi} = \omega/\omega_{gi} = 1 + \delta_{i,1}$

$$1 + \frac{1}{k_\parallel^2 \lambda_D^2} + \frac{k_\perp^2}{k_\parallel^2} \left(1 - \frac{\omega_{pe}^2/\omega_{ge}^2}{v_{hi}^2 - 1} \right) = \frac{\omega_{pi}^2}{\omega_{gi}^2} \frac{1}{v_{hi}^2} \tag{10.137}$$

The $\delta \ll 1$ solution of this equation defines the fundamental ion-cyclotron wave

$$\boxed{\omega_{ic} = \omega_{gi} \left(1 + \tfrac{1}{2} k_\perp^2 r_{gi}^2 \right)} \tag{10.138}$$

This solution is good for nearly perpendicular propagation. If we calculate the energy contained in this particular wave mode, we find that it is $W_{ic} = W_E$. However, for the more general mode, where the assumptions of very small parallel wavenumber have not been made, the spectral energy density becomes a function of $v_{hi} = l(1 + \delta_{i,l})$

$$\boxed{W_{ic} = \frac{\omega_{pi}^2}{\omega_{gi}^2} \left(\frac{v_{hi}^2 W_{\perp E}}{(v_{hi}^2 - 1)^2} + \frac{W_{\parallel E}}{v_{hi}^2} \right)} \tag{10.139}$$

and the electric field energy splits into parallel and perpendicular parts. Hence, depending on the ratio of ion plasma to ion-cyclotron frequencies the energy in the wave can be very different from the electric field energy.

An example of electrostatic ion-cyclotron waves is shown in Fig. 10.10. These waves have been excited during the injection of a heavy ion beam into the ionospheric plasma and show the harmonic structuring of the spectrum up to the 12th harmonic of the ion-cyclotron frequency. The lowest frequency in this figure is the lower-hybrid frequency, which is not at an exact harmonic.

Lower-Hybrid Waves

When the ions are unmagnetized, we can make the approximation of large ion gyroradii, $k_\perp^2 r_{gi}^2 \gg 1$. In such a case the ions follow straight line orbits. It is clear that this approximation is particularly useful for high ion temperatures. Hence, it is complementary to the existence of ion-acoustic waves, which require low ion and high electron temperatures. We therefore use in addition $\eta_e \ll 1$ and the high phase velocity approximation, $\omega/k_\parallel \gg v_{the\parallel}$. The dispersion relation then becomes

$$\epsilon(\omega, \mathbf{k}) = \left(1 - \frac{\omega_{pe}^2}{\omega^2} \right) \frac{k_\parallel^2}{k_\perp^2} - \frac{1}{2k_\perp^2 \lambda_{Di}^2} Z' \left(\frac{\omega}{k v_{thi}} \right) = 0 \tag{10.140}$$

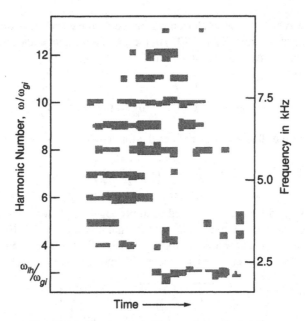

Fig. 10.10. Spectrogram of ion-cyclotron harmonics.

This dispersion relation simplifies for purely perpendicular propagation to

$$1 + \frac{\omega_{pe}^2}{\omega_{ge}^2} - \frac{\omega_{pi}^2}{\omega^2}\left(1 + \frac{3}{2}\frac{k_\perp^2 v_{thi}^2}{\omega^2}\right) = 0 \tag{10.141}$$

which has the approximate solution

$$\omega^2 = \omega_{lh}^2\left[1 + \frac{3}{2}\left(1 + \frac{\omega_{pe}^2}{\omega_{ge}^2}\right)k_\perp^2\lambda_{Di}^2\right] \tag{10.142}$$

This dispersion relation describes perpendicular *lower-hybrid waves*. In the long wave-length domain, $k_\perp^2\lambda_{Di}^2 \ll 1$, this wave becomes a lower-hybrid oscillation. Interestingly, the perpendicular lower-hybrid wave has dispersion similar to that of electron Langmuir waves in an unmagnetized plasma.

Keeping the finite though small parallel wavenumber term in the dielectric response function and expanding the plasma dispersion function for large argument produces the following general dispersion relation for lower-hybrid waves

$$\omega^2 = \omega_{lh}\left[1 + \frac{m_i}{m_e}\frac{k_\parallel^2}{k_\perp^2} + \frac{3k^2}{4k_\perp^2}\left(1 + \frac{\omega_{pe}^2}{\omega_{ge}^2}\right)k^2\lambda_{Di}^2\right] \tag{10.143}$$

We have assumed here that $k_\parallel \ll k_\perp$ in simplifying the right-hand side. This is justified insofar as the second term in the brackets suggests that the ratio of parallel to perpendicular wavenumbers must be of the order of $k_\parallel / k_\perp \approx (m_e/m_i)^{1/2}$ in order to affect the propagation of the waves. As we have mentioned these lower-hybrid waves propagate at a natural plasma resonance and are therefore of immense importance for the interaction between waves and particles. They can be used to heat a plasma and, in cases when the ion temperature is high, contribute to energy losses from the plasma. Some of these effects will be discussed in the following chapters.

Let us finally estimate the energy contained in the perpendicular lower-hybrid mode by using Eq. (10.141). If we multiply by frequency and take the derivative, we find that

$$W_{lh} \approx \left(1 + \frac{\omega_{pe}^2}{\omega_{ge}^2} \right) W_E \tag{10.144}$$

is larger than the electric field energy by a factor which in a dense plasma can be much larger than one. Lower-hybrid waves thus carry very much energy in dense plasmas, while in dilute plasmas their energy is just equal to the electric field energy.

10.6. Electromagnetic Plasma Waves

Electromagnetic wave propagation in a homogeneous plasma embedded into a uniform magnetic field must be treated on the basis of the general dispersion relation (9.50)

$$D(\omega, \mathbf{k}) = 0 \tag{10.145}$$

where $D(\omega, \mathbf{k})$ is the determinant of the full dispersion tensor (9.55)

$$N^2 \left(\frac{\mathbf{kk}}{k^2} - \mathbf{I} \right) + \epsilon(\omega, \mathbf{k}) \tag{10.146}$$

which contains the dielectric tensor (9.54). This tensor has been reduced to the general form in Eq. (10.94), valid for gyrotropic distribution functions. For a Maxwellian plasma the integrations over the parallel and perpendicular velocities can be performed with the help of the plasma dispersion function, Z, and its derivative, Z', using the Weber integrals, and taking advantage of the recursion relations between the Bessel functions. This calculation yields the dielectric tensor

$$\epsilon(\omega, \mathbf{k}) = \mathbf{I} + \sum_s \begin{pmatrix} \epsilon_{s1} & \epsilon_{s2} & \epsilon_{s4} \\ -\epsilon_{s2} & \epsilon_{s1} - \epsilon_{s0} & -\epsilon_{s5} \\ \epsilon_{s4} & \epsilon_{s5} & \epsilon_{s3} \end{pmatrix} \tag{10.147}$$

The components of the tensor inside the sum are given by

$$
\epsilon_{s0} = \frac{2\omega_{ps}^2}{\omega k_\parallel v_{\mathrm{ths}\parallel}} \sum_l \eta_s \Lambda_l'(\eta_s) \left[Z(\zeta_{s,l}) - \frac{k_\parallel v_{\mathrm{ths}\parallel}}{2\omega} A_s Z'(\zeta_{s,l}) \right]
$$

$$
\epsilon_{s1} = \frac{\omega_{ps}^2}{\omega k_\parallel v_{\mathrm{ths}\parallel}} \sum_l \frac{l^2 \Lambda_l(\eta_s)}{\eta_s} \left[Z(\zeta_{s,l}) - \frac{k_\parallel v_{\mathrm{ths}\parallel}}{2\omega} A_s Z'(\zeta_{s,l}) \right]
$$

$$
\epsilon_{s2} = \frac{i\,\mathrm{sgn}(q_s)\omega_{ps}^2}{\omega k_\parallel v_{\mathrm{ths}\parallel}} \sum_l l \Lambda_l'(\eta_s) \left[Z(\zeta_{s,l}) - \frac{k_\parallel v_{\mathrm{ths}\parallel}}{2\omega} A_s Z'(\zeta_{s,l}) \right]
$$

$$
\epsilon_{s3} = -\frac{\omega_{ps}^2}{k_\parallel^2 v_{\mathrm{ths}\parallel}^2} \sum_l \left(1 - \frac{A_s}{A_s+1} \frac{l\omega_{gs}}{\omega} \right) \left(1 + \frac{l\omega_{gs}}{\omega} \right) \Lambda_l(\eta_s) Z'(\zeta_{s,l})
$$

$$
\epsilon_{s4} = \frac{k_\perp}{2k_\parallel} \frac{\omega_{ps}^2}{\omega \omega_{gs}} \sum_l \left(A_s + 1 - \frac{l\omega_{gs}}{\omega} A_s \right) \frac{l\Lambda_l(\eta_s)}{\eta_s} Z'(\zeta_{s,l})
$$

$$
\epsilon_{s5} = -\frac{i\,\mathrm{sgn}(q_s)}{k_\perp k_\parallel} \frac{\omega_{ps}^2}{2\omega \omega_{gs}} \sum_l \left(A_s + 1 - \frac{l\omega_{gs}}{\omega} A_s \right) \Lambda_l'(\eta_s) Z'(\zeta_{s,l})
$$

(10.148)

where $\zeta_{s,l} = (\omega - l\omega_{gs})/k_\parallel v_{\mathrm{ths}\parallel}$, the sum is taken from $l = -\infty$ to $l = +\infty$, and we defined the temperature anisotropy of species s as

$$
A_s = \frac{T_{s\perp}}{T_{s\parallel}} - 1
$$

(10.149)

In an isotropic plasma $A_s = 0$, and many of the above terms disappear.

Weak Damping

Under the assumption of weak wave damping in a magnetized plasma with damping rate $|\gamma(\omega, \mathbf{k})| \ll \omega(\mathbf{k})$, where ω is the real part of the complex frequency, we can find an expression for the damping rate, γ, from the general dispersion relation $D(\omega, \mathbf{k}) = 0$. The procedure is similar to that applied when calculating the weak damping rate of electrostatic waves, with the only difference that now the full dispersion relation replaces the plasma response function. Splitting into real and imaginary parts

$$
D(\omega, \gamma, \mathbf{k}) = D_r(\omega, \gamma, \mathbf{k}) + i D_i(\omega, \gamma, \mathbf{k})
$$

(10.150)

the dispersion relation can be expanded around the real frequency as

$$
D(\omega, \gamma, \mathbf{k}) = D_r(\omega, 0, \mathbf{k}) + i\gamma \frac{\partial D_r(\omega, \gamma, \mathbf{k})}{\partial \omega} \bigg|_{\gamma=0} + i D_i(\omega, 0, \mathbf{k}) = 0 \quad (10.151)
$$

and one finds for weakly damped waves

$$D_r(\omega, 0, \mathbf{k}) = 0 \tag{10.152}$$

$$\gamma(\omega, \mathbf{k}) = -\frac{D_i(\omega, 0, \mathbf{k})}{\partial D_r(\omega, \gamma, \mathbf{k})/\partial\omega|_{\gamma=0}} \tag{10.153}$$

Electromagnetic Modes

High-frequency electromagnetic waves above the electron plasma or electron gyrofrequency, whichever is higher, are only weakly affected by the presence of the plasma. Those waves behave approximately like free space radiation to which we will return in the special case of *auroral kilometric radiation* when treating linear plasma instabilities. In dense cold plasmas, $\omega_{pe}^2 > \omega_{ge}^2$, no electromagnetic wave propagation is allowed between the electron plasma and the electron gyrofrequency. This conclusion is not changed for dense hot plasmas. In dilute cold plasmas, $\omega_{pe}^2 < \omega_{ge}^2$, electromagnetic waves can propagate in different modes at all frequencies, a conclusion which holds as well for hot plasmas, but the lower frequency branches are confined in both cases to the plasma. It is these lower frequency branches which we are interested in, because the specific plasma properties enter only at these lower frequencies. The three electromagnetic wave modes of interest are the whistler mode, the electromagnetic ion-cyclotron or ion whistler mode, and the Alfvén mode. These waves have already been met in the fluid plasma section. Here we show how they emerge from the more precise kinetic treatment.

Whistlers

Electromagnetic waves are transverse waves, satisfying the condition $\mathbf{k} \cdot \delta\mathbf{E} = 0$. Restricting, for simplicity, to parallel propagation, $k_\perp = 0$, and transforming to right- and left-hand circular polarizations

$$\delta E_{R,L} = \delta E_x \mp i\delta E_y \tag{10.154}$$

the dispersion relation for the two parallel propagating electromagnetic modes separates, as in the fluid plasma model treated in Chap. 9

$$\begin{pmatrix} N_\parallel^2 - 1 - \sum_s(\epsilon_{s1} + i\epsilon_{s2}) \\ N_\parallel^2 - 1 - \sum_s(\epsilon_{s1} - i\epsilon_{s2}) \end{pmatrix} \begin{pmatrix} \delta E_R \\ \delta E_L \end{pmatrix} = \begin{pmatrix} 0 \\ 0 \end{pmatrix} \tag{10.155}$$

All other components of the dielectric tensor vanish. For instance, the component ϵ_{s3} drops out, because it does not contribute to the electromagnetic wave for parallel propagation. Also, $\epsilon_{s0} = 0$ because $k_\perp = 0$, so that $\epsilon_{s1} = \epsilon_{xx} = \epsilon_{yy}$, and circular polarization makes the other components vanish. Clearly, the above equation splits into two separate dispersion relations for the R- and L-modes .

$$N_{\|R,L}^2 = 1 + \sum_s (\epsilon_{s1} \pm i\epsilon_{s2}) \tag{10.156}$$

Although all particle components contribute to each of the R- and L-modes, they are independent of each other. In the low-frequency domain the R-mode will become the whistler mode, while the L-mode is the electromagnetic ion-cyclotron wave.

Because of the condition of vanishing perpendicular wavenumber, the argument of the $\Lambda_l(\eta_s)$ function vanishes, $\eta_s = 0$. Since $\Lambda_l(0) = I_l(0)$ and $I_{-l} = I_l \propto (\eta_s/2)^l$, only the $l = \pm 1$ terms in the components of the dielectric tensor (10.147) contribute. This is easily seen from the definition of ϵ_{s1}. To prove this for ϵ_{s2}, one uses the identity

$$\Lambda_l' = -\Lambda_l + I_l' \exp(-\eta_s) \tag{10.157}$$

and the recurrence relation $I_l' = 2(I_{l-1} + I_{l+1})$. Hence, in the sum over l only the terms $l = \pm 1$ are non-zero. This yields for the two surviving components

$$\epsilon_{s1} = \frac{\omega_{ps}^2}{2k_\| v_{ths\|}\omega} \left\{ Z(\zeta_{s,1}) + Z(\zeta_{s,-1}) - \frac{k_\| v_{ths\|}}{2\omega} A_s \left[Z'(\zeta_{s,1}) + Z'(\zeta_{s,-1}) \right] \right\}$$

$$\tag{10.158}$$

$$\epsilon_{s2} = \frac{2i\,\text{sgn}(q_s)\omega_{ps}^2}{k_\| v_{ths\|}\omega} \left\{ Z(\zeta_{s,1}) - Z(\zeta_{s,-1}) - \frac{k_\| v_{ths\|}}{2\omega} A_s \left[Z'(\zeta_{s,1}) - Z'(\zeta_{s,-1}) \right] \right\}$$

Using these expressions in Eq. (10.156), the dispersion relation can be written as

$$N_{\|R,L}^2 = 1 + \sum_s \frac{\omega}{2k_\| v_{ths\|}} \frac{\omega_{ps}^2}{\omega^2} \left\{ \left[1 \mp \text{sgn}(q_s)\right]\tilde{Z}_{s,1} + \left[1 \pm \text{sgn}(q_s)\right] \tilde{Z}_{s,-1} \right\} \tag{10.159}$$

where

$$\tilde{Z}_{s,\pm 1} = Z(\zeta_{s,\pm 1}) - \frac{k_\| v_{ths\|}}{\omega} A_s Z'(\zeta_{s,\pm 1}) \tag{10.160}$$

Let us consider the R-mode. Its dispersion relation from Eq. (10.159) reads

$$N_{\|R}^2 = 1 + \frac{\omega_{pe}^2}{k_\| v_{the\|}\omega} \left[Z(\zeta_{e,1}) - \frac{k_\| v_{the\|}}{2\omega} A_e Z'(\zeta_{e,1}) \right]$$

$$+ \frac{\omega_{pi}^2}{k_\| v_{thi\|}\omega} \left[Z(\zeta_{i,-1}) - \frac{k_\| v_{thi\|}}{2\omega} A_i Z'(\zeta_{i,-1}) \right] \tag{10.161}$$

This equation is still fairly general. The further discussion depends on the approximations introduced. Clearly, the ions have no resonance at the electron cyclotron frequency, because in the whistler range, $\omega \leq \omega_{ge}$

$$\zeta_{i,-1} = (\omega + \omega_{gi})/k_\parallel v_{\text{thi}\parallel} \approx \omega/k_\parallel v_{\text{thi}\parallel} \tag{10.162}$$

is usually much larger than one. This allows to use the large argument expansion for the ion terms in Eq. (10.161), which implies that ion damping can be neglected, because it is proportional to $\exp(-\zeta_{i,1}^2)$. The main damping comes from electrons near the cyclotron resonance frequency. Hence, entirely neglecting the contribution of the ions and assuming that the anisotropy is negligible and that we can use the large argument expansion for the electrons, we obtain

$$N_{\parallel R}^2 = 1 + \frac{\omega_{pe}^2}{\omega(\omega_{ge} - \omega)} + i \frac{\pi^{1/2}\omega}{k_\parallel v_{\text{the}}} \frac{\omega_{pe}^2}{\omega^2} \exp\left[-\left(\frac{\omega_{ge} - \omega}{k_\parallel v_{\text{the}}}\right)^2\right] \tag{10.163}$$

This is the kinetic dispersion relation for parallel propagating whistler waves in a plasma. The first two terms are identical to the fluid dispersion relation (9.125) of whistler waves. But the presence of the imaginary part of the dispersion relation demonstrates that, similar to Landau damping of Langmuir waves, whistlers are also damped due to thermal effects. This damping is particularly strong near the electron-cyclotron frequency, where the argument of the exponential function in Eq. (10.163) vanishes. Far away from $\omega = \omega_{ge}$ the damping vanishes, and whistlers are only weakly damped. Under this assumption we can neglect the imaginary part and obtain the parallel whistler dispersion relation

$$\boxed{\frac{k_\parallel^2 c^2}{\omega_{ge}^2} = \frac{\omega^2}{\omega_{ge}^2}\left[1 + \frac{\omega_{pe}^2}{\omega(\omega_{ge} - \omega)}\right]} \tag{10.164}$$

This equation shows the electron-cyclotron resonance of the fluid waves. But one should remember that it has been derived under the two restricting conditions of large $\zeta_{e,-1} \gg 1$, which at $\omega = \omega_{ge}$ is clearly violated, and the assumption of weak damping, which also does not hold at resonance. Here whistlers become strongly cyclotron damped, feeding their energy into the plasma electrons. Thus electromagnetic wave propagation in the whistler mode ceases at the electron cyclotron frequency.

We can easily estimate the cyclotron damping rate of whistlers sufficiently far away from the resonance, where the approximation of small damping is satisfied so that the dispersion relation (10.164) gives a valid expression for the whistler mode frequency. Assuming that $\gamma \ll \omega$ and using Eq. (10.153), we obtain

$$\gamma \approx -\frac{\pi^{1/2}\omega}{k_\parallel \lambda_D} \frac{\omega_{pe}(\omega_{ge} - \omega)}{\omega_{pe}^2 - 2\omega(\omega_{ge} - \omega)} \exp\left[-\left(\frac{\omega_{ge} - \omega}{k_\parallel v_{\text{the}}}\right)^2\right] \tag{10.165}$$

for the weak damping rate of whistlers sufficiently far below the electron-cyclotron frequency. The above expression shows that whistlers are damped waves, but sufficiently far below the cyclotron frequency this damping is not very strong, because cyclotron resonance does not extend far from ω_{ge}. Moreover, for parallel whistler wavelengths much longer than the electron gyroradius, this damping becomes very small by the exponential factor. Such long-wavelength whistlers propagate practically without any cyclotron damping along the magnetospheric magnetic field lines.

Electromagnetic Ion-Cyclotron Waves

The ion equivalent to whistlers are electromagnetic ion-cyclotron waves. These waves have frequencies far below the electron-cyclotron frequency, $\omega \ll \omega_{ge}$. Hence

$$N_{\parallel L}^2 = 1 + \frac{\omega_{pe}^2}{\omega^2} \frac{\omega}{k_\parallel v_{the}} \left[Z(\zeta_{e,-1}) - \frac{k_\parallel v_{the}}{\omega} A_e Z'(\zeta_{e,-1}) \right]$$
$$+ \frac{\omega_{pi}^2}{\omega^2} \frac{\omega}{k_\parallel v_{thi}} \left[Z(\zeta_{i,1}) - \frac{k_\parallel v_{thi}}{\omega} A_i Z'(\zeta_{i,1}) \right] \tag{10.166}$$

We neglect the anisotropies and expand the dispersion functions in the large argument limit. For the electrons this expansion is obvious because of the large electron-cyclotron frequency, $\zeta_{e,-1} = (\omega + \omega_{ge})/k_\parallel v_{the} \approx \omega_{ge}/k_\parallel v_{the}$. For the ion component it requires that $\omega_{gi} - \omega \gg k_\parallel v_{thi}$. Hence, the parallel wavelength must be much larger than the ion gyroradius. Under these conditions the dispersion relation can be written as

$$N_{\parallel L}^2 = 1 - \frac{\omega_{pe}^2}{\omega \omega_{ge}} + \frac{\omega_{gi}^2}{\omega(\omega_{gi} - \omega)} + i \frac{\pi^{1/2} \omega_{pi}^2}{\omega k_\parallel v_{thi}} \exp \left[-\left(\frac{\omega_{gi} - \omega}{k_\parallel v_{thi}} \right)^2 \right] \tag{10.167}$$

where we have neglected the exponentially small electron damping. The frequency in the electron term can be replaced by the ion gyrofrequency. Moreover, for weak damping the real part of the dispersion relation yields the electromagnetic ion-cyclotron wave

$$\boxed{\frac{k_\parallel^2 c^2}{\omega_{gi}^2} = \frac{\omega^2}{\omega_{gi}^2} \left[1 - \frac{\omega_{pi}^2}{\omega_{gi}^2} \left(1 - \frac{\omega_{gi}^2}{\omega(\omega_{gi} - \omega)} \right) \right]} \tag{10.168}$$

with a resonance at the ion-cyclotron frequency. Since we have neglected the electron term, electromagnetic ion-cyclotron waves are restricted to frequencies below the ion-cyclotron frequency, ω_{gi}. Sometimes the ion-cyclotron dispersion relation is given in the simplified form

$$\frac{k_\parallel^2 v_A^2}{\omega_{gi}^2} = \frac{\omega^2}{\omega_{gi}(\omega_{gi} - \omega)} \tag{10.169}$$

Clearly, for frequencies much below the ion-cyclotron frequency this dispersion relation goes over into the dispersion relation of Alfvén waves. Damping of electromagnetic ion-cyclotron waves is weak. It is given by the same expression as for whistlers, if the electron parameters are replaced by those of the ions.

Kinetic Alfvén Waves

As has been shown in Chap. 9, at extremely low frequencies both whistlers and ion-cyclotron waves become Alfvén waves. In this frequency domain the Alfvén wave has both polarizations, right-hand and left-hand, and can, depending on the amplitudes of the two circularly polarized components, assume any polarization.

These arguments hold for very long wavelengths. The dispersion relation of Alfvén waves in this domain can be obtained simply in the limit of very low frequencies, $\omega \ll \omega_{gi}$. There is, however, a range of wavelengths when the dispersion of the Alfvén wave changes. This happens when the wavelengths become either comparable to the ion gyroradius, $r_{gi} = v_{thi}/\omega_{gi}$, in a hot electron plasma, or to the electron inertial length, c/ω_{pe}, in a cold electron plasma. In these two ranges kinetic effects must be taken into account. The dispersion relations can be obtained from the above general dispersion relations for electron- and ion-cyclotron waves, but it is more instructive and also simpler to derive them separately by using a combined fluid-kinetic approach with the electrons taken to be fluid-like and the ions treated kinetic.

We allow for oblique propagation. Since Alfvén waves are electromagnetic waves, the wave electric potential is zero and the electric field is determined by the wave magnetic potential, $\delta \mathbf{E} = -\partial \mathbf{A}/\partial t$. Gyrotropy implies that the magnetic potential has two components. This is equivalent to introducing two different electric potential components, $\phi_\perp$, $\phi_\parallel$, and to represent the electric field as

$$\delta \mathbf{E} = (\delta E_\perp, \delta E_\parallel) = -(\nabla_\perp \phi_\perp, \nabla_\parallel \phi_\parallel) \tag{10.170}$$

Because no space charges are involved, Poisson's equation yields

$$\nabla_\perp^2 \phi_\perp + \nabla_\parallel^2 \phi_\parallel = \frac{e}{\epsilon_0}(\delta n_i - \delta n_e) = 0 \tag{10.171}$$

and from the only non-vanishing component of Ampère's law one has

$$\nabla_\parallel \nabla_\perp^2 (\phi_\perp - \phi_\parallel) = \mu_0 \frac{\partial (j_{i\parallel} + j_{e\parallel})}{\partial t} \tag{10.172}$$

where the parallel currents are calculated from the ion and electron Vlasov equations.

Let us consider two extreme cases. When the electron temperature is so high that the electron thermal velocity is much larger than the Alfvén velocity, the electrons almost immediately respond to the electric potential of the wave and electron inertia can be neglected. The electrons can be treated as a mass-less fluid of temperature T_e, with their density given by Boltzmann's law. At such low frequencies the ions, which behave as kinetic particles, contribute only through the zero harmonic number, $l = 0$, in the ion term of the dispersion relation. We then find for the ion density as the zero-order moment of the Maxwellian ion distribution, keeping only the term $l = 0$

$$\frac{e\delta n_i}{\epsilon_0} = -\frac{\omega_{pi}^2}{v_{thi}^2}[1 - \Lambda_0(\eta_i)]\phi_\perp + \frac{\omega_{pi}^2}{\omega^2}k_\parallel^2(1 - i\gamma_{li})\Lambda_0(\eta_i)\phi_\parallel \qquad (10.173)$$

and from Boltzmann's equation for the electron density

$$\frac{e\delta n_e}{\epsilon_0} = \frac{\omega_{pe}^2}{v_{the}^2}(1 + \gamma_{le})\phi_\parallel \qquad (10.174)$$

where the Landau damping terms in the low-frequency range are given by

$$\gamma_{li} = 2\pi^{1/2}\beta_i^{-3/2}\exp(-\beta_i^{-1})$$

$$\gamma_{le} = \pi^{1/2}\beta_i^{-1/2}\left(\frac{m_e T_i}{m_i T_e}\right)^{1/2} \qquad (10.175)$$

with $\beta_i = 2v_{thi}^2/v_A^2$. Electron Landau damping of the Alfvén waves is very small, unless the ion temperature is extremely high. In the following we neglect electron Landau damping. Similarly, we obtain for the parallel wave current densities

$$\mu_0 j_{e\parallel} = -\frac{\omega_{pe}^2}{v_{the}^2 c^2}\frac{\omega}{k_\parallel}\phi_\parallel$$

$$\qquad (10.176)$$

$$\mu_0 j_{i\parallel} = \frac{\omega_{pe}^2}{c^2}\frac{k_\parallel}{\omega}\Lambda_0(\eta_i)(1 - i\gamma_{li})\phi_\parallel$$

Substituting these expressions into Poisson's and Ampère's laws and ignoring damping yields

$$\left[\Lambda_0(\eta_i) - \frac{\omega^2}{k_\parallel^2 c_{ia}^2}\right]\left[1 - \frac{\omega^2}{k_\parallel^2 v_A^2}\frac{1 - \Lambda_0(\eta_i)}{\eta_i}\right] = \frac{\omega^2}{k_\parallel^2 v_{thi}^2}[1 - \Lambda_0(\eta_i)] \qquad (10.177)$$

This dispersion relation shows how the Alfvén wave couples to the ion-acoustic wave for perpendicular wavelengths comparable to the ion gyroradius. For a sufficiently warm plasma the Λ_0-term in the first bracket can be ignored

$$\frac{\omega^2}{k_\parallel^2 v_A^2} = \frac{\eta_i}{1 - \Lambda_0(\eta_i)} + \frac{T_e}{T_i}\eta_i \qquad (10.178)$$

Clearly, this relation describes Alfvén waves. When its right-hand side becomes one, for $\eta_i \to 0$, the wave is the usual torsional Alfvén wave. In the more general case of

$\eta_i \neq 0$, the Alfvén wave propagates obliquely to the magnetic field. This wave is known as the *kinetic Alfvén wave*. For not too small ion temperatures the argument of Λ_0 can be assumed small, $\eta_i \ll 1$. The above dispersion relation then simplifies further to

$$\omega_{\text{ka}}^2 = k_\parallel^2 v_A^2 \left[1 + k_\perp^2 r_{gi}^2 \left(\frac{3}{4} + \frac{T_e}{T_i} \right) \right] \qquad (10.179)$$

The kinetic Alfvén wave propagates across the magnetic field. Usually its parallel wavelength is much longer than the ion gyroradius. Therefore we have $k_\parallel \ll k_\perp \propto r_{gi}$, and the perpendicular wavenumber is much larger than the parallel wavenumber. This wave is Landau damped by the ions and, in addition, carries a non-vanishing parallel current. Moreover, it has a non-vanishing parallel electric field component.

When considering the phase velocity of the kinetic Alfvén wave, we find that the parallel phase velocity is increased due to the factor in brackets in Eq. (10.179). This factor is of the order of two. Hence, kinetic Alfvén waves propagate with a parallel phase velocity which is about 50% higher than the normal Alfvén speed. Similarly, the group velocity of the kinetic Alfvén wave is enhanced

$$\left(\frac{\partial \omega_{\text{ka}}}{\partial k} \right)^2 \approx 2 v_A^2 \left(1 + \frac{1}{8} \cos^4 \theta \right) \approx 2 v_A^2 \qquad (10.180)$$

If the plasma is cool, such that $v_{\text{the}} < v_A$, electron inertia must be taken into account and the Boltzmann approximation for the electrons gets invalid. Instead one can use the fluid electron parallel equation, since their perpendicular motion still does not contribute at these low frequencies. The electrons merely perform a drift across the magnetic field in the perpendicular wave electric field. In this case the relevant length is not the ion gyroradius but the electron inertial length, c/ω_{pe}, and the dispersion relation reads

$$\omega_{\text{ska}}^2 = k_\parallel^2 v_A^2 \frac{1 + k_\perp^2 r_{gi}^2}{1 + k_\perp^2 c^2 / \omega_{pe}^2} \qquad (10.181)$$

The kinetic Alfvén wave described by this dispersion relation still contains the thermal effect of the ions, but additionally takes into account the inertia of the electron background plasma, which makes the wave 'heavier'. It effectively increases the mass which enters the Alfvén speed and slows the wave down. If we neglect the ion thermal effect, the wave is usually called *shear kinetic Alfvén wave*.

Figure 10.11 shows the angular dependence of the two kinetic Alfvén wave phase velocities for $k_\perp r_{gi} \approx 1$ and $k_\perp c/\omega_{pe} \approx 1$. In the first case the phase velocity is larger than the Alfvén velocity, in the second it is smaller.

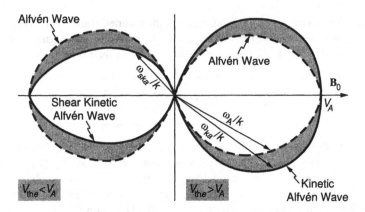

Fig. 10.11. Angular dependence of the phase velocities of the kinetic Alfvén waves.

The shear kinetic Alfvén wave in Eq. (10.181) is important insofar as in a cold plasma its parallel current is transported by the dense electron background and therefore can be rather strong. It is widely believed that these field-aligned Alfvénic currents transport the field-aligned energy along the magnetic field during all disturbances of the magnetosphere, causing the various auroral effects and producing the particle and wave phenomena observed during substorms.

Since the Earth's plasma sheet is a hot plasma region, kinetic Alfvén waves in the plasma sheet propagate in the non-inertial kinetic mode and will not carry bulk currents. The transition to the inertial shear kinetic mode will happen at that position in space where the electron thermal velocity drops below the local Alfvén speed. This becomes possible closer to the Earth, because here the plasma is cooler while the magnetic field strengthens. However, because of the high plasmaspheric density it can actually occur only at auroral latitudes, where the plasma density stays comparably low. The auroral lower magnetosphere is thus the only prospective candidate for the existence of inertial shear kinetic Alfvén waves. Writing the above condition as

$$\frac{v_{the}^2}{v_A^2} = \frac{m_i}{m_e}\beta \tag{10.182}$$

where β is the plasma beta, and assuming that the ions are sufficiently cold so that the plasma energy is carried by the electrons, one finds that inertial shear kinetic Alfvén waves with a dispersion given by Eq. (10.181) can propagate only if the condition

$$\beta < \frac{m_e}{m_i} \tag{10.183}$$

is satisfied. Otherwise, the kinetic Alfvén wave is non-inertial and satisfies the dispersion relation (10.179). Figure 10.12 shows the schematic boundary between the regions

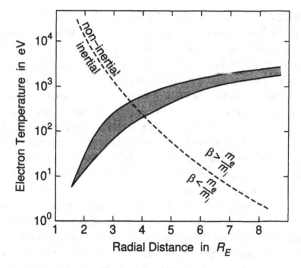

Fig. 10.12. Regimes of kinetic and shear kinetic Alfvén waves.

$\beta < m_e/m_i$ and $\beta > m_e/m_i$ in the auroral magnetosphere as function of the distance from the Earth's surface. For local electron temperatures below this curve the kinetic Alfvén waves are inertia-dominated, above they are non-inertial. We have included also a typical range of auroral region electron temperatures. Close to the Earth the electron temperatures fall into the inertial range while at distances larger than a few R_E the electron temperature rises above the regime of the shear kinetic Alfvén wave.

Concluding Remarks

In the present Chapter we have discussed, from a kinetic theory viewpoint, the main wave modes in a homogeneous magnetized plasma under the condition that these waves are stable. The most important among these modes are the natural plasma resonances at the lower-hybrid, upper-hybrid and cyclotron frequencies, which provide the channels in (ω, k) space along which energy can be transported, once it has been injected into a certain frequency and wavenumber range. For instance, a right-hand polarized electromagnetic wave at a frequency below the electron-cyclotron frequency with a wavelength longer than the electron gyroradius, will necessarily propagate in the whistler mode. On the other hand, it is impossible to inject electromagnetic waves into a dense plasma between the electron cyclotron and the upper-hybrid frequency.

In the last two chapters we dealt with homogeneous plasmas only. Hence, the applicability of the present theory is restricted to cases when the wavelengths are very short in comparison with the characteristics scale of any inhomogeneity. Nevertheless, the range of applicability of the homogeneous theory is rather wide. When the inhomogeneity of the plasma is weak, the wave propagation can be treated as in optics as propagation of rays in a medium where the refraction index and phase velocities change slowly in space. This method is called *ray tracing*. On the other hand, when the changes are very steep, one can use the refraction and transmission theory at discontinuities, where the waves must satisfy the jump conditions derived in Chap. 8. In cases where the scale of the inhomogeneity becomes comparable to the wavelength, inhomogeneous theory must be applied, where one has to use eikonal methods and eigenvalue solutions of the Vlasov-Poisson system. The inhomogeneity does not only introduce a new kind of waves, called *drift waves*, but also changes the properties of the linear waves and allows *linear conversion* of waves from one mode to another.

We finally mention that the plasma dispersion function used in this chapter is based on a Maxwellian distribution function. One can define other plasma dispersion functions by using other distribution functions, as has been done for the kappa distribution (see Sec. 6.3). However, the kappa distribution is not a stable distribution function. For theoretical purposes it is more convenient to approximate unstable distributions as superposition of Maxwellians.

Further Reading

A classic general treatment of plasma waves, but with no emphasis on space physics, is the monograph [5]. The relativistic dielectric tensor applying to radiation belt particles is given in [3] and [6]. A discussion of ion Bernstein modes can be found in [2]. Whistler waves are discussed in [4] and [7]. A brief collection of our knowledge about most aspects of Alfvén waves is contained in [1].

[1] A. Hasegawa and C. Uberoi, *The Alfvén Wave* (U.S. Dept. of Energy, Springfield, 1982).

[2] S. Ichimaru, *Basic Principles of Plasma Physics* (W. A. Benjamin, Reading, 1973).

[3] D. C. Montgomery and D. A. Tidman, *Plasma Kinetic Theory* (McGraw-Hill, New York, 1964).

[4] S. Sazhin, *Whistler-mode Waves in a Hot Plasma* (Cambridge University Press, Cambridge, 1993).

[5] T. H. Stix, *Waves in Plasma* (American Institute of Physics, New York, 1992).

[6] D. G. Swanson, *Plasma Waves* (Academic Press, Boston, 1989).

[7] A. D. M. Walker, *Plasma Waves in the Magnetosphere* (Springer Verlag, Heidelberg, 1993).

Outlook

The present book closes with the linear kinetic theory of waves in a plasma. This is the appropriate point to cut the material and to take a rest. Looking back, we started from the undisturbed single particle motion in electric and magnetic fields. Then, starting from basic kinetic theory, we came to consider a plasma as a mixture of several charged fluids in equilibrium. We showed that under some severe but very practical assumptions it can be described as one single fluid, similar to the conducting fluids in magnetohydrodynamics, but with the important difference that Ohm's law takes a fairly complicated form because of the different mobilities of electrons and ions.

Still assuming equilibrium, we allowed for the existence of discontinuities. We did not ask for the mechanism of their formation, but rather derived the conditions, which are imposed on these plasma discontinuities, by the mere requirement that the fluid and field conservation laws must be satisfied at these boundaries. A large number of important space plasma problems could be discussed using these equilibrium considerations, providing insight into the gross structure of the Earth's environment and into processes like convection and reconnection.

As a natural extension of this equilibrium theory, we considered the propagation of small-amplitude waves, with slight and restorable deviations from equilibrium, in both the fluid and kinetic approximations. We followed a systematic approach, climbing up the ladder from simple fluid theory to the complications of linear kinetic theory. The important result is that a plasma, in all its approximations, allows for a large but limited number of waves to propagate. These waves are eigenmodes of the plasma.

The well-known electromagnetic modes are severely changed when traversing a plasma and, moreover, in some frequency ranges cannot even propagate. In addition, there are a large number of wave modes, which can exist only inside a plasma. All these waves are involved in the energy transport in a plasma and they play a central role in its dynamics and evolution. Their properties have been discussed in detail, but we must admit that we have not been exhaustive and that some important aspects could not be treated. Some are mentioned in the concluding remarks at the end of each chapter, but there remain a number of topics which go far beyond the material treated in this book.

The first important topic concerns the question, how waves can be excited in a plasma. This question is related to the stability of a plasma, the spontaneous, induced, or forced generation of oscillations and waves and their growth to large amplitudes. The second topic is a direct consequence of the former. When plasma waves grow to large amplitudes, linear theory looses validity and becomes irrelevant for the further evolution of the waves. The plasma starts entering the nonlinear stage, saturation processes set in, particle dynamics begins to be affected by the large-amplitude waves, and interaction between the waves themselves modifies the wave modes. The reaction of the plasma to these nonlinear effects may lead to equilibrium states, which are quite different from the linear ones considered in this book.

Nowadays space plasma physics, with the much higher resolution of the measurements in space and time, has grown into this field of instability and nonlinearity and has started to become mature in understanding the nonlinear connections. Even to understand the global effects in space plasma physics, one must now study the nonlinear production of transport coefficients, nonlinear wave dissipation, and the nonlinear acceleration and heating of particles in wave fields. All these processes are important not only for the plasma physics of near-Earth space, but also to solar physics and astrophysics.

The spatial limitations of the present book have not allowed to include these topics here. Instead of shortening the material, we have chosen to include it in a companion text, *Advanced Space Plasma Physics*. Many of the space plasma instabilities, from the macro-scale Rayleigh-Taylor and Kelvin-Helmholtz instabilities to the electrostatic and electromagnetic microinstabilities are discussed therein. The chapters on nonlinear aspects include nonlinear waves, weak and strong turbulence, and a number of special topics like auroral particle acceleration, soliton formation and caviton collapse, anomalous transport, and the theory of collisionless shocks.

Both texts have been written in the spirit to provide physical understanding in addition to mathematical clarity and completeness. In cases where we had to decide for one or the other, we have always chosen the former point of view. This, we hope, will be advantageous for the reader.

A. Some Basics

A.1. Useful Constants

As throughout this book, the values of the constants are given in SI units.

c	velocity of light	$3.00 \cdot 10^8 \, \mathrm{m\,s^{-1}}$
μ_0	free space magnetic permeability	$4\pi \cdot 10^{-7} \, \mathrm{H\,m^{-1}}$
ϵ_0	vacuum dielectric constant	$8.85 \cdot 10^{-12} \, \mathrm{F\,m^{-1}}$
e	electron charge	$1.60 \cdot 10^{-19} \, \mathrm{C}$
m_e	electron mass	$9.11 \cdot 10^{-31} \, \mathrm{kg}$
m_p	proton mass	$1.67 \cdot 10^{-27} \, \mathrm{kg}$
k_B	Boltzmann's constant	$1.38 \cdot 10^{-23} \, \mathrm{J\,K^{-1}}$
R_0	ideal gas constant	$8.31 \, \mathrm{J\,K^{-1}\,mol^{-1}}$
R_E	equatorial radius of Earth	$6.37 \cdot 10^6 \, \mathrm{m}$
M_E	magnetic moment of Earth	$8.05 \cdot 10^{22} \, \mathrm{A\,m^2}$
AU	Sun-Earth distance	$1.50 \cdot 10^{11} \, \mathrm{m}$

A.2. Energy Units

Since thermal energy and temperature are related by $W = k_B T$, Joule and Kelvin are often used interchangeably. Moreover, particle energy is often measured in eV and overall energy budgets are often given in the cgs-unit erg. The four units are related as follows:

1 J	—	$7.24 \cdot 10^{22}$ K	$6.24 \cdot 10^{18}$ eV	10^7 erg
1 K	$1.38 \cdot 10^{-23}$ J	—	$8.62 \cdot 10^{-5}$ eV	$1.38 \cdot 10^{-16}$ erg
1 eV	$1.60 \cdot 10^{-19}$ J	$1.16 \cdot 10^4$ K	—	$1.60 \cdot 10^{-12}$ erg
1 erg	10^{-7} J	$7.24 \cdot 10^{15}$ K	$6.24 \cdot 10^{11}$ eV	—

A.3.   Useful Formulas

The numerical values of all plasma parameters and other important quantities can, of course, be calculated using the formulas given in the main text. However, it is often useful to have a simpler formula, which already incorporates the numerical values of any constants included in the formula as well as possible unit conversion factors. The units are adapted to the typical range of values found in space plasmas.

f_{pe}	electron plasma frequency	$9.0 \cdot 10^3 \sqrt{n}$	in Hz
f_{pi}	proton plasma frequency	$2.1 \cdot 10^2 \sqrt{n}$	in Hz
f_{ge}	electron gyrofrequency	$2.8 \cdot 10^1 B$	in Hz
f_{gi}	proton gyrofrequency	$1.5 \cdot 10^{-2} B$	in Hz
r_{ge}	electron gyroradius	$3.1 \cdot 10^1 \sqrt{T_e}/B$	in km
		$1.1 \cdot 10^2 \sqrt{W_e}/B$	in km
r_{gi}	proton gyroradius	$1.3 \cdot 10^3 \sqrt{T_i}/B$	in km
		$4.6 \cdot 10^3 \sqrt{W_i}/B$	in km
v_{the}	electron thermal speed	$3.9 \cdot 10^3 \sqrt{T_e}$	in km/s
		$1.3 \cdot 10^4 \sqrt{W_e}$	in km/s
v_{thi}	proton thermal speed	$9.1 \cdot 10^1 \sqrt{T_i}$	in km/s
		$3.1 \cdot 10^2 \sqrt{W_i}$	in km/s
v_E	E×B drift speed	$1.0 \cdot 10^3 E/B$	in km/s
v_A	Alfvén speed	$2.2 \cdot 10^1 B/\sqrt{n}$	in km/s
λ_D	Debye length	$6.9 \cdot 10^1 \sqrt{T_e/n}$	in m
		$2.4 \cdot 10^2 \sqrt{W_e/n}$	in m
n	plasma density		in cm^{-3}
T_e & T_i	electron & proton temperature		in 10^6 K
W_e & W_i	electron & proton thermal energy		in keV
B	magnetic field strength		in nT
E	electric field strength		in mV/m

A.4. Vectors and Tensors

The vector and tensor relations given in the following sections are useful in the context of plasma physics. Here, we have denoted scalars by ϕ, vectors by $\mathbf{A}$, and tensors by $\mathbf{T}$.

Dyadic Notation

Second-rank tensors can be written in *dyadic notation* as

$$\mathbf{T} = \mathbf{AB} \tag{A.1}$$

Thus a second-rank tensor can be constructed from two vectors, i.e, first-rank tensors. In component notation the dyadic product reads

$$T_{ij} = A_i B_j \tag{A.2}$$

where i, j represent the components x, y, z. Hence, the dyadic product represents a two-dimensional matrix

$$\mathbf{T} = \begin{pmatrix} A_x B_x & A_x B_y & A_x B_z \\ A_y B_x & A_y B_y & A_y B_z \\ A_z B_x & A_z B_y & A_z B_z \end{pmatrix} \tag{A.3}$$

Exchanging the vectors in the dyadic product $\mathbf{T} = \mathbf{AB}$ results in the transposed dyad

$$\mathbf{T}' = \mathbf{BA} \tag{A.4}$$

For symmetric tensors, dyad and transposed dyad are identical

$$\mathbf{T} = \mathbf{AA} = \mathbf{T}' \tag{A.5}$$

Second-rank tensors or dyadic products of two vectors are the ones most often used. However, at times it is useful to represent scalars and vectors as zero- and first-rank tensors, respectively, and occasionally third-rank tensors are used. The latter are defined as dyadic products of three vectors, like the heat tensor given in Sec. 6.5.

Dyads can be functions of complex quantities like frequency and wavenumber, in which case they become complex functions themselves. For complex tensors it is useful to introduce another kind of transposed tensor, the *Hermitean conjugate tensor*, $\mathbf{T}^\dagger$. This tensor is defined as the transposed of the conjugate complex tensor

$$\mathbf{T}^* = (\mathbf{AB})^* \tag{A.6}$$

so that

$$\mathbf{T}^\dagger = \{(\mathbf{AB})^*\}' \tag{A.7}$$

With this definition, one can show that

$$\mathbf{A}^* \cdot \mathbf{T} \cdot \mathbf{B} = \mathbf{B} \cdot \mathbf{T}^\dagger \cdot \mathbf{A}^* \tag{A.8}$$

Algebraic Relations

The algebraic relations, inner and cross-product, are governed by the following rules:

$$\mathbf{A} \cdot \mathbf{B} = \mathbf{B} \cdot \mathbf{A} \tag{A.9}$$

$$\mathbf{A} \times \mathbf{B} = -\mathbf{B} \times \mathbf{A} \tag{A.10}$$

$$\mathbf{A} \cdot (\mathbf{B} \times \mathbf{C}) = \mathbf{B} \cdot (\mathbf{C} \times \mathbf{A}) = \mathbf{C} \cdot (\mathbf{A} \times \mathbf{B}) \tag{A.11}$$

$$\mathbf{A} \times (\mathbf{B} \times \mathbf{C}) = (\mathbf{A} \cdot \mathbf{C})\mathbf{B} - (\mathbf{A} \cdot \mathbf{B})\mathbf{C} \tag{A.12}$$

$$(\mathbf{A} \times \mathbf{B}) \cdot (\mathbf{C} \times \mathbf{D}) = (\mathbf{A} \cdot \mathbf{C})(\mathbf{B} \cdot \mathbf{D}) - (\mathbf{A} \cdot \mathbf{D})(\mathbf{B} \cdot \mathbf{C}) \tag{A.13}$$

$$\phi\mathbf{T} = \mathbf{T}\phi \tag{A.14}$$

$$\mathbf{CT} = \mathbf{CAB} \neq \mathbf{ACB} \neq \mathbf{ABC} \neq \mathbf{TC} \tag{A.15}$$

$$\mathbf{C} \cdot \mathbf{T} = (\mathbf{C} \cdot \mathbf{A})\mathbf{B} = \mathbf{B}(\mathbf{C} \cdot \mathbf{A}) = \mathbf{T}^t \cdot \mathbf{C} \tag{A.16}$$

$$\mathbf{C} \times \mathbf{T} = (\mathbf{C} \times \mathbf{A})\mathbf{B} = -(\mathbf{A} \times \mathbf{C})\mathbf{B} \tag{A.17}$$

Differential Relations

The vector derivatives tensors obey the following rules:

$$\nabla(\phi\psi) = \phi(\nabla\psi) + \psi(\nabla\phi) \tag{A.18}$$

$$\nabla \cdot (\phi\mathbf{A}) = \mathbf{A} \cdot (\nabla\phi) + \phi(\nabla \cdot \mathbf{A}) \tag{A.19}$$

$$\nabla \times (\phi\mathbf{A}) = \phi(\nabla \times \mathbf{A}) - \mathbf{A} \times (\nabla\phi) \tag{A.20}$$

$$\nabla \cdot (\mathbf{A} \times \mathbf{B}) = \mathbf{B} \cdot (\nabla \times \mathbf{A}) - \mathbf{A} \cdot (\nabla \times \mathbf{B}) \tag{A.21}$$

$$\nabla \times (\mathbf{A} \times \mathbf{B}) = \mathbf{A}(\nabla \cdot \mathbf{B}) - \mathbf{B}(\nabla \cdot \mathbf{A}) + (\mathbf{B} \cdot \nabla)\mathbf{A} - (\mathbf{A} \cdot \nabla)\mathbf{B} \tag{A.22}$$

$$\nabla^2\phi = \nabla \cdot (\nabla\phi) \tag{A.23}$$

$$\nabla^2\mathbf{A} = \nabla(\nabla \cdot \mathbf{A}) - \nabla \times (\nabla \times \mathbf{A}) \tag{A.24}$$

$$\nabla \times (\nabla\phi) = 0 \tag{A.25}$$

$$\nabla \cdot (\nabla \times \mathbf{A}) = 0 \tag{A.26}$$

$$\nabla \cdot (\phi\mathbf{T}) = (\nabla\phi) \cdot \mathbf{T} + \phi(\nabla \cdot \mathbf{T}) \tag{A.27}$$

$$\nabla \cdot \mathbf{T} = \nabla \cdot (\mathbf{AB}) = (\mathbf{A} \cdot \nabla)\mathbf{B} + \mathbf{B}(\nabla \cdot \mathbf{A}) \tag{A.28}$$

$$\nabla \times \mathbf{T} = \nabla \times (\mathbf{AB}) = (\nabla \times \mathbf{A})\mathbf{B} - (\mathbf{A} \times \nabla)\mathbf{B} \tag{A.29}$$

Integral Relations

There are two sets of integral relations for vector derivatives. One set of equations pre-
scribes a transformation from a volume to a surface integral while another set gives the
relation between a surface integral and the integration over the path enclosing the surface. From the equations below, Eqs. (A.31) and (A.34) are of special importance in
electrodynamics and plasma physics. After their inventors, they bear the names *Gauß'
theorem* and *Stokes' theorem*, respectively.

Using the notation $dV = dx\,dy\,dz$ and $d\mathbf{A} = \mathbf{n}\,dA$ with $\mathbf{n}$ the normal vector of a
surface element dA, the volume integrals of the vector derivatives can be rewritten as
integrals over the closed surface, $\mathbf{A}$, bounding the volume, V,

$$\int_V (\nabla\phi)\,dV = \oint_A \phi\,d\mathbf{A} \tag{A.30}$$

$$\int_V (\nabla\cdot\mathbf{Q})\,dV = \oint_A \mathbf{Q}\cdot d\mathbf{A} \tag{A.31}$$

$$\int_V (\nabla\times\mathbf{Q})\,dV = -\oint_A \mathbf{Q}\times d\mathbf{A} \tag{A.32}$$

In a similar way, surface integrals of vector derivatives can be rewritten as integrals
along the closed path, C, enclosing the surface, $\mathbf{A}$,

$$\int_A (\nabla\phi)\times d\mathbf{A} = -\oint_C \phi\,d\mathbf{s} \tag{A.33}$$

$$\int_A (\nabla\times\mathbf{Q})\cdot d\mathbf{A} = \oint_C \mathbf{Q}\cdot d\mathbf{s} \tag{A.34}$$

where $d\mathbf{s} = \mathbf{t}\,ds$, ie., the product of a line element, ds, with its tangent vector, $\mathbf{t}$.

A.5. Some Electrodynamics

Plasma physics builds heavily on knowledge of electrodynamics. In the following we
give a short derivation of Maxwell's equations and other useful formulae.

Maxwell's Equations

Here we give a derivation of Maxwell's equations of electrodynamics, starting from first
principles. Conventionally Maxwell's equations are postulated without derivation. One

can, however, show that very simple assumptions on the continuity of charges and currents suffice to provide a satisfactory basis from which they can be derived axiomatically. Let us assume that there exists a net space charge density

$$\rho = \sum_s q_s n_s \qquad (A.35)$$

where the summation is over the different species (electrons, protons and ions) with different charges, q_s, and densities, n_s. The net space charge is the source of the electric fields. When the charges of a species move with an average velocity, v_s, we also know that they represent currents of density

$$\mathbf{j} = \sum_s q_s n_s \mathbf{v}_s \qquad (A.36)$$

which will become the sources of the magnetic fields.

Since neither charges nor particles can be destroyed in electrodynamics, any local temporal change of the charge density will be due to the divergence of the electric current which transports charges in or out of the volume element. Hence, we have the fundamental charge and current conservation equation

$$\frac{\partial \rho}{\partial t} + \nabla \cdot \mathbf{j} = 0 \qquad (A.37)$$

The second term in this equation is a divergence. Hence, to solve the equation we define the charge density as the divergence of some vector, $\mathbf{E}$,

$$\nabla \cdot \mathbf{E} = \rho / \epsilon_0 \qquad (A.38)$$

where ϵ_0 has been introduced as a dimensional factor. Inserting into Eq. (A.37) we get

$$\nabla \cdot \left(\mathbf{j} + \epsilon_0 \frac{\partial \mathbf{E}}{\partial t} \right) = 0 \qquad (A.39)$$

Since the divergence of the vector in the brackets vanishes, it must be the curl of another vector, $\mathbf{B}$,

$$\nabla \times \mathbf{B} = \mu_0 \mathbf{j} + \epsilon_0 \mu_0 \frac{\partial \mathbf{E}}{\partial t} \qquad (A.40)$$

Here we have introduced another dimensional factor μ_0 to account for the possibly different dimensions of $\mathbf{E}$ and $\mathbf{B}$. By taking the divergence of this expression it is easy to show that it just reproduces the original charge and current continuity equation.

The electric field, $\mathbf{E}$, is generated by the charge density, but we must still ask for the charges generating the magnetic field, $\mathbf{B}$. From observation one knows that such charges, called magnetic monopoles, do not exist. Thus $\mathbf{B}$ must be divergence-free

$$\nabla \cdot \mathbf{B} = 0 \qquad (A.41)$$

Since the time derivative of Eq. (A.41) vanishes, too, we conclude that

$$\frac{\partial \mathbf{B}}{\partial t} = -\nabla \times \mathbf{E} \tag{A.42}$$

must be the curl of a vector. However, this vector can only be the electric field, since no other quantities are involved. The minus sign arises from the experimental *Lenz's rule*.

Poisson's Equation

Poisson's equation is Laplace's equation including a source term. We encounter it in the Poisson law of electrodynamics, i.e., the Maxwell equation (A.38)

$$\nabla \cdot \mathbf{E} = \rho/\epsilon_0 \tag{A.43}$$

when replacing $\mathbf{E}$ by the gradient of the electric potential, $-\nabla\phi$. Then Eq. (A.43) reads

$$\nabla^2 \phi = -\rho/\epsilon_0 \tag{A.44}$$

In vector form with a vector source $\mathbf{Q}$ the same equation becomes

$$\nabla^2 \mathbf{A} = -\mathbf{Q} \tag{A.45}$$

Using the notation $d^3x = d\mathbf{x} = dxdydz$, these equations have the following solutions

$$\phi(\mathbf{x}) = \frac{1}{4\pi\epsilon_0} \int \frac{\rho(\mathbf{x}')}{|\mathbf{x} - \mathbf{x}'|} d^3x' \tag{A.46}$$

$$\mathbf{A}(\mathbf{x}) = \frac{1}{4\pi} \int \frac{\mathbf{Q}(\mathbf{x}')}{|\mathbf{x} - \mathbf{x}'|} d^3x' \tag{A.47}$$

expressing the simple truth that the fields are generated by their sources which are the singularities of the fields, and that the field value in space outside the sources can be obtained simply by integration over all the sources.

Biot-Savart's Law

Integrating Ampère's law, i.e., the Maxwell equation (A.40) for $\partial \mathbf{E}/\partial t = 0$,

$$\nabla \times \mathbf{B} = \mu_0 \mathbf{j} \tag{A.48}$$

with respect to the surface $\mathbf{A}$, which is enclosed by the path C and penetrated by the current $\mathbf{j}$, and using Stokes' theorem (see App. A.4) one obtains

$$\int_A (\nabla \times \mathbf{B}) \cdot d\mathbf{A} = \oint_C \mathbf{B} \cdot d\mathbf{s} = \mu_0 \int_A \mathbf{j} \cdot d\mathbf{A} = \mu_0 I \tag{A.49}$$

where $ds = \mathbf{t}\,ds$ is the product of a line element, ds, with its tangent vector, $\mathbf{t}$, while $d\mathbf{A} = \mathbf{n}\,dA$ is the product of a surface element dA with its normal vector, $\mathbf{n}$. Hence, the line integral along the magnetic field around the current along the boundary C of the surface A is proportional to the total current I flowing through the surface.

For a line current this curve is a circle of radius r, and the above integral yields $2\pi r\, B_\phi$. Hence, the only non-vanishing component of the magnetic field, the azimuthal component, decays radially as $1/r$

$$B_\phi(r) = \frac{\mu_0}{2\pi}\frac{I}{r} \tag{A.50}$$

Taking the curl of Ampère's law and remembering $\nabla \cdot \mathbf{B} = 0$, one finds, with the magnetic disturbance caused by distributed currents

$$\nabla^2 \mathbf{B} = -\mu_0 \nabla \times \mathbf{j} \tag{A.51}$$

This is a vectorial Poisson equation, called *Biot-Savart's law*. According to Eq. (A.47), its solution is

$$\mathbf{B}(\mathbf{x}) = \frac{\mu_0}{4\pi}\nabla \times \int \frac{\mathbf{j}(\mathbf{x}')}{|\mathbf{x} - \mathbf{x}'|}\,d^3x' \tag{A.52}$$

Calculating the curl yields

$$\mathbf{B}(\mathbf{x}) = \frac{\mu_0}{4\pi}\int \frac{\mathbf{j}(\mathbf{x}') \times (\mathbf{x} - \mathbf{x}')}{|\mathbf{x} - \mathbf{x}'|^3}\,d^3x' \tag{A.53}$$

which is Biot-Savart's law in integral form.

Faraday's Law

When integrating the Maxwell equation (A.42) over the surface $\mathbf{A}$ perpendicular to the magnetic field, one obtains

$$\int_A (\nabla \times \mathbf{E}) \cdot d\mathbf{A} = \int_A \frac{\partial \mathbf{B}}{\partial t} \cdot d\mathbf{A} \tag{A.54}$$

The left-hand side can be transformed using Stokes' theorem, while the right-hand side gives the time variation of the magnetic flux

$$\Phi = \int_A \mathbf{B} \cdot d\mathbf{A} \tag{A.55}$$

through the surface. Hence, one finds

$$\frac{d\Phi}{dt} = -\oint_C \mathbf{E} \cdot d\mathbf{s} \tag{A.56}$$

where $d\mathbf{s}$ is a line element of the curve C surrounding the surface A. The change in the magnetic flux is equal to the induced potential difference along C.

A.6. Plasma Entropy

Knowing the ideal gas equation of state, it is possible to derive an expression for the entropy, S, in an ideal gas which is useful to calculate S from measured quantities. The differential of the entropy is defined as the ratio of heat, dQ, and temperature, T, where

$$dQ = dW + pdV \tag{A.57}$$

with W the energy and V the volume. For an ideal gas

$$p = Nk_B T/V = (\gamma - 1)W \tag{A.58}$$

where $n = N/V$ is the number of particle per volume or particle density. From these expressions we find that

$$pdV = Nk_B T dV/V \tag{A.59}$$

or substituting into the above equation for the heat produced

$$dQ = Nk_B T \left[d \ln V + (\gamma - 1)^{-1} d \ln T \right] \tag{A.60}$$

from which we immediately obtain the desired form of the differential entropy

$$dS = Nk_B d \ln \left[V T^{1/(\gamma-1)} \right] \tag{A.61}$$

Performing the integration and remembering that for an ideal gas $Nk_B = R_0 = $ const (see App. A.1), one finds the expression for the entropy in the form

$$S - \tilde{S}_0 = R_0 \ln \left[V T^{1/(\gamma-1)} \right] \tag{A.62}$$

The quantity $\tilde{S}_0$ is an integration constant which is of no importance because only differences in entropy are of interest. Because of the same reason one can replace the volume by the density. This adds only another constant S_0, depending on the number of particles in the volume to the entropy. One then has finally

$$S - S_0 = R_0 \ln \left[\frac{T^{1/(\gamma-1)}}{n} \right] \tag{A.63}$$

as a formula to calculate the entropy change in an ideal gas from the measurements of temperature and density in the gas. With $\gamma = 5/3$ the exponent of T becomes 3/2.

The above equations apply only to isotropic cases. For anisotropic plasmas the corresponding expression for the entropy is

$$S - S_0 = R_0 \ln \left[\frac{1}{n} \left(\frac{\gamma_\| T_\| + 2\gamma_\perp T_\perp}{3} \right)^{3/2} \right] \tag{A.64}$$

where $\gamma_\|$, $\gamma_\perp$ are the adiabatic indices parallel and perpendicular to the magnetic field (see p. 136).

A.7.   Aspects of Analytic Theory

Plasma wave theory makes extensive use of complex variables, since frequency and wave number are complex quantities. Hence, the dispersion relation and all the integrals it contains become complex functions or integrals in the complex plane. Integrals in the complex plane are contour integrals. For their solution special methods have been developed already in the past century. Here, we review the basic methods as far as they apply to plasma theory.

Cauchy-Riemann Equations

A continuous complex function, $f(z) = u(x, y) + iv(x, y)$, in the complex plane, $z = x + iy$, must have a unique derivative with respect to z, $\partial f(z)/\partial z$. Because z itself consists of two variables, x, y, this uniqueness is not obvious. The condition of uniqueness is found by requiring that the derivatives from both sides x and y lead to the same result

$$\frac{\partial f(z)}{\partial z} = \frac{\partial u}{\partial x} + i\frac{\partial v}{\partial y} = \frac{\partial v}{\partial y} - i\frac{\partial u}{\partial y} \tag{A.65}$$

Comparing real and imaginary parts one finds

$$\frac{\partial u}{\partial x} = \frac{\partial v}{\partial y} \qquad \frac{\partial v}{\partial x} = -\frac{\partial u}{\partial y} \tag{A.66}$$

These equations are known as the *Cauchy-Riemann equations*. They constitute the necessary and sufficient conditions for the existence of a unique derivative of $f(z)$ and, hence, the condition for its continuous behavior in the complex plane. Such a function can be non-analytic only at a finite number of isolated points, which are called singular points.

Cauchy's Integral Formula

If $f(z)$ is continuous and analytic in a region, R, of the complex plane which is bounded by a curve, C, and if a is an interior point of R then *Cauchy's integral theorem* states that the integral

$$\oint_C \frac{f(z)dz}{z - a} = 2\pi i f(a) \tag{A.67}$$

is proportional to the value of the function $f(z)$ at the interior point a. Hence, in a more general form the *Cauchy integral formula*

$$F(z) = \frac{1}{2\pi i} \oint_C \frac{f(\zeta)}{\zeta - z}d\zeta \tag{A.68}$$

along a closed curve, C, over an continuous complex function, $f(\zeta)$, defines an analytic function, $F(z)$, in the interior of C. One can then easily define its nth derivative with respect to z as

$$\frac{\partial^n F(z)}{\partial z^n} = \frac{n!}{2\pi i} \oint_C \frac{f(\zeta)}{(\zeta - z)^{n+1}} d\zeta \tag{A.69}$$

so that $F(z)$ possesses all orders of derivatives and can thus be expanded into a Taylor series around any point inside C.

Contour Integration

With the help of these results the Cauchy integral can be used to calculate the value of line integrals which contain singular points. If $z = b$ is a singular point of the function $f(z)$ then in the neighborhood of $z = b$ the function $f(z)$ can be expanded as

$$f(z) = \sum_{n=0}^{\infty} \left[a_n(z-b)^n + \frac{a_{-n}}{(z-b)^n} \right] \tag{A.70}$$

This series is called *Laurent series*. When this expansion has an infinite number of singular terms, the singularity at $z = b$ is called essential and cannot be treated in a simple way. However, when there are only a finite number m of terms which are singular, which is the usual case in physical application, then the function is said to have a pole of order m. In particular, when $m = 1$ the function has a simple pole. In the case of a pole one can define a new function

$$\phi(z) = (z - b)^m f(z) \tag{A.71}$$

for $z \neq b$ so that $\phi(b) = a_{-m}$. This function is analytic at $z = b$. We can now expand this function into a Taylor series around b

$$\phi(z) = (z - b)^m f(z) \tag{A.72}$$
$$= a_{-m} + a_{-m+1}(z - b) + \cdots + a_{-1}(z - b)^{m-1} + a_0(z - b)^m + \cdots$$

Therefore the term a_{-1} is the coefficient of the $(m - 1)$st derivative of $\phi(z)$

$$a_{-1} = \left| \frac{1}{(m - 1)!} \frac{d^{m-1}[(z - b)^m f(z)]}{dz^{m-1}} \right|_{z=b} \tag{A.73}$$

For the special case of a simple pole this residue reduces to the special case $m = 1$

$$a_{-1} = \lim_{z \to b} (z - b) f(z) \tag{A.74}$$

These residua are very useful in calculating complex line integrals. When a function $F(z)$ has a series of poles the line integral over this function turns out to be the sum of all the residua, r_n, of this function at the poles inside the closed contour, C, of integration

$$\oint_C F(z)dz = 2\pi i \sum_n r_n \tag{A.75}$$

where the residua are calculated using the above rules. One first determines the poles of $F(z)$ and their order. Then one expands $F(z)$ around each pole into a Laurent series and determines the coefficient a_{-1} at this pole by simple use of the above formula. The value of the integral is just the sum over all r_n inside C.

If one knows that a function is analytic in one domain of the complex plane but one does not know its behavior at the outside one can use the method of analytic continuation to extend the region of analyticity of the function. Analytic continuation requires investigation of the behavior of the poles during the crossing of the contour C to the outside. If the poles all remain inside C, the continuation is trivial. But if the poles move out of C, one must deform the contour in such a way that the pole remains always on the same side of the new contour. The pole pushes the contour ahead of it. This process is called *analytic continuation*.

Plemelj-Dirac Formula

When the integrand in the complex contour integral has a simple pole but the pole approaches the real axis $\zeta \to \pm x$ either from above or from below that is from the half-plane where the integrand is analytic with the exception of the pole, then the Cauchy integral takes one particularly simple form discovered by Plemelj and Dirac

$$\lim_{\eta \to 0} \int_{-\infty}^{\infty} \frac{f(x)dx}{x - a \pm i\eta} = P \int_{-\infty}^{\infty} \frac{f(x)dx}{x - a} \mp i\pi f(a) \tag{A.76}$$

where a is real and the symbol P designates the principal value of the integral which is defined as the sum of two integrals

$$P \int_{-\infty}^{\infty} \frac{f(x)dx}{x - a} = \lim_{\epsilon \to 0} \left[\int_{-\infty}^{a-\epsilon} \frac{f(x)dx}{x - a} + \int_{a+\epsilon}^{\infty} \frac{f(x)dx}{x - a} \right] \tag{A.77}$$

taken along the real x-axis. The above formula can symbolically be written as

$$\lim_{\eta \to 0} \frac{1}{x \pm i\eta} = \frac{P}{x} \mp i\pi \delta(x) \tag{A.78}$$

As one easily realizes either form of this formula arises from the integration along the real axis and the deformation of the contour into a half-circle around the pole at $x = a \pm i\eta$. Hence, the factor two in front of the imaginary part disappears, and the principal value integral must be included as a "boundary condition" during analytic continuation between the two parts of the complex plane.

Maxwellian Integrals

Maxwellian integrals are integrals over Gaussian functions multiplied by some power of the integrand. They appear frequently in plasma physics where velocity distribution functions are modeled by products or sums of Maxwellians. The basic integral is the definite integral along the full real axis over a Gaussian function

$$\int_{-\infty}^{\infty} \exp(-x^2)\,dx = \sqrt{\pi} \tag{A.79}$$

This integral is closely related to the error function

$$\mathrm{erf}(z) = \frac{2}{\sqrt{\pi}} \int_{0}^{z} \exp(-x^2)\,dx \tag{A.80}$$

The Gaussian function is twice the value of the error function at infinity, with the factor of two a consequence of the symmetry of the integrand. The integral can be calculated using the methods of complex path integration described below. Generalizations of the above integral needed in plasma physics are of the form

$$\int_{0}^{\infty} x^\alpha \exp(-ax^2)\,dx = \begin{cases} \Gamma(l+1/2)/2a^{l+1/2} & \text{for } \alpha = 2l \\ l!/2a^{l+1/2} & \text{for } \alpha = 2l+1 \end{cases} \tag{A.81}$$

where $\Gamma(l+1/2) = [(2l-1)!!/2^l]\sqrt{\pi/a}$, and a has a positive real part to make the integral converging. Because of the asymmetry of the integrand for odd α it is clear that in the second case the integrals from $-\infty$ to $+\infty$ vanish identically, while for even α they are twice the value given above. The above formula can be verified by observing that the basic Maxwellian integral can be reproduced by multiple differentiation with respect to a.

Plasma Dispersion Function

The *plasma dispersion function* $Z(\zeta)$, sometimes also called *Fried-Conte function* because Fried and Conte first tabulated it in 1961, is a special example of an analytic function which is instrumental in plasma physics and particularly in plasma wave theory. It arises from the singular integral over a Maxwellian distribution (see Sec. 6.3)

$$Z(\zeta) = \frac{1}{\pi^{1/2}} \int_{-\infty}^{\infty} \frac{\exp(-z^2)\,dz}{z - \zeta} \tag{A.82}$$

with $\text{Im}\,\zeta > 0$. By differentiation with respect to ζ it is easy to show that $Z(\zeta)$ satisfies the first order differential equation

$$\frac{dZ(\zeta)}{d\zeta} = -2[1 + \zeta Z(\zeta)] \tag{A.83}$$

which can be used also as a recurrence relation to replace derivatives. By analytic continuation one finds some other useful relations

$$\begin{aligned} Z(-\zeta) &= -Z(\zeta) + 2\pi^{1/2}i \exp(-\zeta^2) && \text{for } \text{Im}\,\zeta > 0 \\ \tilde{Z}(\zeta) &= Z(\zeta) - 2\pi^{1/2}\exp(-\zeta^2) && \text{for } \text{Im}\,\zeta < 0 \end{aligned} \tag{A.84}$$

where $\tilde{Z}(\zeta)$ is the analytic continuation of $Z(\zeta)$. The complex conjugate of the plasma dispersion function is

$$[Z(\zeta)]^* = Z(\zeta^*) - 2\pi^{1/2}\exp(-\zeta^2) \tag{A.85}$$

The expansion of $Z(\zeta)$ for small argument $\zeta < 1$ is found by Taylor expansion of the above Cauchy integral and integrating term by term

$$\begin{aligned} Z(\zeta) &= i\pi^{1/2}e^{-\zeta^2} - 2\zeta\left(1 - \frac{2\zeta^2}{3} + \frac{4\zeta^4}{15} - \cdots\right) \\ &= \pi^{1/2}\sum_{n=0}^{\infty}\frac{i^{n+1}\zeta^n}{\Gamma(1+n/2)} \end{aligned} \tag{A.86}$$

The asymptotic expansion for $\zeta \gg 1$ is given by

$$Z(\zeta) = -\frac{1}{\zeta}\left(1 + \frac{1}{2\zeta^2} + \frac{3}{4\zeta^4} + \cdots\right) + \sigma\pi^{1/2}i\exp(-\zeta^2) \tag{A.87}$$

where

$$\sigma = \begin{cases} 0, & \text{Im}\,\zeta > 0 \\ 1, & \text{Im}\,\zeta = 0 \\ 2, & \text{Im}\,\zeta < 0 \end{cases} \tag{A.88}$$

The plasma dispersion function is closely related to a number of other functions. One of them is the error function of a complex argument

$$\text{erf}(z) = \left(1 + \frac{2i}{\pi^{1/2}}\int_0^z e^{t^2}dt\right)e^{-z^2} \tag{A.89}$$

One easily realizes that

$$Z(\zeta) = i\pi^{1/2}\text{erf}(\zeta) \tag{A.90}$$

B. Some Extensions

B.1. Coulomb Logarithm

In Sec. 4.1 we derived the collision frequency in a plasma, where the particles undergo pure Coulomb collisions. Here we present a rigorous derivation.

Rutherford Scattering

Let a charge, q, be scattered from a much heavier charge, q', in such a way that the heavy charge can be considered to be at rest. The Coulomb force F acting on q is then

$$F = \frac{qq'}{4\pi\epsilon_0 r^2} \tag{B.1}$$

with r the instantaneous distance between q and q'. The problem is symmetric (see Fig. B.1) and for equal sign of the charges the path described by the lighter charge (in the rest frame of the heavier charge) will be a hyperbola with its symmetry axis along x. Only the x component of the momentum of the scattered charge, $F\cos\phi/m$, is changed during the collision. With an initial velocity $v_0\cos\phi_0$, we obtain

$$2mv_0\cos\phi_0 = \frac{qq'}{4\pi\epsilon_0} \int\limits_{-\sin\phi_0}^{\sin\phi_0} \frac{dt}{d\phi}\frac{d\sin\phi}{r^2} \tag{B.2}$$

for the total change in momentum. In a central force field the angular momentum

$$mr^2\frac{d\phi}{dt} = mbv_0 \tag{B.3}$$

is conserved (b is the collision or impact parameter defined in Fig. B.1; see also p. 49). Hence, the value of the integral is

$$2mv_0\cos\phi_0 = \frac{qq'\sin\phi_0}{2\pi\epsilon_0 bv_0} \tag{B.4}$$

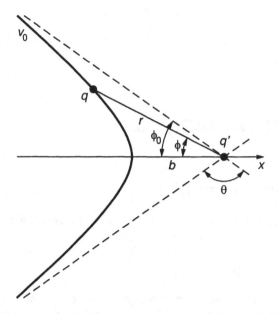

Fig. B.1. Geometry of scattering a charge q at a heavier charge q'.

Since the angle of deflection is $\theta = \pi - 2\phi_0$ and thus $\cot(\theta/2) = \tan\phi_0$, we can rewrite the above equation and find the well-known Rutherford scattering formula

$$\cot(\theta/2) = \frac{4\pi\epsilon_0 m b v_0^2}{qq'} \tag{B.5}$$

Mean Scattering Angle

To calculate the collision frequency one needs the average collisional cross-section with the average taken over all particles incident on q' with their different impact parameters, b. The differential cross-section is defined as

$$d\sigma = 2\pi b\, db \tag{B.6}$$

A stream of particles, all with initial velocity v_0 but different impact parameters, $b_{min} < b < b_{max}$, will be deflected by a mean scattering angle

$$\langle \theta^2(v_0) \rangle = \frac{\int \theta^2(v_0, b)\, d\sigma(b)}{\int d\sigma(b)} \tag{B.7}$$

where the integration in both integrals is over the interval $b_{min} < b < b_{max}$. Taking into account Debye screening, a good approximation for b_{max} is the Debye length. It is more difficult to find the lower limit of the impact parameter. The usual choices are

$$b_{max} = \lambda_D$$
$$b_{min} = qq'/4\pi\epsilon_0 m v_0^2 \tag{B.8}$$

with the latter formula resulting from the balance between initial particle energy and Coulomb energy. The main contribution to the integrals comes then from small-angle scattering of incident particles. Using the small-angle limit of Eq. (B.5)

$$\theta^2(v_0, b) = \frac{4}{v_0^4 b^2} \left(\frac{qq'}{4\pi\epsilon_0 m} \right)^2 \tag{B.9}$$

and performing the integrations one obtains finally

$$\langle \theta^2(v_0) \rangle = \frac{4}{\pi} \frac{b_{min}^2}{b_{max}^2} \ln \left(\frac{b_{max}}{b_{min}} \right) \tag{B.10}$$

where

$$\frac{b_{max}}{b_{min}} = \frac{4\pi\epsilon_0 \lambda_D m v_0^2}{qq'} \tag{B.11}$$

is the ratio of incident energy to Coulomb energy at the distance of one Debye length from the scattering center. For a thermal plasma one can use $m v_0^2 = k_B T_e$ to obtain

$$\frac{b_{max}}{b_{min}} = \frac{4\pi\epsilon_0 \lambda_D k_B T_e}{qq'} \tag{B.12}$$

and setting $q = q' = e$ one can rewrite this expression as

$$\Lambda = \frac{b_{max}}{b_{min}} = 4\pi n \lambda_D^3 \tag{B.13}$$

Λ is within a factor 4π equal to the plasma parameter introduced in Eq. (1.5) which is proportional to the number of particles in a Debye sphere. The logarithm $\ln \Lambda$ of this quantity is called the Coulomb logarithm and has been used in the expression for Spitzer's plasma collision frequency.

B.2. Transport Coefficients

The Coulomb logarithm enters the collision frequencies. The two equations from where this is obvious are the retardation of a particle by collisions

$$\frac{d\mathbf{v}}{dt} = -\nu_c \mathbf{v} \tag{B.14}$$

and the equation describing energy losses during frictional motion

$$\frac{dW_{\text{kins}}}{dt} = \frac{m_s}{2}\frac{dv^2}{dt} = -\nu_W \frac{m_s}{2}v^2 \tag{B.15}$$

Under anisotropic conditions the collision frequencies have parallel and perpendicular components. For two species, α, β, the general expressions for the collision frequencies obtained from a Boltzmann collision integral formulation are

$$\begin{aligned}
\nu_{c\alpha\beta} &= 2(1 + m_\alpha/m_\beta)\nu_{0\alpha\beta}x_\beta^2\psi(x_\beta) \\
\nu_{W\alpha\beta} &= 2\nu_{c\alpha\beta} - \nu_{\perp\alpha\beta} - \nu_{\parallel\alpha\beta} \\
\nu_{\perp\alpha\beta} &= 2\nu_{0\alpha\beta}[\phi(x_\beta) - \psi(x_\beta)] \\
\nu_{\parallel\alpha\beta} &= 2\nu_{0\alpha\beta}\psi(x_\beta) \\
\nu_{0\alpha\beta} &= (q_\alpha^2 q_\beta^2 n_\beta/4\pi\epsilon_0^2 m_\alpha^2 |v_\alpha - v_\beta|^3)\ln\Lambda_{\alpha\beta}
\end{aligned} \tag{B.16}$$

where the two functions of $x_\beta = |v_\alpha - v_\beta|/\sqrt{2}v_{\text{th}\beta}$ are given by

$$\begin{aligned}
\phi(x) &= \frac{2}{\sqrt{\pi}}\int_0^x dt\,\exp(-t^2) \\
\psi(x) &= \frac{1}{2x^2}\left[\phi(x) - x\frac{d\phi(x)}{dx}\right]
\end{aligned} \tag{B.17}$$

The Coulomb logarithm for electron-electron collisions has the value

$$\ln\Lambda_{ee} = \begin{cases} 16.0 - 0.5\ln n_e + 1.5\ln T_e & T_e \leq 7\cdot 10^4\,\text{K} \\ 21.6 - 0.5\ln n_e + \ln T_e & T_e \geq 7\cdot 10^4\,\text{K} \end{cases} \tag{B.18}$$

while we have for electron-ion collisions $\ln\Lambda = \ln\Lambda_{ei} = \ln\Lambda_{ie}$ from Eq. (B.13) with the numerical values

$$\ln\Lambda_{ei} = \begin{cases} 16.0 - 0.5\ln n_e + 1.5\ln T_e - \ln Z_i & T_e \leq 1.4\cdot 10^5\,\text{K} \\ 21.6 - 0.5\ln n_e + \ln T_e & T_e \geq 1.4\cdot 10^5\,\text{K} \end{cases} \tag{B.19}$$

In both cases, the density is measured in cm^{-3} and the temperature in K. Temperature equilibrium between electrons and ions is reached in a time

$$dT_e/dt = \nu_{\text{eqei}}(T_i - T_e) \tag{B.20}$$

where the equilibrium collision frequency is determined as

$$\nu_{\text{eqei}} = \frac{e^2 q_i^2 n_e \ln\Lambda_{ei}}{3(2\pi)^{1/2}\pi m_e m_i \epsilon_0^2 (v_{\text{the}}^2 + v_{\text{thi}}^2)^{3/2}} \tag{B.21}$$

Characteristic collision times for the electrons and ions are

$$\tau_e \approx 2.8 \cdot 10^5 \, T_e^{3/2}/(n_e \ln \Lambda_{ei})$$
$$\tau_i \approx 1.7 \cdot 10^7 \, T_i^{3/2}/(n_e \ln \Lambda_{ei})$$

(B.22)

Again, the density is measured in cm^{-3}, and the temperature in K. The momentum transfer rate, $\mathbf{R}_{ei} = -\mathbf{R}_{ie}$, first used in Eq. (7.42) can be written as

$$\mathbf{R}_{ei} = en_e \left(\frac{\mathbf{j}_\parallel}{\sigma_\parallel} + \frac{\mathbf{j}_\perp}{\sigma_\perp} - \frac{0.7 \nabla_\parallel T_e}{e} - \frac{3}{2} \frac{\mathbf{B} \times \nabla T_e}{eB\omega_{ge}\tau_e} \right)$$

(B.23)

where the parallel and perpendicular conductivities are

$$\sigma_\parallel = 2\sigma_\perp$$
$$\sigma_\perp = \epsilon_0 \omega_{pe}^2 \tau_e$$

(B.24)

The electron and ion heat fluxes are given by

$$\mathbf{q}_e = \frac{0.7 T_e}{e} \mathbf{j}_\parallel + \frac{3 T_e}{2eB\omega_{ge}\tau_e} \mathbf{B} \times \mathbf{j}_\perp - \kappa_{e\parallel} \nabla_\parallel T_e - \kappa_{e\perp} \nabla_\perp T_e - \frac{5 n_e T_e}{2 m_e B \omega_{ge}} \mathbf{B} \times \nabla T_e$$

$$\mathbf{q}_i = -\kappa_{i\parallel} \nabla_\parallel T_i - \kappa_{i\perp} \nabla_\perp T_i + \frac{5 n_i T_i}{2 m_i B \omega_{gi}} \mathbf{B} \times \nabla T_i$$

(B.25)

and the thermal conductivities entering these expressions are for $\omega_{gs}\tau_s \gg 1$

$$\kappa_{e\parallel} = 3.16 \, n_e T_e \tau_e / m_e$$
$$\kappa_{i\parallel} = 3.90 \, n_i T_i \tau_i / m_i$$
$$\kappa_{e\perp} = 4.66 \, n_e T_e / (m_e \omega_{ge}^2 \tau_e)$$
$$\kappa_{i\perp} = 2.00 \, n_i T_i / (m_i \omega_{gi}^2 \tau_i)$$

(B.26)

B.3. Geomagnetic Indices

Magnetic indices are derived from ground-based magnetograms and are meant to quantify disturbed states of the Earth's plasma environment. While planetary range indices like Kp describe only the overall disturbance level and will not be described here, two indices quantify the disturbance and the dissipation of energy of a certain element of the geo-plasma space. These are the *Dst index*, which was introduced in Sec. 3.5 and quantifies the ring current, and the *AE index*, which gives a measure of the auroral electrojets and the substorm activity (see Secs. 5.5 through 5.7).

AE Index

The auroral electrojet indices AE, AU, and AL were introduced as a measure of global auroral electrojet activity. The present auroral indices are based on 1-min readings of the northward H component trace from twelve auroral zone observatories located between about 65° and 70° magnetic latitude with a longitudinal spacing of $10 - 40°$.

For each of the twelve observatories, the readings of the H component are referenced to a quiet day level, H_0. The base value H_0 for the month under consideration is calculated as the average over all the readings from the five most quiet days in that month. The data of all twelve observatories are then plotted as a function of universal time. The upper and lower envelopes are defined as AU and AL, while AE is defined as the separation between the upper and lower envelopes

$$AU(t) = \max_{i=1,12} \left\{ H(t) - H_0 \right\}_i$$

$$AL(t) = \min_{i=1,12} \left\{ H(t) - H_0 \right\}_i \qquad \text{(B.27)}$$

$$AE(t) = AU(t) - AL(t)$$

where t is universal time. AU and AL are thought to represent the maximum eastward and westward electrojet current, respectively. AE represents the total maximum electrojet current and is most often used.

The main uncertainties of the AE index stem form the use of the H component, from longitudinal gaps in the distribution of the twelve observatories, from the small latitudinal range covered by these magnetic stations, and from the effects of strong local field-aligned currents.

At some magnetic observatories the angle between the local magnetic H component and the global eccentric dipole north-south direction is greater than 30°. Since the electrojets tend to flow along the auroral oval, i.e., perpendicular to the global eccentric dipole north-south direction rather than perpendicular to the local H direction, these observatories tend to underestimate the electrojet current.

The most severe longitudinal gaps in the AE observatory coverage are in Siberia, but also in western Canada and the Atlantic sector gaps spanning more than two hours of local time exist. Substorm current wedges associated with weak or moderate substorms may cover less than two hours of local time and can easily be missed by the twelve-station AE network.

Probably more severe is the small latitudinal range covered by the AE observatories. During times of very weak activity, when the interplanetary magnetic field is northward directed and convection ceases (see Sec. 5.2), the auroral oval contracts northward and the electrojets tend to flow poleward of 70° latitude. In this situation, the AE network, with all its stations south of 70°, will not detect the maximum disturbances.

The first three uncertainties of the standard AE index can, in principle, be avoided by the use of eccentric dipole coordinates and by including more stations in the network. However, the last uncertainty cannot be overcome even by an ideal AE index. For regions east or west of strong local field-aligned currents, a significant part of the north-south component of the magnetic disturbance stems from the field-aligned currents. This effect is most pronounced behind the head of the westward traveling surge, where strong field-aligned currents flow upward (see Sec. 5.7). Here, the southward perturbation due to the westward flowing ionospheric current can be reduced by up to 30% by the northward magnetic field associated with the upward field-aligned current to the west.

The other two electrojet indices have the same uncertainties as the AE index, but in addition are influenced by azimuthally uniform non-electrojet fields like that of the ring current, which cancel out in AE.

Dst Index

The ring current index Dst was introduced as a measure of the ring current magnetic field and thus its total energy, as described in Eq. (3.34). Since the westward ring current causes a reduction of the terrestrial dipole field, Dst is typically negative. During a magnetic storm, a typical Dst trace looks like the magnetogram in Fig. 3.10. The present Dst index is based on hourly averages of the northward horizontal H component recorded at four low-latitude observatories, Honolulu, San Juan, Hermanus, and Kakioka. All four observatories are $20 - 30°$ away from the dipole equator to minimize equatorial electrojet effects (see Sec. 4.5) and are about evenly distributed in local time.

At each observatory a magnetic perturbation amplitude is calculated by subtracting from the hourly averaged measured H component a quiet time reference level, $H_0(t')$, and the Sq field, $H_{sq}(t')$ (see Sec. 4.5), which both vary with local time, t'. All four magnetic disturbances are then averaged to further reduce local time effects and multiplied with the averages of the cosines of the observatories' dipole latitudes, Λ_i, to obtain the value of the ring current field at the dipole equator

$$Dst(t) = \frac{1}{16} \left[\sum_{i=1}^{4} \cos \Lambda_i \right] \cdot \left[\sum_{i=1}^{4} \{ H(t) - H_0(t') - H_{sq}(t') \}_i \right] \qquad (B.28)$$

where t is the universal time.

In contrast to the AE index, where most of the uncertainties lie in the uneven and too widely spaced station network, the Dst uncertainties are mainly caused by magnetic contributions of sources other than the ring current to the H component measured at the four Dst observatories, namely the magnetopause current, the partial ring current, and the substorm current wedge.

The magnetopause current (see Sec. 8.6) is regulated by the solar wind pressure. Its magnetic perturbation peaks around noon. A typical magnetopause contribution is included in the quiet time reference level, but dynamic variations of the solar wind pressure can yield positive or negative deviations from this average. In particular, sudden changes of the solar wind pressure associated with an interplanetary shock front can change the local time average of the magnetic perturbation due to the magnetopause current by typically $10 - 40$ nT.

The westward partial ring current (see Sec. 3.3) prevails in the afternoon sector and its closure via field-aligned currents causes a southward magnetic disturbance around the dusk meridian. The substorm current wedge (see Sec. 5.7) dominates in the midnight and early morning sector. Here, again mainly the field-aligned currents create a significant northward magnetic disturbance.

The combined effects of these three variable current systems yield an uncertainty of the quiet time reference level. Experimentally, one has found that the uncertainty maximizes around noon, where it may reach up to 50% of a typical Dst value of 50 nT. On the nightside, the uncertainty is relatively small, around $5 - 10$ nT. Hence, solar wind pressure changes are the dominant source of uncertainties in the Dst index.

B.4. Liouville Approach

The derivation of the Vlasov equation from the Klimontovich-Dupree equation given in Chap. 6 is only one way of finding the kinetic equations of a plasma. Historically, one has gone a different route, starting from the Liouville equation of statistical mechanics and descending from it along the construction of reduced distribution functions to the Vlasov equation. The two approaches are entirely equivalent. We have chosen the more simple treatment in the main text, but since the Liouville equation approach is slightly more rigorous we will, for completeness, sketch it here.

Liouville Equation

The Liouville approach starts from the assumption that each particle with index i spans its own six-dimensional phase space. Hence, the total phase space of all, say n, particles has $6n$ dimensions numbered by the index i of the particle to which the individual phase space belongs, $\mathbf{x}_1, \mathbf{v}_1, \ldots, \mathbf{x}_i, \mathbf{v}_i, \ldots, \mathbf{x}_n, \mathbf{v}_n$. The interaction between all these particles is contained in the total n-particle Hamiltonian $\mathcal{H}_n(\mathbf{x}_1, \mathbf{v}_1, \ldots, \mathbf{x}_i, \mathbf{v}_i, \ldots, \mathbf{x}_n, \mathbf{v}_n, t)$, which is a function of $6n + 1$ coordinates.

As done in Sec. 6.1, one can define an exact phase space distribution function of all coordinates and time, $\mathcal{F}_n(\mathbf{x}_1, \mathbf{v}_1, \ldots, \mathbf{x}_i, \mathbf{v}_i, \ldots, \mathbf{x}_n, \mathbf{v}_n, t)$, which is again conserved during the dynamic evolution of the plasma. This conservation can be expressed as the total time derivative of $\mathcal{F}_n$, taken with respect to all $6n$ coordinates which is the same as writing the Poisson bracket of $\mathcal{F}_n$ with the n-particle Hamiltonian $\mathcal{H}_n$

$$\frac{\partial \mathcal{F}_n}{\partial t} = [\mathcal{H}_n, \mathcal{F}_n] \tag{B.29}$$

This equation is the *Liouville equation* for the exact phase space distribution function $\mathcal{F}_n$. As with the Klimontovich-Dupree equation, this equation cannot be solved directly without knowing all the orbits of all the particles under all of their interactions. Instead solutions are found by taking moments of this equation.

Reduced Distribution Functions

The moments of the distribution function are defined as integrals over some of the microscopic individual phase spaces of particles contributing to $\mathcal{F}_n$. Such an integration, for instance, with respect to the whole individual phase space of particle i, gets rid of the coordinates of this particle and thus of its individuality. Its contribution to $\mathcal{F}_n$ and its dynamics is smeared out to all the remaining particles as an average effect on the distribution function. The loss of individuality is no problem in plasma physics because all electrons are indistinguishable as are all ions of the same kind.

Performing $6(n - 1)$ such consecutive integrations, the individuality of all particles will be destroyed and one is left with a reduced distribution function $\mathcal{F}_1(\mathbf{x}_1, \mathbf{v}_1, t)$ which depends merely on the phase space coordinates of one particle, which can be any electron or ion in the plasma. The distribution function $\mathcal{F}_1$ describes all electrons or ions equally well, distinguishing them only with respect to their velocities and positions. Hence, at a given position $\mathbf{x}_1$ there can be many particles of same velocity $\mathbf{v}_1$ at time t, and the function $\mathcal{F}_1(\mathbf{x}_1, \mathbf{v}_1, t)$ gives the probability density of finding these particles at this place.

Accordingly, the equation describing the dynamic evolution of $\mathcal{F}_1$ is obtained by performing $6(n - 1)$ integrations on Eq. (B.29). This procedure sounds easy, but it introduces a serious difficulty insofar as the Hamiltonian is a nonlinear function, and thus the integrations over subspaces of the phase space produce non-vanishing average terms in each step. In the last step, the integration over the coordinates of index 2, one obtains an equation which contains a large number of correlation terms, which depend on all the higher order reduced distribution functions.

BBGKY Hierarchy

Hence, the moment procedure does not close, but produces a whole hierarchy of evolution equations for the reduced distribution functions. To obtain the lowest one, one must solve that for the former and so on. This hierarchy is called *BBGKY hierarchy* after the initials of their inventors. It shows that the statistical character of a plasma is, in principle, very complicated and cannot be resolved entirely without some assumptions. These assumptions are that at some stage in the hierarchy, one simply neglects the higher order correlations and their contribution to the reduced distribution function.

At the lowest level, neglecting all the higher order correlations, the continuity or kinetic equation for $\mathcal{F}_1$ becomes

$$\frac{\partial \mathcal{F}_1}{\partial t} = [\mathcal{H}_1, \mathcal{F}_1] \tag{B.30}$$

Expanding the Poisson bracket and replacing $f(\mathbf{x}, \mathbf{v}, t) = \mathcal{F}_1(\mathbf{x}_1, \mathbf{v}_1, t)$, i.e., dropping the index 1, since the lowest level phase space distribution function depends only on the space and velocity coordinates of one particle, one obtains

$$\frac{\partial f}{\partial t} + \mathbf{v} \cdot \nabla_{\mathbf{x}} f + \frac{q}{m}(\mathbf{E} + \mathbf{v} \times \mathbf{B}) \cdot \nabla_{\mathbf{v}} f = 0 \tag{B.31}$$

This is exactly the same equation as the Vlasov equation (6.20). The method of reduced distribution functions and the Liouville equation thus lead to the same result for the kinetic equation of a plasma if all the correlations between the particles as well as the collisions are neglected.

B.5. Clemmow-Mullaly-Allis Diagram

Investigation of wave propagation even in a cold plasma has turned out to be a complicated play with the dispersive properties of the different wave modes. There is, however, a simple way of classifying the different types of waves in dependence on the plasma parameters, the so-called CMA-diagram in Fig. B.2.

For this diagram one chooses the ratios of cyclotron, ω_{ge}^2/ω^2, and plasma frequencies, ω_{pe}^2/ω^2, to the wave frequency as rectangular axes in a plane. In this coordinate system the resonances and cut-offs are represented as particular curves. For example, the resonance frequency of the R-mode, $\omega_{R,res} = \omega_{ge}$ does not depend on the plasma density and in this system becomes a horizontal line valid for all ratios ω_{pe}^2/ω^2. Similarly, in a pure charge-compensated electron plasma the cut-off $\epsilon_3 = 0$ is a constant vertical line cutting the abscissa at $\omega_{pe}^2/\omega^2 = 1$ and valid for all ratios ω_{ge}^2/ω^2, while the resonance $\epsilon_1 = 0$ connects the point $\omega_{ge}^2/\omega^2 = 1$ on the ordinate with the point $\omega_{pe}^2/\omega^2 = 1$ on the abscissa. The cut-off $\epsilon_L = 0$ can be written as $1 - \omega_{pe}^2/\omega^2 = \omega_{ge}^2/\omega^2$ and describes a parabola with its minimum at $\omega_{pe}^2/\omega^2 = 1$.

The two straight lines, $\epsilon_R = \infty$ and $\epsilon_3 = 0$, divide the CMA-plane into four regions which are further subdivided by the remaining resonances and cut-offs at different angles. The total number of separate regions obtained is eight. Now, in each of these regions one can draw a polar plot of the phase velocity of the different waves normalized to the velocity of light, ω/kc, as a function of the angle θ between the magnetic field

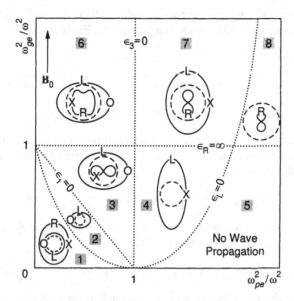

Fig. B.2. CMA diagram for a cold electron plasma.

(parallel to the ordinate) and the direction of the wave vectors. The dashed circles denote $v_{ph} = \omega/k = c$. One observes the different connections between the L-, R-, O-, and X-modes in the different regions of the diagram as well as the changes of the polarization properties from region to region. In some regions no propagation is possible due to reflection from this domain.

B.6. Magnetized Dielectric Tensor

In a hot plasma of arbitrary temperature the calculation of the dielectric tensor, $\epsilon(\omega, \mathbf{k})$, entering the dispersion relation Eq. (9.55) demands solving the linearized Vlasov equation (10.84). This requires explicit calculation of the current integral over all particle species, s, in Eq. (10.92)

$$\delta \mathbf{j}(\mathbf{k}, \omega) = -\sum_s \frac{\epsilon_0 \omega_{ps}^2}{n_0 \omega} \int_0^\infty \int_0^{2\pi} \int_{-\infty}^\infty v_\perp dv_\perp d\psi dv_\parallel$$

$$\delta \mathbf{E}(\mathbf{k}, \omega) \cdot \int_0^\infty d\tau \, e^{-i\varphi(\tau)} \{ \mathbf{v} \, \mathbf{k}(\tau) + [\omega - \mathbf{k} \cdot \mathbf{v}(\tau)] \mathbf{l} \} \cdot \frac{\partial f_0[\mathbf{v}(\tau)]}{\partial \mathbf{v}(\tau)} \quad \text{(B.32)}$$

under the additional assumption of gyrotropy, i.e., that there is no difference between the two directions transverse to the ambient magnetic field, $\mathbf{B}_0$. In this case we can write the wave vector as

$$\mathbf{k} = k_\perp \hat{\mathbf{e}}_\perp + k_\| \hat{\mathbf{e}}_\| \tag{B.33}$$

Similarly, we decompose the velocity into transverse and parallel components

$$\mathbf{v} = (v_\perp \cos\psi, \, v_\perp \sin\psi, \, v_\|) \tag{B.34}$$

The motion of particles in uniform magnetic fields consists of a parallel translation with speed $v_\| = \text{const}$ and a gyration (see Chap. 2). A particle starting with initial phase angle ψ has at time τ the velocity

$$\mathbf{v}(\tau) = [v_\perp \cos(\omega_g \tau + \psi), \, v_\perp \sin(\omega_g \tau + \psi), \, v_\|] \tag{B.35}$$

where $\tau = t' - t$ is the difference between observation time, t', and initial time, t. Correspondingly its position is, after integrating the velocity

$$\mathbf{x}(\tau) = \mathbf{x} + \omega_g^{-1} [v_\perp \sin(\omega_g \tau + \psi), \, -v_\perp \cos(\omega_g \tau + \psi), \, v_\| \tau] \tag{B.36}$$

It is assumed that the wave has so small amplitude that the path of the particle is not distorted. With the help of these expressions we can now calculate the derivative of the distribution function appearing in Eq. (B.32)

$$\frac{\partial f_0[\mathbf{v}(\tau)]}{\partial \mathbf{v}(\tau)} = \left[\frac{\partial f_0}{\partial v_\perp} \cos(\omega_g \tau + \psi), \, \frac{\partial f_0}{\partial v_\perp} \sin(\omega_g \tau + \psi), \, \frac{\partial f_0}{\partial v_\|} \right] \tag{B.37}$$

Moreover, using the above representation of the position vector, the phase function of the plane wave, $\varphi(\tau) = -\omega\tau + \mathbf{k} \cdot [\mathbf{x} - \mathbf{x}(\tau)]$, can be rewritten as

$$\varphi(\tau) = (k_\| v_\| - \omega)\tau + \xi[\sin(\omega_g \tau + \psi) - \sin\psi] \tag{B.38}$$

where $\xi = k_\perp v_\perp / \omega_g$. Inserting for the exponent of the exponential factor in Eq. (B.32) lets the exponential depend on the two sin-functions. Fortunately we can make use of the following addition theorem of Bessel functions, $J_l(\xi)$

$$\sum_{l=-\infty}^{l=\infty} J_l(x) \exp(-il\phi) = \exp(-ix \sin\phi) \tag{B.39}$$

to decompose the exponential into a sum over the product of Bessel functions and simple exponentials which turn out to be trigonometric functions as

$$\sum_{l,l'=-\infty}^{l,l'=\infty} J_l J_{l'} \exp\{-i[(k_\| v_\| + l\omega_g - \omega)\tau + (l - l')\psi]\} = \exp[-i\varphi(\tau)] \tag{B.40}$$

Vice versa, we write the trigonometric functions in Eq. (B.37) as exponentials. In addition, observing that

$$-i[\mathbf{k} \cdot \mathbf{v}(\tau) - \omega] \exp[-i\varphi(\tau)] = \frac{d}{d\tau} \exp[-i\varphi(\tau)] \tag{B.41}$$

allows to simplify the scalar product in Eq. (B.32). We realize that the factor of $\delta\mathbf{E}(\omega, \mathbf{k})$ in Eq. (B.32) is the linear conductivity. Inserting it into Eq. (9.54), we obtain

$$\epsilon(\omega, \mathbf{k}) = \left(1 - \sum_s \frac{\omega_{ps}^2}{\omega^2}\right)\mathbf{I} - \sum_s \frac{\omega_{ps}^2}{n_{0s}\omega^2} \int v_\perp dv_\perp dv_\parallel d\psi$$

$$\int_0^\infty d\tau \, e^{-i\varphi(\tau)} \mathbf{v} \left(i\mathbf{v}(\tau)\mathbf{k} - \mathbf{I}\frac{\partial}{\partial\tau}\right) \cdot \frac{\partial f_{0s}[\mathbf{v}(\tau)]}{\partial\mathbf{v}(\tau)} \tag{B.42}$$

The integration over τ can now be performed with the help of the above decompositions into sums. It splits into a number of separate integrations over exponentials of different l, l', which can subsequently be simplified using the orthogonality conditions of trigonometric functions, when performing the sum over l' and when ultimately performing the integration over the velocity phase angle, ψ, in the interval 0 to 2π. Since Eq. (B.42) contains a tensor, the final result also becomes a tensor

$$\epsilon(\omega, \mathbf{k}) = \left(1 - \sum_s \frac{\omega_{ps}^2}{\omega^2}\right)\mathbf{I} - \sum_s \sum_{l=-\infty}^{l=+\infty} \frac{2\pi \omega_{ps}^2}{n_{0s}\omega^2}$$

$$\int v_\perp dv_\perp dv_\parallel \left(k_\parallel \frac{\partial f_{0s}}{\partial v_\parallel} + \frac{l\omega_{gs}}{v_\perp}\right) \frac{\mathbf{S}_{ls}(v_\parallel, v_\perp)}{k_\parallel v_\parallel + l\omega_{gs} - \omega} \tag{B.43}$$

where we have introduced

$$\mathbf{S}_{ls} = \begin{bmatrix} \dfrac{l^2\omega_{gs}^2}{k_\perp^2} J_l^2 & \dfrac{ilv_\perp\omega_{gs}}{k_\perp} J_l J_l' & \dfrac{lv_\parallel\omega_{gs}}{k_\perp} J_l^2 \\[3mm] -\dfrac{ilv_\perp\omega_{gs}}{k_\perp} J_l J_l' & v_\perp^2 J_l'^2 & -iv_\parallel v_\perp J_l J_l' \\[3mm] \dfrac{lv_\parallel\omega_{gs}}{k_\perp} J_l^2 & iv_\parallel v_\perp J_l J_l' & v_\parallel^2 J_l^2 \end{bmatrix} \tag{B.44}$$

with $J_l'(\xi) = dJ_l(\xi)/d\xi$. In the above calculation we have used

$$\int_0^{2\pi} d\psi \, \exp[i(l-l')\psi] = 2\pi \delta_{l,l'}$$

$$\int_0^{2\pi} d\psi \, \cos(\psi + \phi) \exp[i(l+l')\psi] = \pi \left[e^{i\phi} \delta_{l+1,l'} + e^{-i\phi} \delta_{l-1,l'} \right]$$

(B.45)

and the following recursion formulas and sums

$$J_l(z) = \frac{z}{2l} \left[J_{l-1}(z) + J_{l+1}(z) \right]$$

$$J_l'(z) = \frac{1}{2} \left[J_{l-1}(z) - J_{l+1}(z) \right]$$

$$\sum_{l=-\infty}^{\infty} J_l^2(z) = 1$$

(B.46)

$$\sum_{l=-\infty}^{\infty} l J_l^2(z) = 0$$

$$\sum_{l=-\infty}^{\infty} \frac{J_l(z) J_{l-r}(z)}{q-l} = \frac{(-1)^r \pi}{\sin \pi q} J_{r-q}(z) J_q(z)$$

Weber Integrals

The following so-called Weber integrals

$$\int_0^{\infty} x \, dx \, J_0(px) e^{-q^2 x^2} = (2q^2)^{-1} \exp\left(\frac{-p^2}{4q^2} \right)$$

$$\int_{0_.}^{\infty} x \, dx \, J_l(px) J_l(rx) e^{-q^2 x^2} = (2q^2)^{-1} \exp\left(\frac{-p^2 + r^2}{4q^2} \right) I_l \left(\frac{pr}{2q^2} \right)$$

(B.47)

are used for calculating the magnetized plasma response function for longitudinal waves.

Index